PROGRESS IN COLLOID & POLYMER SCIENCE
Editors: H.-G. Kilian (Ulm) and G. Lagaly (Kiel)

Volume 88 (1992)

Advances in Colloid Structures

Guest Editors:
J. Sjöblom (Bergen), B. Lindman (Lund),
and P. Stenius (Helsinki)

Steinkopff Verlag · Darmstadt
Springer-Verlag · New York

ISBN 3-7985-0912-3 (FRG)
ISBN 0-387-91409-9 (USA)
ISSN 0340-255 X

Chemistry editor: Dr. Maria Magdalene Nabbe; English editor: James Willis; Production: Holger Frey.

Printed in Germany.

Type-Setting: Mac Millan Ltd., Bangalore, India
Printing: betz-druck gmbh, D-6100 Darmstadt 12

Preface

This volume of Progress in Colloid and Polymer Science consists of some selected papers presented at the XI Scandinavian Symposium on Surface Chemistry, held at the University of Bergen, Department of Chemistry, June 26—28, 1991. These Scandinavian Symposia have been organized about every other year since 1962.

The purpose of these symposia is to present the latest developments in surface and colloid chemistry from both fundamental and applied points of view. In the beginning, these symposia were almost entirely Scandinavian, but they now have a truly international character. At the Bergen Symposium 120 delegates from 14 different countries participated. In addition to about 50 oral presentations, 20 poster contributions were presented. The memory of the Scandinavian pioneer in surface and colloid chemistry, Prof. Per Ekwall, who died December 19, 1990, was honored at the symposium by a special pictorial presentation compiled by Dr. Krister Fontell.

The XI Scandinavian Symposium on Surface Chemistry was organized around the topics:

— Surfactants and macromolecules in solution and on surfaces;
— Surface forces, ceramics, and composites;
— Functionalized colloids; and
— Development of technical equipment in surface chemistry.

The Advisory Board consisted of Drs. K. S. Birdi and N. Krog of Denmark; I. Danielson and J. Rosenholm of Finland; A. Bjørseth, P.Gramme, B. Hafskjold, H. Hoiland, S. K. Ratkje, P. Read, O. Smidsrød, T. A. Torp and J. Ugelstad of Norway; M. Almgren, K. Holmberg, K. Larsson, G. Lindblom and I. Lundström of Sweden.

Financial support was provided by Berol Nobel Ab, Dipl. ing. Houm A/S, KSV Instruments Ltd. Norsk Hydro A/S, Statoil A/S, International Association of Colloid and Interface Scientists (IACIS), Norwegian Research Council for Science and the Humanites (NAVF), and VISTA.

On behalf of the organizing committee, we would like to thank all the participants and lectures for making the symposium a success, all the contributors to this proceedings volume, all the contributing sponsors and, especially, the conference secretariat (lead by Dr. A. M. Blokhus) for all the practical arrangements.

The next meeting in this series will be held in Helsinki in 1994. We look forward to meeting all of you then.

Johan Sjöblom
Björn Lindman
Per Stenius

Contents

Progress in Colloid & Polymer Science Progr Colloid Polym Sci 88:1–7 (1992)

Phase behavior of a system of nonionic polymer, ionic surfactant, and water

K. Zhang[1]), G. Karlström[2]), and B. Lindman[1])

[1]) Physical Chemistry
[2]) Theoretical Chemistry, Chemical Center, University of Lund, Sweden

Abstract: The phase diagrams for mixtures of ethyl(hydroxyethyl) cellulose and sodium dodecyl sulfate in water and in 0.03% aqueous NaCl solutions have been determined for the water rich part of the system. The obtained phase diagram is characterized by a two-phase region in the water-rich corner with tielines almost parallel to the water-polymer axis. It is demonstrated that the ratio of polymer to surfactant has a critical effect on the phase behavior. Model calculations based on the Flory–Huggins polymer theory are presented. The close agreement between theoretical and experimental phase behavior gives strong support to a molecular interpretation of the observed phase behavior, implying that there is a relatively strong attraction between the polymer and the ionic surfactant.

Key words: Polymer; surfactant; interaction; phase diagram; phase behavior

Introduction

The study of aqueous systems containing both a polymer and a surfactant is a research area which has received increased interest in recent years. Much of this interest originates from industrial applications since polymers and surfactants are extensively used together in technical processes and in products. Frequently, systems which utilize both polymers and surfactants are used. There is, however, also a more fundamental interest in such systems, both from a physico-chemical and a biochemical point of view.

The behavior of systems containing either a surfactant or a polymer has also, in recent years been much better elucidated. Particularly the surfactant-water systems may show several different phases, both of solution and liquid crystalline type [1].

Whereas our understanding of simple polymer solutions or of simple surfactant-water systems is good today, our understanding of aqueous polymer-surfactant systems is rather limited, despite the large interest that such systems have received in recent years [2–23]. Much of the interest in these studies has been on dilute systems, in particular on how physico-chemical properties of a polymer solution change as a function of the concentration of added surfactant.

In our previous work [18–21], we have inter alia studied the interaction between a nonionic polymer, ethyl (hydroxyethyl)-cellulose (EHEC), and sodium dodecyl sulfate (SDS) in aqueous solutions, with and without added electrolyte, and found a strong influence on the phase behavior of only minor amounts of salt (0.03% by weight). The experimental data presented in the previous work only cover part of the phase diagrams, whereas model calculations [21], which are of the Flory–Huggins type, and are presented to rationalize the behavior, present full three-component phase diagrams. The simple EHEC-water system shows the peculiarity (which it has in common with many other systems which contain molecules with ethyleneoxide (EO) groups) that the miscibility decreases with increasing temperature. This means that if a solution is heated, at some temperature the solution becomes turbid and phase separates into two phases. (The temperature where the solution becomes turbid is called the cloud point (CP)). As a consequence the phase behavior of the aqueous polymer-surfactant system may be quite complex. There is thus a need for as complete experimental phase diagrams as possible in order to check the predictions of the theoretical model.

Experimental section

Materials

EHEC powder (Bermocoll CST 103, Batch 3587) was supplied by Berol Nobel AB, Sweden. The average degree of substitution of ethyl and ethylene oxide groups is 1.5 and 0.7, respectively. The number average molecular weight is 80 000. The EHEC solution was dialyzed against pure water (Millipore water, USA) for 5 days to remove the salt (impurity from the manufacturing). All other chemicals were of analytical grade and used without further purification. SDS was supplied by BDH, England. NaCl and dichlormethane were obtained from Merck, FRG.

Samples were prepared by weighing the components. EHEC, SDS, and NaCl concentrations are expressed in percent by weight.

Methods

Boundaries of the two-phase regions in the phase diagrams were acquired by cloud-point (CP) measurements. The cloud points were determined visually. CP was taken as the temperature where the first visible sign of clouding appeared on heating. The heating rate was approximately 1 °C per min.

The tielines were determined by analysis of the phase compositions. The samples were left to be equilibrated at the required temperature for 10 days and separated into two phases. The supernatant phase is solution-like and water-rich, while the other phase is solid-like and polymer-rich.

The SDS content was determined spectrophotometrically [24, 25]. A methylene blue-dodecyl sulphate complex was formed in aqueous solution and extracted to a dichlormethane phase. The absorbance of the organic phase was measured at 655 nm on a Nova spectrophotometer (LKB, England).

The chloride content was determined by titration with mercuric (II) nitrate in the presence of diphenylcarbazon as indicator [26]. The endpoint was detected when a violet color appeared.

The EHEC content in the precipitate was determined by dissolving the precipitate and dialyzing the solution against Millipore water for 2 days to eliminate SDS and NaCl. Then the solution was freeze-dried and the solid weighed.

Theoretical modeling

The purpose of this section is to present the theoretical model and its physical background and limitations. Apart from a brief outline of the formalism, there are four basic problems that will be covered in this section: i) the origin of the clouding phenomenon, ii) the modeling of the surfactant by the Flory–Huggins theory, iii) the incorporation of long-range electrostatic interactions due to charged particles, and iv) the effect of salt in the system.

The formalism used to model the observed phase behavior is based on the Flory–Huggins theory, but it is additionally assumed that each polymer segment (for the EHEC polymer) may exist in one of two different types of conformations (vide infra) [27–29]. Each type of conformation interacts differently with the other entities in the system. The surfactant molecules are modeled as another polymer and the degree of polymerization is used as a parameter, which is fitted in order to obtain a phase behavior resembling the experimental one.

One of the standard ways to model a polymer solution is to use the Flory–Huggins (F–H) theory. Within this model the mixing energy can be written

$$U = N\Phi_1 \Phi_2 w_{12} , \qquad (1)$$

where w_{12} is the effective interaction parameter between a polymer segment and a neighboring solvent molecule. The w's used in this work differ from the ordinary χ-parameters used in F–H theory, since the w's do not have the explicit temperature dependence present in the χ-parameters. N is the total number of cells (solvent molecules + polymer segments in the system) and Φ_1 and Φ_2 are the volume fractions of solvent and solute. The effective interaction parameter w_{12} is related to the direct interaction parameters w'_{ab} according to

$$w_{12} = w'_{12} - 1/2(w'_{11} + w'_{22}) . \qquad (2)$$

The F–H entropy expression is

$$S = -kN(\Phi_1 \ln \Phi_1 + \Phi_2/M \ln \Phi_2) , \qquad (3)$$

where k is Boltzmann's constant and M is the degree of polymerization of the polymer. Equations (1) and (3) can easily be modified to include a two-conformational description of the polymer segments. Thus, each polymer segment can exist in either a polar conformation, which interacts with one set of interaction parameters, or in a less polar conformation, which interacts with another set of interaction parameters [27]. The probability of a polar conformation will be denoted P in the following. One then obtains

$$\begin{aligned} U = N(&\Phi_1\Phi_2(Pw_{1p} + (1-P)w_{1u}) \\ &+ \Phi_2\Phi_2(P(1-P)w_{pu} \\ &+ (1-P)^2 w_{uu}/2))\ , \end{aligned} \tag{1'}$$

and

$$\begin{aligned} S = &-Nk(\Phi_1 \ln \Phi_1 + \Phi_2/M \ln \Phi_2 \\ &+ \Phi_2(P\ln P + (1-P)\ln(1-P)/f) \end{aligned} \tag{3'}$$

Subscripts u and p in Eq. (1′) refer to the less polar and polar conformations of the polymer. The parameter f appearing in (3′) is a measure of the size of phase space corresponding to the less polar conformations relative to that of the polar conformations. Physically, this means that there may be several conformations of less polar type and only a few or one of polar type; f measures the ratio between these numbers. It is further worth noting that Eq. (1′) contains an interaction parameter w_{uu} which is diagonal in indices. This term cannot be removed since it is necessary in order to describe the equilibrium between the polar and the less polar conformations of the polymer. Equations (1′) and (3′) can be generalized to include the effect of another polymer. One obtains,

$$\begin{aligned} U = U' &+ \Phi_1\Phi_3 w_{13} \\ &+ \Phi_2\Phi_3(Pw_{p3} + (1-P)w_{u3}) \end{aligned} \tag{1''}$$

$$S = S' - Nk\,\Phi_3/M_3 \ln \Phi_3\ . \tag{3''}$$

U' and S' in these equations are U and S according to Eqs. (1′) and (3′). These are the equations that we will work with in our theoretical modeling.

There are, at present, basically three different explanations for the decreased solubility of EO-containing molecules at increased temperatures. Two of these models have been given a mathematical formulation and can thus be used for modeling purposes. The third model explains the behavior as a consequence of water structuring around the polymer [30]. A weakness with such a model is that it is hard to test, except by changing the solvent. However, recent investigation of EO-containing molecules in formamide have shown a similar phase behavior for these compounds in formamide as in water [31]. Thus, this model seems less likely. In the other two models the effective interaction between the solvent and the solute becomes temperature-dependent, either through hydrogen bonding [32] or through changes in the polymer conformation [27]. The mathematical formulations of these models are very similar and it is probably possible to use both models to rationalize the phase behavior that we will discuss below. We will, however, use the polymer conformation model since this model can also explain a set of additional observations, such as the rapid change in dielectric constant in a set of EO-containing compounds with temperature [28, 29]. In this model [27] it is assumed that each segment of the polymer chain may exist in either a polar conformation, which interacts favorably with water, or as one of many less polar conformations which interacts unfavorably with water. The polar conformation is energetically favored, but the less polar conformations are entropically favored, since they are more numerous. This means that at higher temperatures the probability that a segment should exist in a less polar conformation will increase, and this will lead to a more repulsive interaction between the polymer and water at higher temperatures.

The system which we are studying is built up from water, polymer, and surfactant molecules. The surfactant molecules may exist as either monomer units or as micelles. In this model, which is crude, we will forget about the surfactant monomers and only model the effect of the micelles. This does not mean that the monomers lack importance for the phase behavior, but rather that we focus on what we believe is the main effect.

The micelles are big entities compared to, for example, the water molecules, and they are charged. This means that they interact strongly with oppositely charged molecules, and that the interaction contribution to the free energy expression is more important than the entropy for almost all concentrations. We have chosen to treat the micelles as polymer molecules. In doing so, we reduce the entropy contribution from the micelles compared to treating the individual monomer molecules as free molecules, but increase it compared to treating the micelles as one molecule. Since it is difficult to define a degree of polymerization, we will use this parameter as a variable in the calculations. (The actual values used are shown in the relevant figure captions). The important point is, however, that for all concentrations except for very dilute ones the influence of this entropy term is small. For very dilute systems the entropy expression used is as reasonable for the micelles as it is for the polymer molecules.

The third point that we will discuss here is the treatment of long-range electrostatics in a Flory–Huggins-type model. The essential assumption in the derivation of the Flory–Huggins free energy expression is that the interactions are short-range and proportional to the volume fractions of the treated species. Obviously, this is not true for the long-range electrostatics, but, nevertheless, it is always possible to expand the part of the total free energy (which is not included in the entropy expression used) around a suitable point in the concentration space in a polynomial of second degree in the volume fractions. This neither means that the interactions present in the system are short-range nor that screening phenomena are not present, it is only a mathematical expansion of the free energy.

The final issue that we set out to discuss in this section is how to include the effect of salt in the model. In principle, there are two ways of doing this. In the first approach, we consider the effect of salt on the interaction between the polymer and the micelles. Salt will obviously screen the electrostatic interaction between different micelles and thus facilitate the binding of the micelles to the polymer. This effect can be obtained by making the interaction between the polymer and the surfactant system more attractive [4, 10, 14, 17]. It is, however, more straightforward to describe salt as part of the solvent and, thus, it shall not affect the interaction parameter between the surfactant and the polymer. (Obviously, the phase behavior along the polymer-surfactant axis must be independent of the solvent.) One may further note that salt at the concentrations that are relevant in this work (0.03%) does not affect the pure polymer-solvent system, and that we, consequently, should not change the polymer-solvent interaction energies. In this model the experimentally observed cloudpoint depression observed when salt is added is obtained if the interaction between the solvent and the surfactant is made less attractive in the presence of salt [21]. In fact, this is a consequence of the fact that there is less entropy to gain by mixing the ions counter to the micelles with the solvent, when these ions are already present in the solvent.

Results and discussion

In Fig. 1, we present the experimentally obtained phase diagrams for the EHEC-SDS-water system at three different temperatures. Apart from the complication that there obviously exists a distribution of EHEC molecules, both with respect to molecular weight and degree of chemical modification, this is a true three-component system. Tielines are almost parallel to the water-EHEC axis. The extension of the two-phase area in the surfactant dimension is somewhat smaller than was previously assumed and, consequently, we have slightly adjusted the interaction parameters previously used [21] in the theoretical modeling. This latter observation is probably partly due to the fact that a different (more hydrophobic) EHEC polymer is used in the present work. It should further be noted that the dashed part of the phase diagram is very uncertain since equilibrium in this region is very slow due to the high viscosity. In Figs. 2a and b, we present the corresponding theoretical phase diagrams. (The parameters used are listed in the figure.) It should be noted that the clouding temperature of both the experimental and the theoretical EHEC-water system is 30 °C, and that 2 °C below this temperature no two-phase region is observed, either experimentally or theoretically at any surfactant concentration. In Fig. 3a, we show the experimentally observed phase diagrams obtained at three different temperatures when 0.03% salt is added to the system. Obviously, it is an approximation to treat

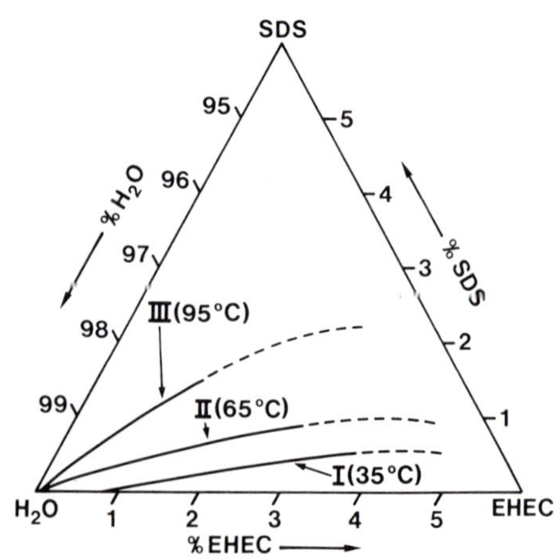

Fig. 1. Experimentally obtained phase diagrams at 35 °, 65 °, and 95 °C for the water-EHEC-SDS system. Dashed part of the curve is uncertain due to slow equilibration. Above the curve there is a one-phase solution, while below there is a separation into two phases

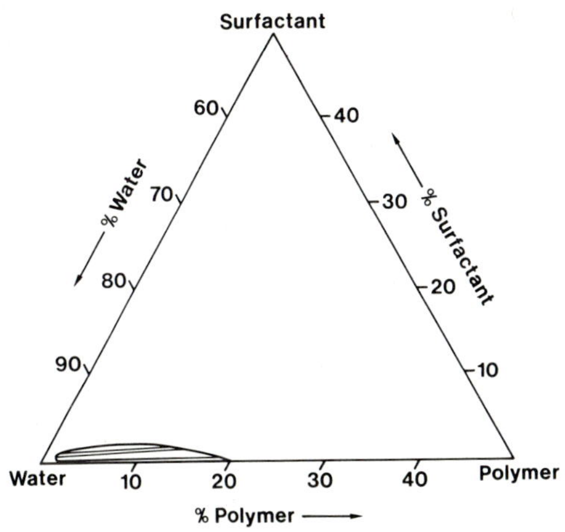

Fig. 2a. Theoretically calculated phase diagrams (with tielines) using Eqs. (1″) and (3″) for 33 °C. The following parameters have been used. $w_{1p} = 562.6$ J/mol, $w_{1u} = 9209.0$ J/mol, $w_{pu} = 5491.0$ J/mol, $w_{uu} = 8793.0$ J/mol, $w_{13} = -5402.5$ J/mol, $w_{p3} = -8039.0$ J/mol, $w_{u3} = -907.6$ J/mol. The degree of polymerization was 2000 for the polymer and 100 for the surfactant

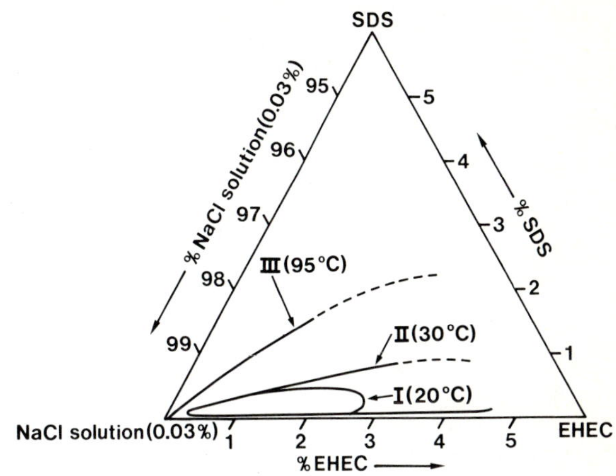

Fig. 3a. Experimentally obtained phase diagrams at 20 °, 30 °, and 95 °C for the water + salt (0.03%)-EHEC-SDS system. The dashed part of curve is very uncertain due to slow equilibration. At 20 ° and 30 °C there is a closed two-phase region encircled by the homogeneous solution phase, while at 95 °C the two-phase region extends to the polymer-water axis

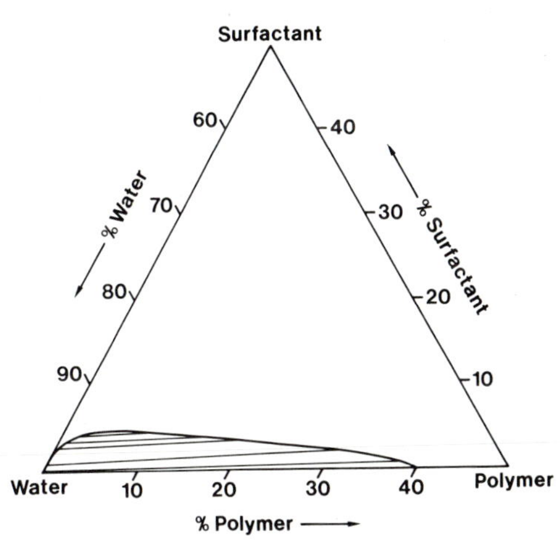

Fig. 2b. As in Fig. 2a, but the temperature is 49 °C

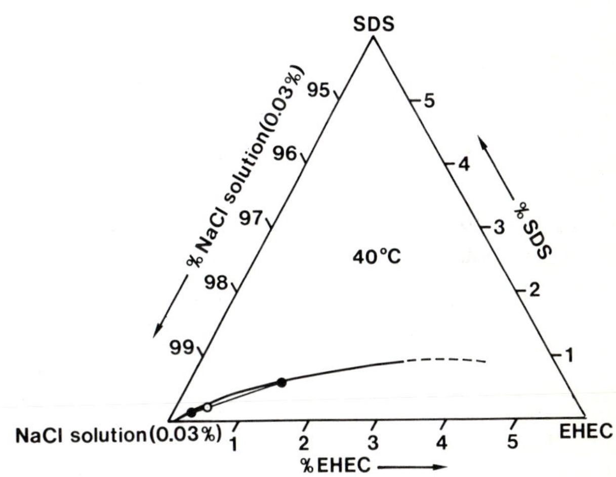

Fig. 3b. As in Fig. 3a, but the temperature is 40 °C. One tieline is indicated

the salt-water mixture as one component, but if the salt concentration is approximately the same in the different phases the approximation is reasonably good. In Fig. 3b, we show one tie line determined at 40 °C. The experimentally obtained tieline deviates significantly more from the water-EHEC axis in this case than in the pure water system. The corresponding theoretical diagrams are shown in Figs. 4a, b, and c. These theoretical diagrams have been obtained by decreasing the attraction between the surfactant and the solvent (water + 0.03% salt) by 475.4 J/mole. In general terms, one may say that the agreement between the experimental and theoretical phase diagrams is semi-quantitative, but they differ at low surfactant and polymer concentrations. It is likely that this is a consequence of the crudeness of the model, particularly of the entropy expression connected with the surfactant molecules, as discussed above. One must also keep in mind that effects such as saturation of a polymer

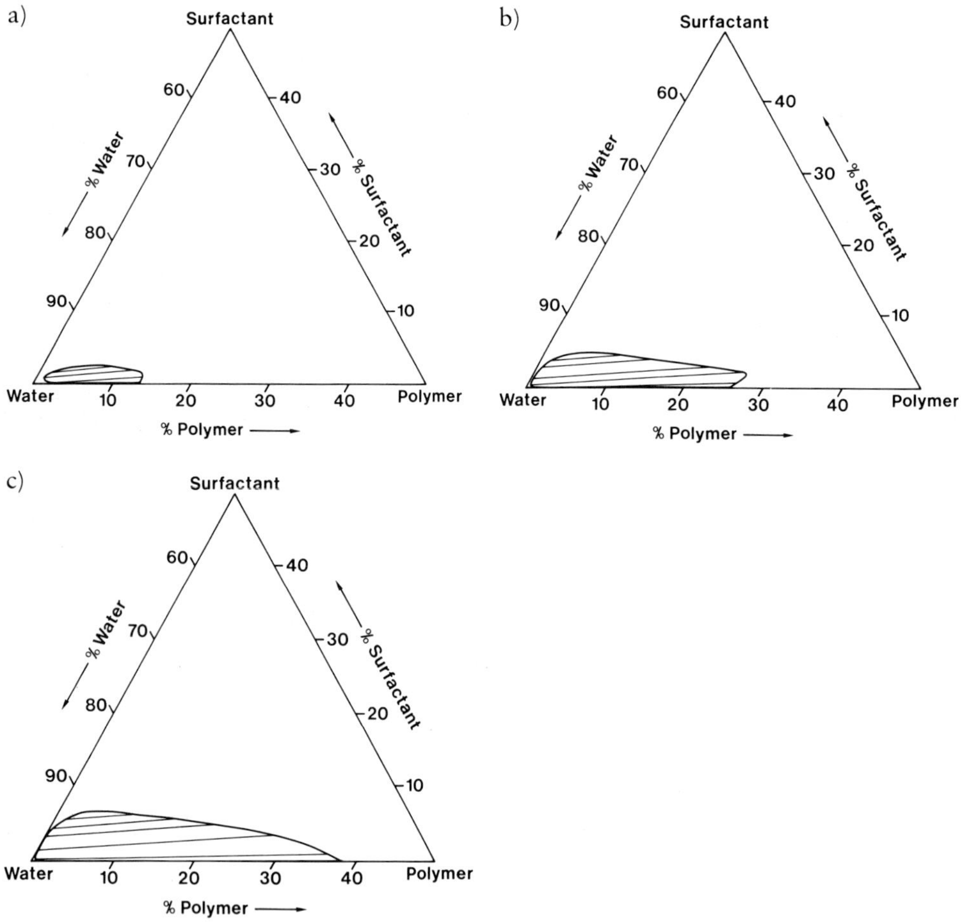

Fig. 4a, b, c. Phase diagram calculated as in Fig. 2 but with $w_{13} = -4927.1$ J/mol for 17 °, 33 °, and 49 °C

with micelles cannot be treated in a mean field model such as the Flory–Huggins model. It is also likely that, in this region, and, especially, close to the clouding temperature, the size of the micelles bound to the polymer is strongly affected by the presence of the polymer, an aspect not considered by the model calculations.

In the previous work [21], we explained the observed phase behavior as a consequence of a strong attraction between the polymer and the surfactant. This will induce a large cloudpoint depression. If, however, the additive prefers to interact with the polar conformation of the polymer and, moreover (and this is most important), it prefers to interact with both the solvent and the different conformers relative to itself, then first a decrease and then an increase in CP are predicted by the model when larger amounts of surfactant are added to the system.

There is, however, one main difference between the theoretically predicted phase diagram and the experimental one, and that is the shape of the upper part of the two phase boundary. In the theoretical phase diagram this is a curved line, whereas, experimentally, it is a fairly straight line up to a polymer concentration of 3–5%, depending on temperature. This indicates that a polymer molecule at a given temperature can, at the most, bind a fixed amount of surfactant micelles before free micelles can start to form. Such a detailed description is not possible to include the crude theoretical model.

Acknowledgement

This work was supported by a grant from the Swedish Board of Technical Development. The work has benefitted from discussions with Dr. A. Carlsson, and part of the experimental work was performed by Ingegerd Lind.

The EHEC polymer was kindly provided by Lars Andersson at Berol Nobel AB.

References

1. Tiddy GJT (1980) Physics Report 57:1
2. Hall DG (1985) J Chem Soc Farady Trans 1 81:885
3. Hayakawa K, Kwak JCT (1991) In: Rubingh D, Holland PM (eds) Cationic Surfactants: Physical Chemistry. Surfactant Series, Dekker cop. New York, p 189
4. Ruckenstein E, Huber G, Hoffmann H (1987) Langmuir 3:382
5. Hayakawa K, Ohta J, Maeda T, Satake I, Kwak JCT (1987) Langmuir 3:37
6. Shirahama K, Masaki T, Takashima K (1985) In: Dubin (ed) Microdomains in Polymer Solution. Plenum Publishing Corporation, p 299
7. Delville A, Laszlo P, Schyns R (1986) Biophysical Chemistry 24:121
8. Tondre C (1985) J Phys Chem 89:5102
9. Oakes J (1974) J Chem Soc Farady Trans 1 70:2200
10. Goddard ED (1986) Colloids and Surfaces 19:255
11. Goddard ED (1986) Colloids and Surfaces 19:301
12. Satake I, Yang JT (1976) Biopolymers 15:2263
13. Satake I, Hayakawa K, Komaki M, Maeda T (1984) Bulletin of the Chemical Society of Japan 57:2995
14. Saito S (1987) In: Schick MJ (ed) Nonionic Surfactants: Physical Chemistry. Surfactants Series 15:881
15. Dubin PL, Thé SS, McQuigg DW, Gan CH (1988) Langmuir 5:89
16. Skerjanc J, Kogej K, Vesnaver G (1988) J Phys Chem 92:6382
17. Brackman JC, Van Os NM, Engberts JBFN (1988) Langmuir 4:1266
18. Carlsson A, Karlström G, Lindman B (1986) Langmuir 2:536
19. Carlsson A, Karlström G, Lindman B, Stenberg O (1988) Colloid and Polym Sci 266:1031
20. Carlsson A, Karlström G, Lindman B (1989) J Phys Chem 93:3673
21. Karlström G, Carlsson A, Lindman B (1990) J Phys Chem 94:5005
22. Thalberg K, Lindman B, Karlström G (1991) J Phys Chem 95:6004
23. Cabane B (1977) J Phys Chem 81:1639
24. Shirahama K (1974) Colloid Polymer Sci 252:978
25. Longwell J, Maniece WD (1955) Analyst 80:167
26. Vogel AI (ed) (1962) Quantitative Inorganic Analysis 3rd. Longman, London, p 275
27. Karlström G (1985) J Phys Chem 89:4957
28. Viti V, Zampetti P (1973) Chem Phys 2:223; Viti V, Indovina PL, Podo F, Radics L, Nemety G (1974) Mol Phys 27:541
29. Podo F, Nemety G, Indovina PL, Radics L, Viti V (1974) Mol Phys 27:521
30. Kjellander R, Florin E (1981) J Chem Soc Farady Trans 1 77:2053
31. Samii AA, Lindman B, Karlström G (1990) Prog Colloid Polym Sci 82:280; Samii AA, Karlström G, Lindman B (1991) Langmuir 7:1067; Wärnheim T, Sjöberg M (1989) Colloid Interface Sci 131:402.
32. Goldstein RE (1984) J Chem Phys 80:5340; Hirschfelder J, Stevenson D, Eyring H (1937) J Chem Phys 5:896

Authors' address:

G. Karlström
Theoretical Chemistry
Chemical Center
University of Lund
P.O.B. 124
S-221 00 Lund
Sweden

Effect of electrolyte on the depletion and structural forces in a micellar system*)

P. Kékicheff and P. Richetti[1])

Department of Applied Mathematics, Research School of Physical Sciences, Australian National University, Canberra, Australia
[1]) C.R.P.P.-C.N.R.S., Château Brivazac, Avenue Schweitzer, Pessac, France

Abstract: The force as a function of separation is measured between two mica surfaces coated with adsorbed bilayers of cetyltrimethylammonium bromide (CTAB) and immersed in aqueous ionic micellar solutions of CTAB at different ionic strengths. At short separations, the interaction profile is a double-layer repulsion, arising solely from the dissociated counterions, free CTA^+, and added salt. At larger separations, the repulsion is reduced and a secondary minimum in the interaction potential due to the depletion of the micelles from the diffuse double-layer is observed. Finally, at further separations, structural effects superimpose an oscillatory force profile. Addition of electrolyte reduces the thickness of the diffuse double-layers. In addition to the shift of the position of the secondary minimum towards smaller separations, the depletion and structural contributions are depressed progressively as the hard-sphere effective volume of micelles decreases. Finally, at high ionic strength, the structural effects are completely removed, but not the depletion one, as the micelles become highly anisotropic.

Key words: Direct micelles; depletion; structural forces; surface force apparatus

Introduction

The question of how a solvent (simple liquid, complex fluid, polymer solution, supermolecular fluid, etc.) mediates the interaction between a pair of macroparticles is a central problem of colloid science. When the separation between the bodies is large compared to the other length scales in the system the interaction will be characterized by bulk solvent parameters, and treatments of the particle interactions are frequently based on the assumption that the suspending medium can be treated as a continuum [1–3]. Such a continuum model must eventually break down at shorter separations between the particles as the discrete molecular nature of the solvent medium becomes apparent.

Specifically, the finite size of the objects constituting the suspending medium results in a density profile that is modified in the vicinity of solid surfaces. For a nearly spherical molecular solvent, the density profile is an oscillating function of the distance normal to the solid surface and converging to the bulk density after several diameters [4]. For non-adsorbing polymers in free solution a solid surface will be surrounded by a layer of different density in which the configurational restrictions create a solvent region that is depleted in polymer [5]. In both cases (inter alia) the overlap of such interfacial layers leads to a disjoining pressure, giving rise to additional interactions between surfaces. In pure molecular solvents, this excess force is often referred to as the solvation force [6]. The

*) This work was presented at the XI Scandinavian Symposium on Surface Chemistry, Bergen, Norway (26–28 June, 1991)

interaction exhibits spatially decaying oscillations between attraction and repulsion as a result of the tendency of the solvent molecules to pack in layers near the surfaces, with a periodicity close to a molecular diameter and with a range of a few solvent diameters [7, 8]. For non-adsorbing polymer solutions, the osmotic pressure vanishes in the interlayer of completely depleted macromolecules, so that the disjoining pressure is negative, and results in an attractive interaction between surfaces over a range of 1 gyration diameter [5, 9–11].

In order to simplify matters, let us now restrict ourselves to the case of two hard-sphere solute particles (both with diameter d_1) immersed in a solvent of smaller hard-spheres (all with diameter d_2). As mentioned above, structural and depletion contributions to the overall interaction can be interpreted in terms of a break in the isotropy of the solvent density profile when the interfacial layers of the two large spherical particles overlap. According to Attard [12], these effects, mediated by the particles (2) of the solvent, give rise to an interaction between two solutes particles (1), related to the density n_c of solvent spheres contacting one of the solute spheres:

$$f(r) = -\frac{1}{2}\pi k_B T(d_1 + d_2)^2 \times \int_0^{\pi} d\theta \sin\theta \cos\theta \, n_c(\theta, r) \,, \quad (1)$$

where the angle θ positions the solvent spheres at contact relative to the axis linking the centers of the two solute particles, a distance r apart. Note that the contact density n_c is a function both of position around the solute particles and of separation. The major difficulty is to determine $n_c(\theta, r)$, and in most cases it is possible only by making approximations or performing numerical calculations [12]. For sufficiently concentrated solutions of spheres (2) the force given by Eq. (1) has an oscillatory profile with a period close to d_2 and with a magnitude that increases as the contact position of the (1) surfaces is approached. On the other hand, the Asakura–Oosawa expression [5] for the depletion-associated attractive contribution to the interaction can be derived [12] from the general expression given by Eq. (1). Indeed, in the limit of high solute dilution, with non interacting solvent particles, the contact density n_c can be considered as equal to the bulk density, so that its variation is only a function of θ. Hence, $n_c(\theta) = n_{\mathrm{bulk}}$ when $\cos\theta < r/(d_1 + d_2)$ and zero otherwise, and the force between the two particles (1) becomes purely attractive when $D = r - d_1 < d_2$ and a function of the volume fraction $\Phi_2 = \pi d_2^3 n_2/6$ (n_2 is the number density of particles (2)):

$$f(r) = -\frac{3}{2}[(d_1 + d_2)^2 - r^2] k_B T \frac{\Phi_2}{d_2^3} \qquad d_1 < r < d_1 + d_2 \quad (2)$$

The analysis given above considers only pure, excluded volume interactions between particles. It does not include any long-range repulsion that may provide a barrier to the expulsion and packing of the solvent particles (2) in close proximity to the solute surfaces (1). If such barriers exist, e.g., electrical double-layer repulsion, then they would clearly modify the extent and magnitude of the depletion and structural contributions to the overall interaction. The range of these latter interactions may be generalized from the situation of a system of uncharged particles and walls. In this description, the charged walls of the solute particles (1) and the charged particles (2) of the suspending medium behave as a soft system with effective hard-sphere diameters $d_1' = d_1 + 2\delta$ and $d_2' = d_2 + 2\varepsilon$, where δ and ε account for the supply of the electric field around each particle (1) and (2), respectively. The double-layer interaction and depletion/structural forces are considered as purely additive interactions – the repulsive electrostatic barrier shifts the occurrence of these latter contributions towards larger separations between the surfaces of the solute particles (1), in comparison to the situation of a pure hard-sphere system. The assumption of additivity, at least for depletion and electrostatic interactions, is supported by the experimental observation that macromolecules which behave as co-ions do not participate in the double-layer interaction and are expelled from the interlayer between two charged surfaces [13, 14].

Recently, such a simple analysis has enabled us to interpret the measured interaction [15] in aqueous micellar solutions using a surface force apparatus [16, 17]. In these experiments, the large solute particles (1) are modeled by the two mica surfaces of the device, while micelles represent the small dispersed particles (2). The two crossed cylinders (mean radius R) of molecularly smooth mica surfaces are geometrically equivalent to two spheres of radius $2R$ approaching each other when the surface separation D is much smaller than the radii

($D \ll R$). As both the walls (adsorbed bilayer on each mica surface [15]) and the particles (micelles) are charged, and are surrounded by a diffuse outer region of ions and solvent characterized by "thickness" δ and ε, respectively, they behave as a soft system with effective hard-sphere diameters $d'_1 = d_1 + 2\delta$ ($d_1 = 4R$, see above) and $d'_2 = d_2 + 2\varepsilon$. In the limit where $d_1 = 4R \gg d_2, \delta$, and ε, the attractive depletion contribution (Eq. (2) with $r = d_1 + D$) to the total interaction is linear in separation D:

$$f(D)/R = -2\pi(d'_2 + 2\delta - D)\,\alpha n_2 k_B T\,, \quad (3)$$

where the osmotic coefficient $\alpha(\Phi_2)$ is introduced to account for the interactions between the micelles. Equation 3 is an Asakura–Oosawa-like [5] approximation, assuming that the enhancement of the contact density in the wedges that remain between the two curved walls outside the depletion overlap is considered as negligible. This assumption is reasonable as long as the structural effects are weak around the boundaries [12]. Nevertheless, the same structural contribution adds a concentration-dependent constant to Eq. (3) since the free energy is no longer necessarily zero when the depletion attraction first occurs at $D = d'_2 + 2\delta$. Note that this surface separation defines the new range of the interaction for the soft system.

In the charged micellar system outlined above, at least three interactions that contribute to the total interactive potential are sensitive to the ionic strength: the double-layer interaction, the depletion, and the structural contributions. Indeed, these two latter interactions are both affected by the means of the effective hard-sphere diameters which alter the effective volume fractions and, consequently, the contact density $n_c(\theta, D)$. This prompted an extension of our previous study [15] with the investigation of the influence of free added salt on the interactions in micellar solutions. We report preliminary results here.

The paper is organized as follows. In section 2, we specify the experimental conditions under which the results have been obtained. In section 3, we first review our previous results with micellar solutions free of salt, focusing on the evolution of the force profile when the surfactant concentration is increased. Then, we describe how the interaction profile at a fixed surfactant concentration is changed when the ionic strength is increased. In section 4, we discuss the salting-out action of the electrolyte on each of contributions of the overall interaction. Finally, we comment on some consequences for the stability of colloids in presence of smaller macromolecule suspensions.

Experimental

Cetyltrimethylammonium bromide (CTAB, or *n*-Hexadecyltrimethylammonium bromide $CH_3(CH_2)_{15}N^+{-}(CH_3)_3Br^-$) was purchased from Kodak and was used without further purification. Water was distilled twice. Potassium bromide, used to change the ionic strength of the solution, was purchased from Pronalys (analar grade) and was roasted at 500 °C for 12 h before use. The range studied extended up to 0.1 M. The ionic strength was varied by adding concentrated salt solutions of known molarity into the apparatus and mixing to the desired concentration. All the experiments were carried out at 25 °C above the Krafft temperature of the solutions.

The latest version of the surface force apparatus was used [17]. The surface separation is controlled by the expansion of a piezoelectric crystal. The instrument allows the force F between the surfaces (of mean radius of curvature R) to be measured to within 10^{-7} N and the surface separation D to within 0.2 nm by an optical-interference technique. Since the force is measured with a cantilever spring, the mechanical system reaches an unstable region whenever the gradient of the force $\partial F/\partial D$ is larger than the spring constant [16]. At the distances where such instabilities occur the surfaces will jump to the next mechanically stable region.

Results

Comparison of the contact position at equilibrium in micellar solutions with that of bare mica indicates that both mica surfaces have an adsorbed CTAB bilayer, the thickness of which is independent of salt concentration and in direct agreement with previous observations [13, 15].

Without added salt

Direct force-distance measurements at three different surfactant concentrations are reported in Fig. 1. The force F is normalized by the radius of curvature R ($R \approx 1$–2 cm) of the surfaces. In the absence of added salt, the force profile may exhibit

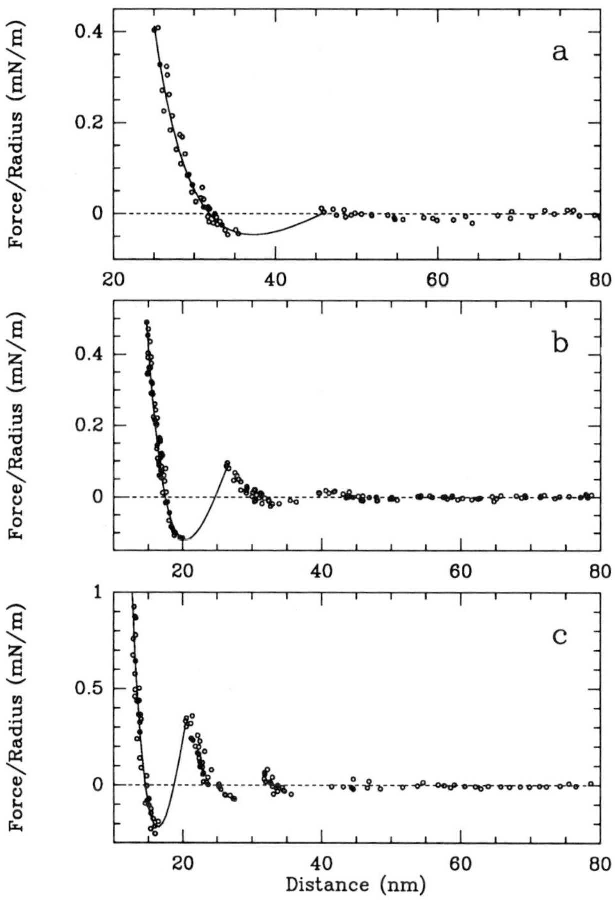

Fig. 1. a) Measured force F (normalized by the mean radius of curvature R of the surfaces) as a function of surface separation, D, between crossed mica cylinders, each coated with an absorbed bilayer of CTAB (thickness about 3.2 nm), and immersed in a micellar solution of CTAB (volume fraction $\Phi_2 \approx 0.009$) in the absence of salt. The attractive minimum around 38 nm is due to depletion of micelles, while the exponential repulsion at smaller separations arises from double-layer forces due to the dissociated counterions and free CTA^+ ions only. The solid line is the best-fit numerical solution (Table 1) to a force-distance profile sum of a decaying exponential and a linear function of the separation over the range 6.3 nm (bilayer-bilayer contact) and 46 nm ($\approx$end of the secondary minimum). b) Same as in a) but for $\Phi_2 \approx 0.03$. Note the occurrence of one oscillation at large separations beyond the depletion minimum. This structural effect is not described by the numerical fit. c) Same as in a–b) but for $\Phi_2 \approx 0.073$ where two oscillations due to structural forces turn up.

three distinct regimes, depending on the surfactant concentration. At small surface separations D the interaction is always exponentially repulsive as expected for a double-layer interaction (Fig. 1a), while at large separations a weak attraction characterized by a shallow well, followed by an oscillatory profile at further separations are revealed (Fig. 1c). The resolution of the surface force apparatus allows unambiguous identification of the last two contributions only at micellar fraction larger than $\Phi_2 \approx 0.007$ (about 25 times the critical micelle concentration, CMC) for the attractive minimum (Fig. 1a) and $\Phi_2 \approx 0.015$ for the oscillatory profile (Fig. 1b). Note the occurrcnce of intrinsic unstable regimes, without data, when $\partial F/\partial D$ is larger than the stiffness of the cantilever spring [16]. According to the interpretation of our previous study [15] and summarized in section 1, the highly charged micelles (effective diameter d'_2) between the two charged walls (adsorbed bilayer on each mica of effective diameter $d'_1 = 4R + 2\delta$) behave as co-ions expelled from the region of close double-layer overlap. The resulting depletion mechanism (attractive force) is in competition with the repulsive double-layer interaction in a similar fashion to the way that DLVO theory considers the summation of electrostatic and dispersion forces. From this competition emerges a secondary minimum in the interaction potential, as shown in Fig. 1.

The solid lines in Fig. 1 represent the numerical fit to the experimental measured data using a Levenberg–Marquard method [18] for a force-distance profile described by a law $f(D)/R = A_1 \exp(-\kappa D) + A_2 D + A_3$ for $D \leq d'_2 + 2\delta$, where the electrostatics is treated by a linearized Poisson–Boltzmann contribution, the depletion attraction is spatially linear since $d'_2 \ll R$ (Eq. (3)), and the van der Waals contribution is neglected at these separations. Table 1 gives the parameters of such a fit, which is a three-independent parameter fit since the magnitude of the force where the surfaces jump inwards (secondary minimum) is imposed on the calculation [15].

At high surfactant concentration (Fig. 1c) the oscillatory profile, which is superimposed at large separations, is due to structural contributions to the overall interaction; they reflect some particle packing effects favoring an integral number of particle layers between the two interacting walls, as it occurs in a pure molecular solvent, despite the dynamic structure and short lifetime of the micelles.

With added salt

Direct force-distance measurements at different added 1:1 electrolyte up to 0.1 M are reported in

Table 1. Parameters of the fits (Fig. 1–2) and decay-screening length calculated with Eq. (4).

Φ_2	κ_0^{-1} (eqn. 4) (nm)	κ^{-1} (nm)	A_2 (10^5 N/m^2)	A_3 (mN/m)
0.009[a])	4.9	4.8	0.10096	– 0.4701
0.030[a])	2.9	3.3	0.63439	– 1.6240
0.073[a])	1.9	2.4	2.58577	– 5.0810
0.030[b])	2.1	2.0_6	0.26425	– 0.62285

[a]) [KBr] = 0; [b]) [KBr] = 0.02 M

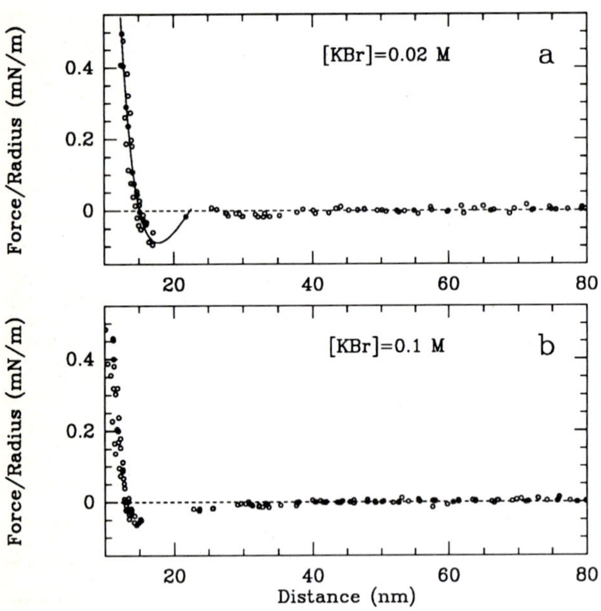

Fig. 2. Same as in Fig. 1 but in presence of potassium bromide a) [KBr] = 0.02 M. Note the nearly complete disappearance of the structural oscillation and the shift of the secondary minimum towards smaller surface separations b) [KBr] = 0.10 M. The depletion attraction is now the only additional contribution to the double-layer repulsion. No fit has been attempted as the micelles are highly anisotropic and polydisperse at this high ionic strength (see text).

Fig. 2. The surfactant concentration was kept constant at 3.05 wt.%, about 100 times the CMC, corresponding to a micellar volume fraction Φ_2 of about 0.03 in the absence of added salt.

At low ionic strength (added salt concentration below 0.01 M), the force-distance profile presents the same main features as without salt illustrated by Fig. 1b. It is only at about 0.02 M of added salt that the change in ionic strength becomes sufficiently significant to alter the force profile (Fig. 2). As expected, the range of the double-layer repulsion progressively decreases as the salt screens the electrostatic interaction and, consequently, the position of the secondary minimum moves towards shorter separations. Simultaneously, the depth of the secondary minimum decreases slowly. But the main feature is the disappearance of the structural oscillations (Fig. 2). Note that for 0.1 M salt, experimental limitations in accuracy preclude the observation of an eventual repulsive maximum before the secondary minimum.

Discussion

A number of different interactions contribute to the overall force-distance profile observed.

Electrostatic contribution

Direct micelles are generally believed to consist of a liquid-like core of monomer hydrocarbon chains, surrounded by a region which contains charged head groups plus some neutralizing counterions. A diffuse outer region of ions and solvent surrounds each micelle. Under appropriate conditions and, particularly, in absence of salt (or at low ionic strength), an approximately constant micellar size can be assumed (the polydispersity is neglected here) and the system can be modeled as a suspension of spherical particles of diameter d_2, each with a net charge Q distributed uniformly over its surface. These particles diffuse in an electrolyte which is characterized by a Debye inverse screening length κ_0.

In these highly asymmetric electrolytes the measured double-layer decay lengths κ_0^{-1} appear to be due entirely to the free surfactant ions CTA^+, and both added salt plus dissociated counterions (bromide) released by the surfactant and the micelles, with no apparent contribution from charged micelles as observed before [13–15]. This decay-length is well described by the following equation [13]:

$$\kappa_0^2 = \frac{4\pi q^2}{\varepsilon k_B T}\left\{2\rho_{cmc} + \frac{(\rho_s - \rho_{cmc})}{N} Q + \rho_{K^+} + \rho_{Br^-}\right\}, \quad (4)$$

where ρ_s is the surfactant concentration, ρ_{cmc} that at the CMC, N the mean micellar aggregation number, Q the mean net effective charge of the micelle, ρ_{K^+} the concentration of potassium ions, and ρ_{Br^-} that of bromide brought in excess by the external salt.

Addition of an electrolyte in a solution usually compresses the diffuse double-layer (i.e., decreases the Debye screening length). In a micellar solution, addition of an external salt may also influence the monomer-micelle equilibrium and the binding equilibrium of the counterions to the surfactant (change of micellar ionization fraction). As a result, the self-assembled aggregates may change their size (growth of the micelle) and/or their shape (increasing anisotropy). The increase in the ionic strength can lead to the so-called sphere-to-rod transition where the quasi-spherical micelles turn into rod-like (spherocylindrical) aggregates [19–23].

Although the CMC of CTAB, which is 9×10^{-4} M in water, is shifted to lower values with electrolyte concentration [24], the change in the relative contribution of the free CTA^+ in solution to the Debye screening length (Eq. (4)) is negligible as the total surfactant concentration is here 7.5×10^{-2} M.

For CTAB with no added salt, different authors [19, 25–28] claim N values in the range 75–105, with an average value of 90. Taking this value and knowing the molecular weight $M = 364.5$ and the partial specific volume of the surfactant [26] $\bar{v} = 0.98$ cm^3/g, the diameter of the "dry" spherical micelle is $d_2 = (6NM\bar{v}/\pi N_{Av})^{1/3} \approx 4.7$ nm. At low ionic strength (0.01–0.04), despite the fact that the mean micellar aggregation number increases (N is reported [28–29] to be constant over this range and approximately 50% larger than the value at zero salt concentration), the micellar volume fraction is grossly unchanged since the anisotropy of the micelles remains very low over this salt concentration [28]. However, at higher ionic strength, the micelles grow very rapidly and become highly anisotropic [23, 28, 29]. In sodium bromide solutions the threshold salt concentration inducing the sphere-rod transition of CTAB micelle is between 0.04 and 0.075 M [28], and the mean micellar aggregation number N becomes a very rapidly increasing function of the salt concentration. Nevertheless, it is emphasized that the physical parameter important to a change of the Debye screening length due to the relative contribution of the dissociated counterions from the micelles (Eq. (4)) is not the mean aggregation number N, but the fractional ionization Q/N of the micelles. Different authors [28–30] report that the binding equilibrium is actually very little affected by a change in the ionic strength, at least to an added salt concentration below 0.1–0.2 M over similar CTAB concentration solutions ($Q/N \approx \text{cst} \approx 0.22$ [28] or 0.16–0.21 [30]). Despite the fact that these studies concern sodium bromide and not potassium bromide as used here, we have assumed that the same feature should occur for potassium bromide solutions as the observations of Candau et al. [29] in 0.1 M KBr are very similar to those of Dorshow et al. [28] in NaBr. Thus, the main change in the Debye screening length appears principally due to the contribution of the external added salt. Table 1 indicates a good agreement between the values of the Debye screening length calculated by Eq. (4) (κ_0^{-1}) with the decay-lengths extracted from the fits to the experimental data (κ^{-1}) at different surfactant and salt concentrations.

van der Waals attraction

A typical Hamaker constant of about 1.5×10^{-20} J for the interaction between mica surfaces coated with adsorbed bilayers across the aqueous micellar solutions [13, 15] indicates that the attractive dispersion forces cannot account for the occurrence of a secondary minimum in the interaction potential, at least for the salt concentrations [KBr] = 0 and 0.02 M. Indeed, for their positions of 20 nm (Fig. 1b) and 17 nm (Fig. 2a) the dispersion concentration would be only 0.014 and 0.023 mN/m, respectively. On the other hand, at high salt concentration (Fig. 2b) the same contribution would be about one-third of the minimum depth.

Attractive depletion interaction

Because the highly charged micelles between two charged walls behave as co-ions expelled from the region of close double-layer overlap, a depletion mechanism must occur [15]. As we have previously demonstrated, Eq. (3) is a good approximation to describe the results in the absence of salt, as long as the surfactant concentration is not too high. Nevertheless, several terms in Eq. (3) depend on the electrolyte concentration.

As a first analysis, let us assume that the aggregation number and the bare micellar radius are constant irrespective of the ionic strength. The added salt shrinks the diffuse double-layers and, consequently, the effective hard-sphere diameters and the corresponding volume fraction are reduced as well. Firstly, addition of salt must reduce the range of the depletion attraction ($d_2' + 2\delta$). This range is roughly given by the separation at the last stable point before the inward jump (due to the mechanical instability of the device) towards the secondary minimum. The decrease of the depletion range is indeed quite clear at 0.02 M (Fig. 2a) compared to the free salt solution (Fig. 1b). At higher salt concentration (Fig. 2b), the micelles are no longer spherical [28–30] and such a simple analysis cannot be made.

Secondly, the osmotic coefficient α is a function of the effective hard-sphere volume fraction Φ_2'. Indeed, $\alpha(\Phi) = P_{osm}/nk_BT = (1 + \Phi + \Phi^2 - \Phi^3)/(1 - \Phi)^3$, according to the Carnahan–Starling [31] approximation, and by use of the geometric limit of planar walls (since $d_1 = 4R \gg d_2$). From the fitting procedure [15] of the experimental data, the osmotic coefficient α is estimated to 4.7 ± 0.1 from $A_2 = 2\pi\alpha n_2 k_B T$ and using an aggregation number of 90 and a micelle diameter around 4.7 nm (see above). Note that we have estimated the uncertainty in the value of the osmotic coefficient to be due only to the scatter of the experimental data, and have not taken into account the variation in the estimates of the aggregation number N according to different authors [19, 25–28]. At [KBr] = 0.02 M the osmotic coefficient is $\alpha = 2.0 \pm 0.5$. The poor estimate arises from the lack of experimental data around the position of the inward jump before the secondary minimum (Fig. 2a). The latter value of α assumes the same mean aggregation number N despite the addition of salt. Now, with a growth of 50% for the micelles (the maximum increase in N according to literature [28–30]) α is then equal to 3.0 ± 0.7. The value for this osmotic coefficient (α between 2.0 and 3.0) compares well with the value obtained for a micellar solution without added salt at a volume fraction $\Phi_2 = 0.009$ where similar magnitude of the structural contribution was observed (Fig. 1a and [15]). In conclusion, from the experimental data a decrease of the osmotic coefficient is qualitatively observable, as expected when the salt concentration is increased. At [KBr] = 0.1 M and at these surfactant concentrations the micelles are no longer spherical, but are large, elongated, and probably flexible [21–23]. These aggregates should overlap, forming a transient network similar to an entangled polymer solution (semi-dilute range) [23, 29]. As mentioned above, the van der Waals interaction is not solely responsible for the secondary minimum (Fig. 2b). Our results suggest that, even in this particular situation with elongated micelles, an important depletion contribution remains. Note that with the same technique and a real non-adsorbing polymer solution, the depletion interaction has been claimed to be so weak as to be unmeasurable, under the resolution of the surface force apparatus [32].

Structural contribution

These structural forces, which have also recently been observed in an uncharged reversed micellar system [33], reflect some particle-packing effects favoring an integral number of particle layers between the two interacting walls, as it occurs in a pure solvent, despite the dynamic structure and short lifetime of such aggregates.

At low ionic strength ([KBr] < 0.02 M), one oscillation is clearly observed in the force-distance profile (Fig. 1b). The corresponding period must be almost equal to the effective hard-sphere diameter of the micelles d_2'. Measured from the two successive maxima in the force-distance profile, we find $d_2' \approx 12.9$ nm. This value appears a little too large as compared to the expected effective diameter deduced from the osmotic coefficient. Indeed, $\alpha = 4.7$ corresponds to an effective volume fraction $\Phi_2' \approx 0.33$ and $d_2' \approx 11$ nm.

Remarkably, when the ionic strength is increased ([KBr] > 0.02 M), the structural oscillations in the force-distance profile disappear (Fig. 2a). This occurs at electrolyte concentrations well below the threshold of the reported sphere-rod transition for the micelles [28], and is a complementary manifestation of the reduction of the osmotic coefficient and, conversely, of the effective hard-sphere volume fraction. However, in the case of higher surfactant concentrations than that reported here at very high ionic strength, in addition to a reduction of the effective volume fraction, other explanations could be given. First, the micelles which have become very anisotropic are also polydisperse [28–30]. In such a situation where particles of different size coexist, it is known that the structural contribution to the interaction due to some packing effect is much reduced [34]. Secondly, as mentioned above, these elongated large micelles are flexible [21–23], and it must be recalled that the surfactant molecules retain a very high mobility in the micelles. This, added to the high conformational mobility of the surfactant alkyl chains makes it possible for the chain packing to readjust very rapidly when a fold occurs in an initially linear part of such an elongated micelle. Finally, the particular experimental geometry of crossed cylinders [16] may complicate the effect.

Concluding remarks

In summary, at low ionic strength the observed deviation from the pure double-layer repulsion is due to the depletion of the micelles in the double-layer overlap. The modeling force, which is in fair agreement with the measured profile, is based on the simple assumption that all the contributions to the interaction are purely additive. Direct force measurements between charged CTAB bilayers in micellar solutions at large volume fractions indicate that depletion and structural contributions superimpose on top of classical double-layer interactions. At high ionic strength the situation is more complicated, as the micelles can no longer be considered as spherical with a low polydispersity. The force profile has thus only been interpreted qualitatively. This difficulty can be avoided in order to complete the present preliminary study by using surfactants of shorter chain length that form spherical micellar structure up to high ionic strength. We have chosen the surfactant dodecyltrimethylammonium bromide (DTAB) which gives this property (the study is underway).

Nevertheless, the case of very anisotropic micelles is of most interest. At these surfactant and salt concentrations, these large, elongated micelles of high flexibility should overlap, forming a transient network similar to an entangled polymer solution (semi-dilute range) [23, 29]. Such an analogy between a solution of giant one-dimensional micelles and a polymer solution could be used to shed light on the behavior of these complex systems, where these living polymers would behave as long linear chain polymers that can break and recombine reversibly. In view of the present results, these large, elongated, flexible micelles seem to be good systems to directly measure depletion interactions, both in dilute and semi-dilute regimes. A direct comparison with non-adsorbing polymer solutions would then be possible. In a forthcoming paper we will report a complete study in the semi-dilute regime where the depletion forces have been undoubtedly observed.

Finally, our results underline the consequences that addition of surfactant micelles can have in a colloid suspension. For instance, the role of emulsifier in causing flocculation and creaming of emulsions has been observed since 1952 [35], but no really plausible explanation was offered before 1983 [36]. Our results confirm the depletion mechanism proposed by different authors in the creaming of oil-in-water emulsions [36, 37] and to account for flocculation of polystyrene latex by surfactant [38]. Thus, just as addition of salt may destabilize a charged suspension, addition of charged micelles may induce flocculation, reflecting the occurrence of a secondary minimum in the interaction potential between two colloids. The addition of an oscillatory potential to the depletion minimum confirms the interpretation of ordered micelle structuring in thin films formed from surfactant solutions, where the stepwise thinning of liquid foam films (so-called stratification) is due to a layer-by-layer destruction of a structure of micelles formed in the film [39]. Because it turns out that the main parameter governing the oscillatory potential is the effective volume of the particles, one can suggest the possibility of several metastable states for a dispersion of particles in relatively low concentrated micellar solutions at different surfactant concentration and ionic strength, leading to the presence of reentrant phase transitions. At some concentrations the repulsive barrier preceding the depletion minimum might be high enough to prevent any destabilization, while at higher concentrations the preceding well, due to structural effects, might be deep enough to favor a new flocculation. In the presence of ionic surfactant a simple fashion to disperse again large charged colloid particles would be to add some salt. It is a quite unusual suggestion in reference to the DLVO theory. Nevertheless, such an observation has already been made by Bibette et al. [40]. The salting-out effect would let down the energy barrier, keeping the particles trapped together by reducing the depletion strength as well the formation of the first macromolecular layer, as shown in Fig. 2a.

Acknowledgments

Fruitful discussions with J. Bibette, C. J. Drummond, and J. L. Parker have been appreciated. P. Richetti thanks the Groupement de Recherche no. 936 for interest in this work.

References

1. Derjaguin BV, Landau L (1941) Acta Phys Chim USSR 14:633
2. Verwey EJW, Overbeek JThG, (1947) Theory of the Stability of Lyophobic Colloids. Elsevier, Amsterdam
3. Mahanty J, Ninham BW (1976) Dispersion Forces. Academic Press, New York
4. see for example: Tarazona P (1985) Phys Rev A31:2672; Kjellander R, Sarman S (1988) Chem Phys Lett 149:102
5. Asakura S, Oosawa F (1954) J Chem Phys 22:1255; (1958) J Polym Sci 33:183
6. Derjaguin BV, Churaev NV (1974) J Colloid Interface Sci 49:249; Marčelja S, Radic N (1976) Chem Phys Lett 42:129; Le Neveu DM, Rand RP, Parsegian VA (1976) Nature (London) 259:601
7. Mitchell DJ, Ninham BW, Pailthorpe BA (1977) Chem Phys Lett 51:257; (1978) J Chem Soc Faraday Trans 2 74:1098, 1116; van Megen W, Snook IK (1979) J Chem Soc Faraday Trans 2 75:1095; Snook IK, van Megen WJ (1979) J Chem Phys 72:2907
8. Horn RG, Israelachvili JN (1981) J Chem Phys 75:1400; Christenson HK (1983) J Chem Phys 78:6906; Christenson HK, Horn RG (1983) Chem Phys Lett 98:45
9. Vrij A (1976) Pure Appl Chem 48:471; Vincent B, Luckham PF, Waite FA (1980) J Colloid Interface Sci 73:508; Sperry PR, Hopfenberg HB, Thomas NL (1981) J Colloid Interface Sci 82:62; De Hek H, Vrij A (1981) J Colloid Interface Sci 84:409; Sperry PR (1984) J Colloid Interface Sci 99:97
10. Joanny JF, Liebler L, de Gennes PG (1979) J Polymer Sci Polymer Phys 17:1073; Feigin RI, Napper DH (1980) J Colloid Interface Sci 74:567; (1980) 75:525
11. Gast AP, Hall CK, Russel WB (1983) J Colloid Interface Sci 96:251; (1983) Faraday Discuss Chem Soc 76:189
12. Attard P (1989) J Chem Phys 91:3072, 3083
13. Pashley RM, Ninham BW (1987) J Phys Chem 91:2902; Pashley RM, McGuiggan PM, Horn RG, Ninham BW (1988) J Colloid Interface Sci 126:569
14. Marra J, Hair ML (1989) J Colloid Interface Sci 128:511
15. Richetti P, Kékicheff P (1992) Phys Rev Lett 68:1951
16. Israelachvili JN, Adams GE (1978) J Chem Soc Faraday Trans 1 74:975
17. Parker JL, Christenson HK, Ninham BW (1989) Rev Sci Instrum 60:3135
18. In Numerical Recipes Cambridge University Press (1986)
19. Reiss-Husson F, Luzzati V (1964) J Phys Chem 68:3504
20. Mazer NA, Benedek GB, Carey MC (1976) J Phys Chem 80:1075; Young CY, Missel PJ, Mazer NA, Benedek GB, Carey MC (1978) ibid 82:1375; Missel PJ, Mazer NA, Benedek GB, Carey MC (1983) ibid 87:1264
21. Porte G, Appell J, Poggi Y (1980) J Phys Chem 84:3105; Porte G, Appell J (1981) ibid 85:2511; Appell J, Porte G, Poggi Y (1982) J Colloid Interface Sci 87:492; Appell J, Porte G (1983) J Phys Lett (Paris) 44:L-689
22. Ikeda S, Ozeki S, Tsunoda MA (1980) J Colloid Interface Sci 73:27; Ozeki S, Ikeda S (1980) ibid, 77:219; Ikeda S, Hayashi S, Imae T (1981) J Phys Chem 85:106; Imae T, Ikeda S (1984) Colloid Polym Sci 262:497
23. Imae T, Kamiya K, Ikeda S (1985) J Colloid Interface Sci 108:215
24. Mukerjee P, Mysels KJ (1971) Natl Stand Ref Data Ser (U.S. Natl. Bur. Stand., Washington, DC), no. 36
25. Tartar HV (1959) J Colloid Interface Sci 14:115
26. Ekwall P, Mandell L, Solyom P (1971) J Colloid Interface Sci 35:519
27. Dorrance RC, Hunter TF (1974) J Chem Soc Faraday Trans 1 70:1572; Aikawa M, Yekta A, Turro N (1979) J Chem Phys Lett 68:285; Lianos P, Zana R (1981) J Colloid Interface Sci 84:100
28. Dorshow R, Briggs J, Bunton CA, Nicoll DF, (1982) J Phys Chem 86:2388; Briggs J, Dorshow RB, Bunton CA, Nicoll DF (1982) J Chem Phys 76:775; Dorshow R, Bunton CA, Nicoll DF (1983) J Phys Chem 87:1409
29. Candau SJ, Hirsch E, Zana R (1984) J Phys (Paris) 45:149
30. Larsen JW, Tepley LB (1974) J Colloid Interface Sci 49:113; Zana R (1980) J Colloid Interface Sci 78:330
31. Carnahan NF, Starling KF (1970) J Chem Phys 53:600
32. Luckham PF, Klein J (1985) Macromolecules 18:721
33. Parker JL, Richetti P, Kékicheff P, Sarman S (1992) Phys Rev Lett 68:1955
34. Van Beurten VP, Vrij A (1981) J Chem Phys 74:2744
35. Cockbain EG (1952) Trans Faraday Soc 48:185; Higuchi W, Okada R, Lemberger A (1962) J Pharm Sci 51:683; Becher P (1965) In: Emulsions; Theory and Practice. Rheinhold: New York

36. Fairhurst D, Aronson MP, Gum ML, Goddard ED (1983) Colloids Surf 7:153; Aronson MP (1989) Langmuir 5:494; Aronson MP (1991) In: Emulsions: A Fundamental and Practical Approach. NATO Advanced Research Workshop, Bergen, Norway
37. Bibette J, Roux D, Nallet F (1990) Phys Rev Lett 65:2470
38. Ma C (1987) Colloids Surfaces 28:1
39. Nikolov AD, Wasan DT (1989) J Colloid Interface Sci 133:1; Nikolov AD, Kralchevsky PA, Ivanov IB, Wasan DT (1989) ibid 133:13; Nikolov AD, Wasan DT, Denkov ND, Kralchevsky PA, Ivanov IB (1990) Prog Colloid Polym Sci 82:87
40. Bibette J, Roux D, Pouligny B (1992) J Phys II France 2:401

Authors' address:

P. Kékicheff
Department of Applied Mathematics
Australian National University
G.P.O. Box 4
Canberra ACT 2601
Australia

P. Richetti
C.R.P.P.-C.N.R.S.
Château Brivazac
Avenue Schweitzer
F-33600 Pessac
France

Progress in Colloid & Polymer Science Progr Colloid Polym Sci 88:18–22 (1992)

Phase behavior of alkanolammonium carboxylates

T. Wärnheim and A. Jönsson[1])

Karlshamns Oils & Fats AB, Karlshamn, Sweden
[1]) Institute for Surface Chemistry, Stockholm, Sweden

Abstract: Partial binary phase diagrams for different alkanolammonium carboxylates with water have been determined. The alkanolamines employed were monoethanolamine, triethanolamine and triisopropanolamine, while the fatty acid has been varied with respect to chain length (C8-C22) and saturation.

The phase diagrams show two features that distinguish them from ordinary soap-water phase diagrams. First, multi-phase regions and miscibility gaps occur at low concentrations of surfactant, in particular, for triethanolammonium carboxylates. Secondly, the stability regions of the liquid crystalline phases are different, and in general smaller, compared to those formed by the corresponding alkali carboxylates. It is suggested that the first feature is due to that alkanolamines are weak bases and, thus, that any surfactant aggregate comprises a few percent of unhydrolyzed acid, while the second feature is attributed to the larger size of the counterion compared to alkali ions, which tends to destabilize ordered structures.

Key words: Alkanolamine; phase diagram; alkanolammonium carboxylate

Introduction

Formulations in cosmetics, as well as in technical areas, very often contain alkanolamines together with fatty acids or soaps [1]. Although ordinary soaps, i.e., alkali carboxylates, remain a closely studied model system from almost any point of view in surfactant science, the effects of added alkanolamines or of alkanolammonium carboxylates have been comparatively little studied, in spite of the importance of their applications.

As for investigations of phase equilibria, a starting point for further studies, phase diagrams for alkanolammonium carboxylates and additives have rarely been presented in the literature. The only systematic efforts seems to be phase diagrams of monoethanolammonium oleate with water and other components, studied by Friberg and coworkers [2–4].

We have thus recently investigated the effect of varying alkyl chain length and molecular structure of the alkanolamine in phase diagrams for a number of alkanolammonium carboxylates with water [5, 6]. In this paper, a brief review and an extension of the work is reported.

Material and methods

Chemicals: Monoethanolamine (MEA) (p.a., BDH), triethanolamine (TEA) p.a., Merck), triisopropanolamine (TPA) (95%, Aldrich), octanoic acid (C8) (puriss, Fluka), dodecanoic acid (C12) puriss, Fluka), tetradecanoic acid (C14) (puriss, Fluka), hexadecanoic acid (C16) (99%, Kebo), octadecanoic acid (C18) (puriss, Merck), cis-9-octadecenoic acid (oleic acid) (C18 : 1) (99.5% + , Fluka), eicosanoic acid (C20) (puriss, Fluka), cis-13-docosenoic acid (erucic acid) (C22 : 1) (90%, Riedel-de-Haën) were used as received. Maximum water content of the alkanolamines was less than 0.5%. Alkanolammonium carboxylates, denoted as, for example, TEA-C8 for triethanolammonium octanoate, were prepared by mixing either molten or dissolved components in a 1 : 1 molar ratio. Melting points, whenever checked, were comparable to literature values. IR were employed in order to ensure that no formation of fatty amides occurred with MEA. Water was twice distilled.

Phase diagrams: Phase diagrams were determined as described in previous work [5, 6].

Critical micelle concentrations (CMCs): CMCs were determined from surface-tension measurements using the du Noüy ringmethod [7], or from density measurements using an Anton Paar densitometer.

Counterion association: Determinations of the counterion association to micelles was performed using the PGSE FT NMR-method, as described by Stilbs [8].

Results and discussion

The phase diagrams for the different monoethanolammonium carboxylates with water are shown in Fig. 1. MEA-C8 (Fig. 1a) forms a continuous solution phase L. A hexagonal phase E is stable between 54 and 79 wt% surfactant at 20 °C; it decreases in stability with temperature and finally melts at 31 °C. MEA-C12 (Fig. 1b) reveals a phase behavior more reminiscent of normal soaps: a hexagonal phase E precipitates at 28 wt% surfactant at 20 °C while a lamellar phase D is stable at higher concentrations, approaching 60 wt%. The temperature stability of the liquid crystalline phases is limited, however; they both melt at a maximum 90 °C. MEA-C18 : 1 shows a large miscibility gap with an equilibrium between a dilute aqueous phase and a lamellar phase (Fig. 1c). Finally, the phase diagram of MEA-C22 : 1 (Fig. 1d) is very similar to MEA-C18 : 1 although the temperature stability of the lamellar phase has increased.

An extensive number of triethanolammonium carboxylates was investigated. TEA-C8 (Fig. 2a) forms no liquid crystalline phases with water at any concentration in the investigated temperature

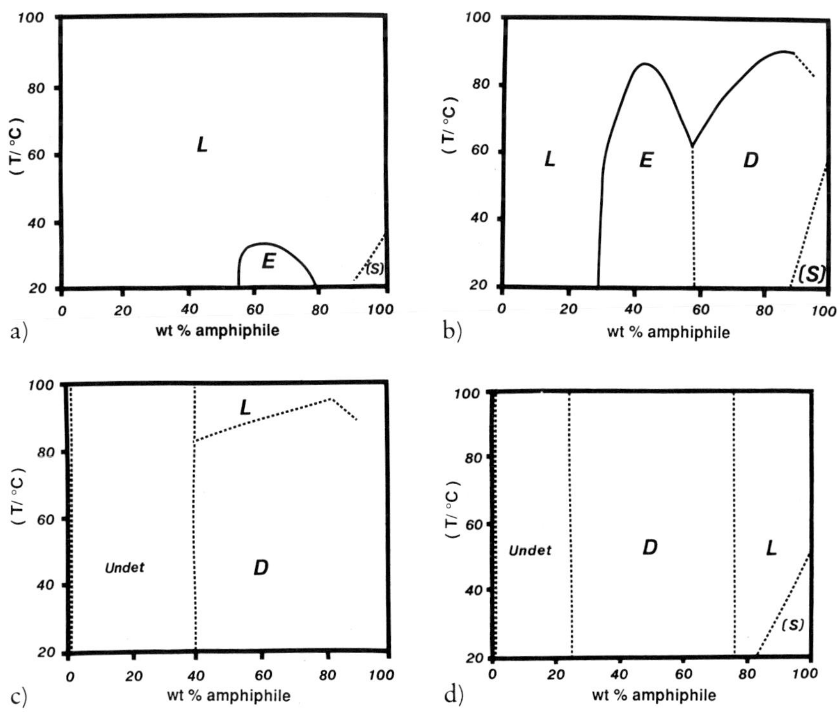

Fig. 1. Partial phase diagrams for the systems monoethanolammonium carboxylate (MEA-CX) and water at varying temperature. Notations in this and in the following phase diagrams are according to Ekwall [9]. D denotes a lamellar phase, E a hexagonal phase, L a solution phase, and (s) solid surfactant. The regions denoted "undetermined" in the phase diagrams comprise multi-phase areas, as discussed in detail in the text. a) Monoethanolammonium octanoate (MEA-C8)–water. (Redrawn from [6].) b) Monoethanolammonium dodecanoate (MEA-C12)–water. (Redrawn from [6].) c) Monoethanolammonium oleate (MEA-C18:1)–water. (Redrawn from [6].) d) Monoethanolammonium erucate (MEA-C22:1)–water

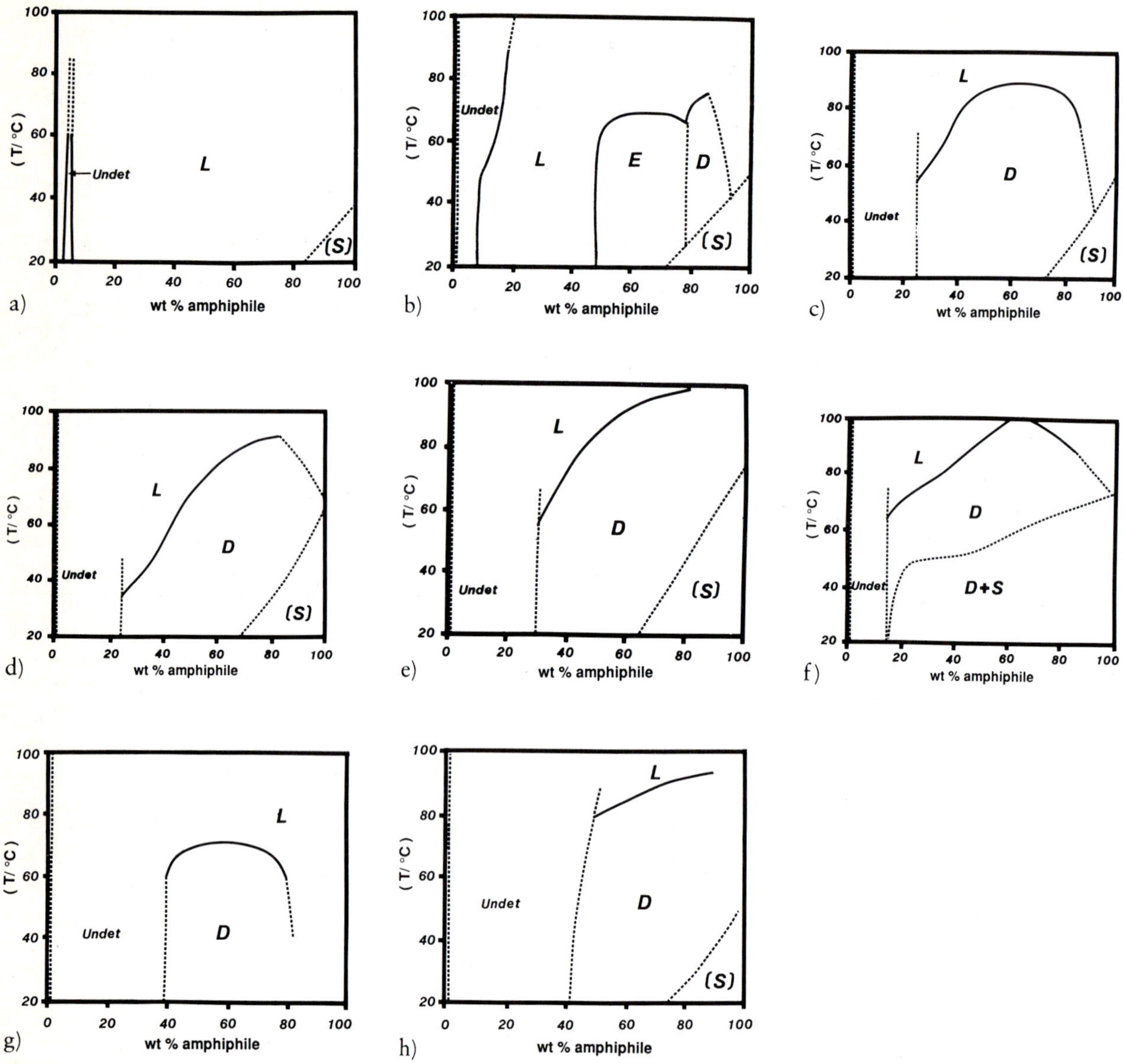

Fig. 2. Partial phase diagrams for the systems triethanolammonium carboxylate (TEA-CX) and water at varying temperature. Notations as in Fig. 1a. a) Triethanolammonium octanoate (TEA-C8)–water. (Redrawn from [6].) b) Triethanolammonium dodecanoate (TEA-C12)–water. (Redrawn from [6].) c) Triethanolammonium tetradecanoate (TEA-C14)–water. d) Triethanolammonium hexadecanoate (TEA-C16) water. e) Triethanolammonium octadecanoate (TEA-C18)–water. f) Triethanolammonium eicosanoate (TEA-C20)–water. g) Triethanolammonium oleate (TEA-C18:1)–water. (Redrawn from [6].) h) Triethanolammonium erucate (TEA-C22:1)–water. (Redrawn from [5])

interval, in contrast to, for example, sodium or potassium octanoate [9]. Another conspicuous feature is the miscibility gap in the aqueous region between 3 and 6 wt% surfactant at 20 °C. The miscibility gap is more extensive with TEA-C12 (Fig. 2b); it occurs between 0.1 and 7 wt% surfactant at 20 °C. It increases with increasing temperature; in fact, a solution of 10 wt% TEA-C12 is stable at room temperature, but phase-separates with increasing temperature, in analogy with the clouding of nonionic ethoxylated surfactants. A hexagonal and a lamellar phase are stable in the TEA-C12 system, although the existence regions are smaller compared to, for example, MEA-C12.

Further increasing the chain length leads to a more extensive miscibility gap and destabilizes the hexagonal phase. For TEA-C14 (Fig. 2c), TEA-C16 (Fig. 2d), TEA-C18 (Fig. 2e) and TEA-C20 (Fig. 2f) with water, respectively, there is only a lamellar phase D. The lamellar phase increases somewhat in temperature stability and there is also a gradual increase in the melting-point curve of the solid surfactant. The derivatives of unsaturated fatty acids, TEA-C18 : 1 (Fig. 2g) and TEA-C22 : 1 (Fig. 2h) show similar phase diagrams with water: miscibility gaps and lamellar phases stable at higher surfactant contents.

Finally, the phase diagram of TPA-C12 has been determined (Fig. 3). These systems shows a miscibility gap at low concentrations of surfactant, at 20 °C between 0.1–6 wt%, while a hexagonal phase E precipitates at higher concentration of surfactant and is stable between 38–73 wt%, but melts below 40 °C.

Let us consider the different features in the phase diagrams of Figs. 1–3.

There is a clear effect of the size of the counterion at constant alkyl chain length of the carboxylate. For MEA-C8 (Fig. 1a) and TEA-C8 (Fig. 2a) it is clear that an increase in counterion size causes the hexagonal phase E to disappear. The same trend for the liquid crystalline phases is even more evident when comparing MEA-C12 (Fig. 1b), TEA-C12 (Fig. 2b) and TPA-C12 (Fig. 3). The extensions of the liquid crystalline regions decrease with increasing counterion size, and the preferred curvature of the aggregate increases only an E-phase occurs with TPA-C12. As already pointed out, the preference for an increased curvature of the surfactant aggregates in octanoate and dodecanoate systems can be noted also in comparison with the corresponding alkali carboxylates [9].

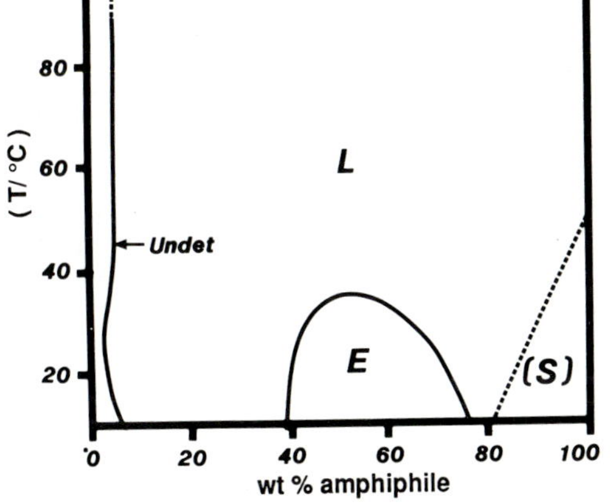

Fig. 3. Partial phase diagram for the system triisopropanolammonium dodecanoate (TPA-C12) and water at varying temperature. Notations as in Fig. 1

The effect of increasing the alkyl chain length is dramatic comparing TEA-C8 with TEA-C12 or even TEA-C14 (Figs. 2a–c), but less so comparing e.g. TEA-C14 with TEA-C20 (Figs. 2c–f). The first series shows drastic qualitative differences in the phase sequences, while the latter only shows increments in the extension regions and temperature stability of the lamellar liquid crystalline phase (as well for the crystalline surfactant) with increasing alkyl chain length.

Miscibility gaps occur already with the short-chain fatty acids for TEA and TPA as counterion, and for a sufficiently long alkyl chain, also for MEA. Such miscibility gaps do not occur in binary soap-water systems. However, a closer look at this feature gives a plausible explanation. All employed alkanolamines are weak bases, with pKa equal to 8 (TEA), 7.9 (TPA), and 9.5 (MEA), respectively [6]. A simple estimation gives for dilute solutions 3 mole% (TEA-systems) and 1 mole% (MEA-systems), respectively, of acid and base present in uncharged form [6]. Thus, for the aggregated systems, there is a proportion of solubilized fatty acids in soap micelles; at too high ratios of acid, another phase will precipitate (Figs. 1c, d, 2, and 3).

A closer investigation of the TEA-C12-system (Fig. 2b) is consistent with this explanation. At low concentrations of surfactant in the multiphase region, samples contain crystals dispersed in a solution phase. The crystals melt at 44 °C and are clearly pure dodecanoic acid. At higher concentrations of surfactant there is an equilibrium between a solution phase and a lamellar phase. Further increasing the concentration above 7 wt% gives an isotropic solution phase. This is analogous to the behavior of ternary phase diagrams for soap-water-polar solubilizate, and at low ratios for polar solubilizate-soap [9]. These complexities also clearly show that we do not have a true binary system simply from considering the phase rule. A final proof is that the miscibility gaps of the TEA-C8 and TEA-C12 systems can be shifted and made to disappear by increasing pH; a phase diagram for 0.1 M NaOH (aq) and TEA-C8 has no miscibility gap. Thus, the miscibility gaps are due to the fact that alkanolamines are weak bases.

Also, the preference of lamellar phases in all long-chained systems (Figs. 1c, d, 2c–h) can be explained from considerations of the partitioning between charged-uncharged amphiphiles in the aggregate: for long alkyl chain lengths in normal soaps the hexagonal phase is not very stable and the lamellar phase precipitates when adding minute amounts of polar solubilizate [9].

Comparing the CMCs of alkanolammonium carboxylates with normal soaps, it should be noted that the latter are considerably higher: the CMCs of octanoates are 0.03 M (TEA-C8), 0.08 M (MEA-C8) and 0.35 M (NaC8) [10], respectively. This is due to the proportion of uncharged amphiphiles in the aggregates; the decrease in CMC is consistent with the calculated proportion of acid in the aggregate [6].

Let us now consider the effect of the ion size. It should be recognized that this factor should work in the opposite direction compared to the acid-base equilibrium. An increase in counterion size will (all other factors being equal) make the screening of the charges at the aggregate surface less efficient, promoting aggregates with high curvature and, in general, making aggregation less favorable. The comparison of the series of C8 and C12 surfactants reveals that, at high concentrations of amphiphile, the effects from the ion size become predominant. The liquid crystals disappear with TEA-C8 (Fig. 2a) and, almost, with MEA-C8 (Fig. 1a); the increase in preferred curvature for the series MEA-C12, TEA-C12, and TPA-C12 clearly shows the influence of counter ion size (Figs. 1b, 2b, 3).

There is also a previously noted effect from the ion size: a decrease in Krafft-point with the bulky alkanolamines compared to alkali soaps (compare, for example, the Krafft-point of NaC16: around 50 °C, and TEA-C16: below room-temperature); this has been pointed out by Shinoda [11].

Finally, it should be noted that organic counterions may associate to the surfactant aggregates due to hydrophobic interactions. This has been systematically investigated by Stilbs and coworkers [12] for cationic surfactants with carboxylate counterions, and anionic surfactants with alkylammonium counterions. For surfactants where an appreciable hydrophobic interaction occurs, the counterion association β determined from NMR self-diffusion measurements will be $\gg 0.5$, approaching 1.0 for strongly interacting systems. A similar measurement on the present systems, TEA-C12 and MEA-C12, gives β in the order of 0.53 and 0.60, respectively. Also, a consideration of the shape of the alkanolamines with hydrophilic groups at all directions of the molecule also suggests that the hydrophobic interaction is of minor importance in these systems.

The main conclusions from this study are that alkanolammonium carboxylate-water systems have a phase behavior that deviates considerably from ordinary soap-water systems. Some features, miscibility gaps, decreases in CMC, and the destabilization of the hexagonal phase for the systems with long alkyl chains can be attributed to the presence of a few mole % of fatty acid in the soap aggregates. A general destabilization of liquid crystalline phases and an increase in Krafft-point compared to the alkali soaps are attributed to the larger size of the alkanolammonium counterions.

Acknowledgement

This work was financially supported by Karlshamns Research Foundation. Marie Sjöberg is thanked for performing the self-diffusion measurements.

References

1. Falbe J (1988) Surfactants in Consumer Products. Springer Verlag, Berlin
2. Friberg S, Liang P, Lockwood FE, Tadros M (1984) J Phys Chem 88:1045
3. Friberg S, Wohn C, Greene B, van Gilder R (1984) J Coll Interf Sci 101:593
4. Friberg S, Osborne DW (1986) J Am Oil Chem Soc 63:123
5. Bokström J, Jönsson A, Malmvik AC, Wärnheim T (1990) J Am Oil Chem Soc 67:733
6. Wärnheim T, Jönsson A (1990) J Coll Interf Sci 138:314
7. Padday JF (1969) in "Surface and Colloid Science" vol 1, Wiley-Interscience, New York
8. Stilbs P (1982) J Coll Interf Sci 87:385
9. Ekwall P (1975) Adv Liq Cryst 1:1
10. Mukerjee P, Mysels KJ (1971) "Critical Micelle Concentrations" Nat Stand Ref Data Ser, NBS, Washington
11. Kunieda H, Shinoda K (1976) J Phys Chem 80:2468
12. a) Jansson M, Stilbs P (1987) J Phys Chem 91:113
    b) Jansson M, Jönsson A, Li P, Stilbs P (1991) Coll Surfaces 59:387

Authors' address:

T. Wärnheim
Karlshamns Oils & Fats AB
R & D
S-374 82 Karlshamn
Sweden

Surface potentials in concentrated isotropic micellar systems

C. J. Drummond[1]) and B. S. Murray[2])

[1]) Department of Applied Mathematics, Research School of Physical Sciences, The Australian National University, Canberra, Australia
[2]) School of Chemistry, The University of Melbourne, Parkville, Victoria, Australia

Abstract: The solvatochromic acid-base indicator 2,6-diphenyl-4-(2,4-6-triphenyl-1-pyridinio)phenoxide, $E_T(30)$, is employed to determine the effective surface potentials of micelles (Ψ_0^{expt}) in relatively concentrated aqueous dodecyltrimethylammonium chloride (DTAC) systems (i.e. up to 40 wt% DTAC). Ψ_0^{expt} decreases with increasing DTAC concentration. This is attributed to a concomitant increase in the amount of dissociated chloride ions in solution when the DTAC micellar concentration is increased. The effective surface potentials are compared with the electrostatic surface potentials obtained in relatively dilute micellar DTAC/NaCl systems and with the electrostatic surface potentials calculated from the non-linearized Poisson–Boltzmann (P–B) equation and a cell model (Ψ_0^{cell}). The P–B cell model, which incorporates a Langmuir adsorption constant in order to categorize the association of Cl^- to the surfactant headgroups, adequately describes Ψ_0^{expt} up to 7 wt% DTAC. Beyond 7 wt%, Ψ_0^{expt} and Ψ_0^{cell} progressively diverge. The divergence is primarily considered to be a result of the fact that the P–B cell model neglects the ion-ion correlations and the finite size of the ions.

Key words: Micelles; electrostatic surface potentials; counter-ion association/dissociation; concentrated surfactant system; dodecyltrimethylammonium chloride (DTAC)

Introduction

A theoretical thermodynamic model that incorporates an electrostatic term calculated from the Poisson–Boltzmann (P–B) equation (inter alia) goes close to quantitatively describing the phase behaviour in a number of test binary and ternary surfactant/water systems [1–3]. Nevertheless, for many self-assembled surfactant aggregates the experimentally measured electrostatic surface potentials are significantly less than those calculated with the P–B treatment when it is assumed that there is no electro-neutralization of the headgroup charge by counter-ion association [4–8]. A representative example is the dialkyldimethylammonium bromide self-assembly system where surface charge densities derived from surface potentials, measured with both the surface force apparatus [4] and solvatochromic acid-base indicators [7], and zeta potentials [8] are in quantitative agreement. When compared with the P–B treatment, each of the three independent methods for determining the electrostatic potential indicate that $93 \pm 4\%$ of the quaternary ammonium headgroups are electrically neutralized by counter-ion association. In charged surfactant self-assembly systems the degree of counter-ion association/dissociation may depend on the particular headgroup, the interfacial curvature and additives. For most concentrated self-assembled charged surfactant systems, where there is no extraneous electrolyte, it is generally considered that the counter-ion association/dissociation dictates the electrostatic properties of the aggregates and, hence, the surfactant phase behaviour. The degree of counter-ion association/dissociation cannot be predicted a priori. Therefore,

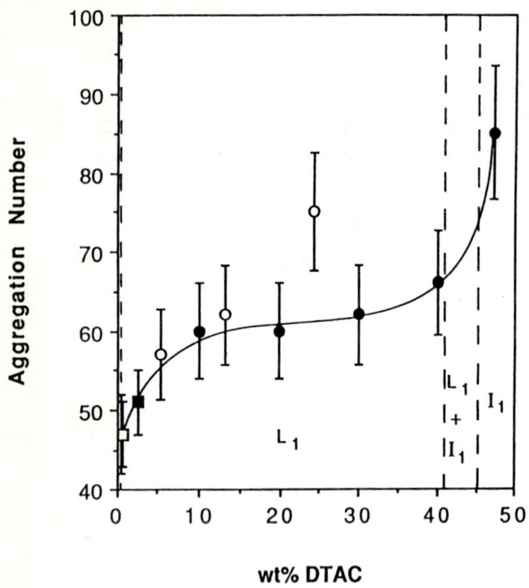

Fig. 1. Micellar aggregation number (N_{agg}) as a function of weight percent DTAC in the binary DTAC/water system at 25 °C. The N_{agg} values represented by ■, ○, □ and ● were taken from [12], [13], [14] and [15], respectively. L_1 denotes an isotropic micellar phase and I_1 denotes a cubic phase [11]

experimental investigations of counter-ion association/dissociation in concentrated self-assembled charged surfactant systems are warranted [9, 10].

The dodecylytrimethylammonium chloride (DTAC)/water system has an isotropic micellar phase at 25 °C that extends from 0.5 (the CMC) to ca. 41 wt% DTAC [11]. Within this range, as shown in Fig. 1, the micellar aggregation number increases only slightly with increasing DTAC concentration [12–15]. The DTAC micelles can be considered as spheres. In essence, these characteristics make the DTAC/water system ideal for investigating the effect of counter-ion association/dissociation on the electrostatic properties of the aggregates in concentrated surfactant systems.

In the present work, the solvatochromic acid-base indicator 2,6-diphenyl-4-(2,4,6-triphenyl-1-pyridinio)phenoxide, $E_T(30)$, has been employed to determine the effective surface potentials of micelles in relatively concentrated DTAC systems. The effective surface potentials are compared with the electrostatic surface potentials calculated from the non-linearized P–B equation and a cell model. To our knowledge, this is the first report of a comparison between experimental surface potentials and the predictions of a P–B cell model for concentrated micellar systems. The effective potentials are also compared with electrostatic potentials obtained in DTAC/NaCl systems where the surfactant concentrations is relatively dilute [5].

Experimental section

Reagents

DTAC was supplied by Tokyo Kasei Kogyo Co., Ltd. and was vacuum desiccated prior to use. $E_T(30)$ was obtained from Aldrich Chemical Co. and was used as received. The water (conductivity $< 0.9\ \mu S\ cm^{-1}$ and air/water surface tension $= 72.0\ mN\ m^{-1}$ at 25 °C) was prepared from deionised water that had been passed through a still in which the water was distilled once from alkaline potassium permanganate and then distilled again. All other reagents were either analytical grade or spectroscopic grade and were not further purified.

Effective surface potential measurements

The solvatochromic acid-base indicator $E_T(30)$ was employed to measure the effective micellar surface potentials (Ψ_0^{expt}). The effective surface potentials were determined by using the relationship

$$\Psi_0^{expt} = 2.303\,(kT/e)\,(pK_a^0 - pK_a^{obs})\ , \qquad (1)$$

where k, T, e, pK_a^0 and pK_a^{obs} are the Boltzmann constant, the absolute temperature, the elementary electrostatic charge, the intrinsic pK_a of $E_T(30)$ residing at the aqueous/surfactant interface in the absence of any electrostatic field, and the apparent pK_a of $E_T(30)$ residing at the charged aqueous/surfactant interface, respectively. A complete description of the procedure is given in [16]. In the present work, the pH titrations were performed with a Radiometer PHM83 Autocal pH-meter and a combined electrode. UV/visible spectra for the pK_a^{obs} determination were measured on a Hewlett Packard 8451A diode array spectrophotometer. λ_{max} values for the phenoxide form of $E_T(30)$ were measured on a Varian Cary 219 spectrophotometer. Effective dielectric constants for the micellar interfacial microenvironment were derived from the λ_{max} values [16]. All measurements were made at 25 °C.

Results and discussion

The experimentally measured effective surface potentials are plotted as a function of wt% DTAC in Fig. 2.

Maeda and Satake [17] have recently shown, with an ion(surfactant)-selective electrode study, that above the CMC the Cl^- activity increases and the DTA^+ activity decreases as a function of [DTAC]. Their study was limited to DTAC concentrations less than 3 wt%. The same technique has also been employed by Kale et al. [18] to investigate the analogous dodecyltrimethylammonium bromide (DTAB) system. They found that above the CMC the Br^- activity increases and the DTA^+ activity decreases as a function of [DTAB] at least up to ca. 18 wt% DTAB (the self-imposed limit in their study). As shown in Fig. 3, in relatively dilute DTAC solutions the micellar surface potential monotonically decreases as the [Cl^-] in solution increases [5]. It has been inferred from surface force apparatus studies [19–22] that above the CMC, micelles and their associated counterions do not contribute to the Debye length. In view of these previous findings, it seems reasonable to

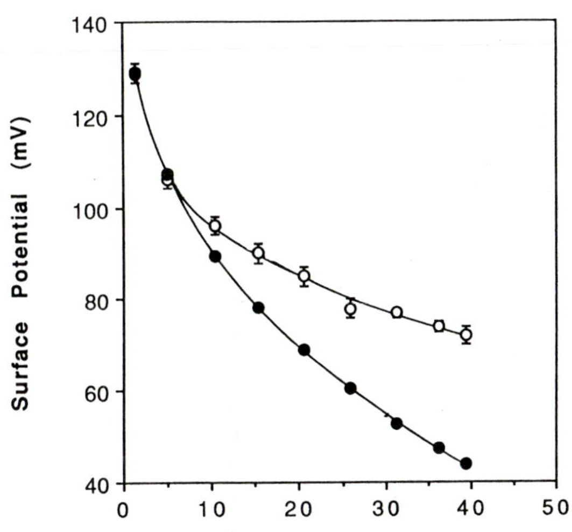

Fig. 2. Experimental (○) and cell model (●) micellar surface potentials as a function of weight percent DTAC in the DTAC/water system at 25 °C

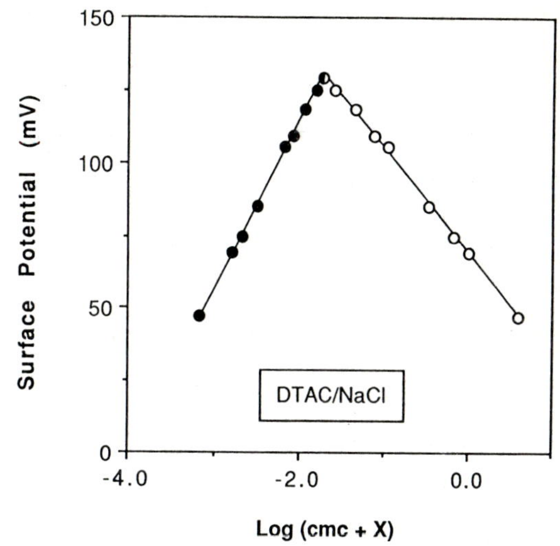

Fig. 3. Micellar surface potential in the aqueous DTAC/NaCl system at 25 °C as a function of log(CMC + X) where $X = 0$ for the ● data and X = [NaCl] in $mol\,dm^{-3}$ for the ○ data. [DTAC] is constant at $0.05\ mol\,dm^{-3}$ (= 1.32 wt%). This plot was derived from information contained in [5] and [29]

consider the hypothesis that in the concentrated DTAC systems the dissociated counter-ions govern the electrostatic properties of the micelles. In other words, it is considered that the concentration of micelle dissociated counter-ions and the concentration of 1 : 1 surfactant monomer dictate the overall electrostatic screening.

The variation of Ψ_0 with increasing volume fraction Φ of the micellar phase has been theoretically treated by using the cell model which has been described in earlier work [9, 10]. The volume V of an individual micelle core was calculated from N_{agg} and the equation of Tanford [23] for the volume of a saturated hydrocarbon chain v, i.e.,

$$v = 27.4 + 26.9 n_c \,, \tag{2}$$

where n_c is the number of methylene groups in the chain. Since $V = (4/3)\pi R^3$, the effective spherical micellar radius R may be calculated, along with ϕ from the concentration of DTAC for each composition. The density of the surfactant phase is taken as $0.913\ g\,cm^{-3}$ [24] and that of the aqueous phase as that of water, i.e. $1.0\ g\,cm^{-3}$. Each micelle may be considered to be surrounded by a shell of aqueous phase of radius R_c, given by

$$(R/R_c)^3 = \phi \,. \tag{3}$$

Table 1. Experimental and cell model data for the aqueous DTAC system

wt%	ϕ	Ψ_0^{expt} (mV)	N_{agg}	R (nm)	Ψ_0^{cell} (mV)	α	R_c (nm)	Cl^-_{tot} (mM)	$n^+_{R_c}$ (mM)	$n^-_{R_c}$ (mM)
1.32	0.014	129 ± 2	48	1.59	129	0.73	6.52	50.6	23.0	34.0
5.03	0.055	106 ± 2	55	1.66	107	0.67	4.36	186.5	25.5	82.2
10.42	0.114	96 ± 2	58	1.69	89	0.65	3.48	398.6	26.8	172.0
15.29	0.168	90 ± 2	60	1.71	78	0.64	3.10	609.9	27.3	273.9
20.48	0.224	85 ± 2	61	1.72	69	0.63	2.83	861.8	27.5	410.7
25.82	0.283	78 ± 2	62	1.73	61	0.62	2.64	1154.2	27.4	586.8
31.34	0.343	77 ± 1	62	1.73	53	0.61	2.47	1508.3	27.2	823.0
36.24	0.397	74 ± 1	64	1.75	47	0.60	2.38	1859.1	26.9	1069.8
39.45	0.432	72 ± 2	66	1.77	44	0.59	2.34	2113.0	26.7	1256.7

Thus R_c is a time-averaged quantity such that $2(R_c - R)$ is the mean maximum separation between micelle surfaces and at R_c the local electrostatic potential Ψ will have its lowest value throughout the system. The N_{agg}, R and R_c values for each experimental ϕ are listed in Table 1. The Poisson–Boltzmann equation is used to describe the spatial variation of Ψ and is solved numerically between R and R_c since the boundary conditions are evident, i.e.,

$$[d\Psi/dr]_{r=R} = (-\sigma_0/\varepsilon) , \tag{4}$$

and

$$[d\Psi/dr]_{r=R_c} = 0 , \tag{5}$$

where σ_0 is the surface charge density, ε is the permittivity of the solvent and r is the radial distance from the centre of the micelle.

As stated earlier, it is necessary to allow for neutralization of σ_0 to some extent, via counter-ion association, for calculated potentials to take on experimental values. In the cell model the association of the counter-ions, Cl^- in this case, is described via a Langmuir adsorption constant, K_d, given by

$$K_d = (k_d/k_a) = (n_{ci}(R)\sigma_0)/(\sigma_{max} - \sigma_0) , \tag{6}$$

where k_d and k_a are the rate constants for the dissociation and association of Cl^- with the DTA^+ headgroups, respectively. The term $n_{ci}(R)$ is the concentration of counter-ions at $r = R$ and $\sigma_{max} = eN_{agg}/4\pi R^2$, the maximum possible surface charge density. The value of σ_0 is usually expressed via the degree of dissociation α, where $\alpha = \sigma_0/\sigma_{max}$.

It is important to note that the numerical solution to the above equations yields the potential difference between R and R_c, Ψ_0^{cell}, and the potential at R_c is not necessarily the same as the potential far away from the same particle (i.e. same σ_0 and R) in bulk electrolyte, where there will be an equal concentration of positive and negative ions. Also, in order to solve the equations, K_d must be known, which cannot be measured independently. However, for the purposes of modelling the phase volume dependence of Ψ_0 it is reasonable to take the value of Ψ_0 measured via the solvatochromic acid-base indicator Ψ_0^{expt} for a dilute system, assume that this is the correct surface potential, and then vary K_d in the cell model until a value of σ_0 is obtained such that Ψ_0^{cell} is equal to Ψ_0^{expt}. For the most dilute system studied, $\phi = 0.0145$, the value of K_d required to fit Ψ_0^{cell} to Ψ_0^{expt} was 9.8 mol dm^{-3}, and the corresponding value of $\alpha = 0.726$. This value of K_d was therefore used in the cell model to calculate potentials for each value of ϕ. The value of the CMC was assumed to be $0.0193 \text{ mol dm}^{-3}$ throughout. The CMC decreases as the [DTAC] increases (vide infra), however, the assumption of a constant CMC introduces negligible errors in the Ψ_0^{cell} calculations. Table 1 gives the values of Ψ_0^{cell} calculated along with the corresponding values of α. Table 1 also gives the concentrations of DTA^+ and Cl^- ions at R_c (i.e. $n^+_{R_c}$ and $n^-_{R_c}$, respectively) and the total concentration of Cl^- ions in the shell of volume $(4/3)\pi(R_c^3 - R^3)$ (i.e. Cl^-_{tot}). In Fig. 2 the values of Ψ_0^{cell} are plotted against wt% DTAC. It is clear that the agreement between Ψ_0^{cell} and Ψ_0^{expt} becomes quite poor past about 7 wt% DTAC (i.e. $\phi \approx 0.08$).

A number of checks were made to confirm that the Ψ_0^{cell} values were valid. As expected, when the same K_d value and 1:1 [electrolyte] are imposed, as ϕ gets close to zero there is little difference between

Ψ_0^{cell} and Ψ_0 calculated with the approximate analytical solution to the spherical P-B equation derived by Ohshima et al. [25]. Ψ_0^{cell} values as a function of ϕ are also identical to those obtained by Bell and Dunning [26] when the same R, N_{agg} and $\alpha(=1)$ are employed.

In the above analysis it is tacitly assumed that the values of the measured surface potentials are the actual electrostatic surface potentials (i.e. $(\Psi_R - \Psi_{R_c})$) and that the value of K_d, obtained as a fit to the low ϕ data, is constant throughout. With regard to the latter assumption, the value of K_d might be expected to vary if the chemical environment of the surfactant headgroups varies and this should be reflected in a change in the value of ε_{eff}. Figure 4 indicates that ε_{eff} decreases with increasing ϕ; this change may be a direct consequence of the increasing $[Cl^-]$ [5]. The values of K_d required to fit Ψ_0^{cell} to Ψ_0^{expt} are shown in Fig. 5, along with the corresponding values of α. It is seen that K_d must increase markedly and also that the extrapolated values of α must be greater than 1.0 in order to fit the Ψ_0^{expt} values in the highest ϕ systems, which is unrealistic. It is also noted that a decrease in ε_{eff} should lead to a decrease in K_d rather than an increase. The above factors when taken together suggest that a variation in K_d is not primarily responsible for the divergence of Ψ_0^{cell} and Ψ_0^{expt}.

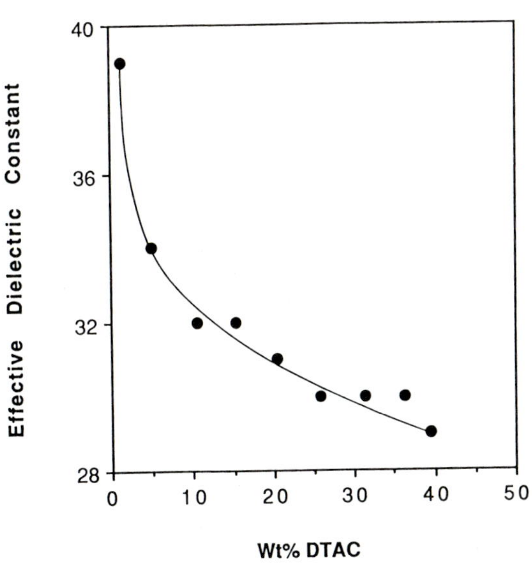

Fig. 4. Effective interfacial dielectric constant as a function of weight percent DTAC in the DTAC/water system at 25 °C

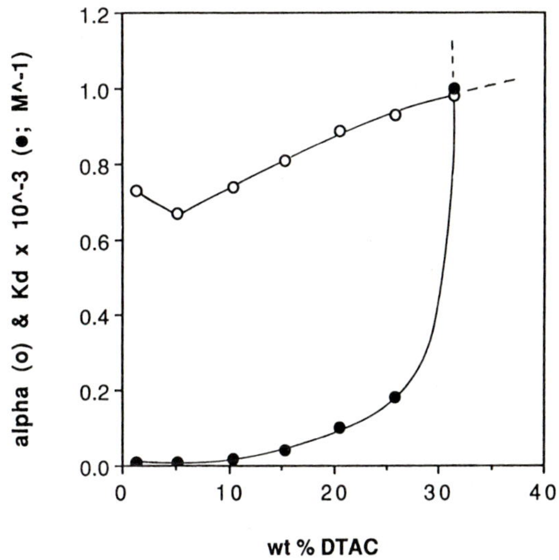

Fig. 5. K_d and α values required to fit Ψ_0^{cell} to Ψ_0^{expt}. When [DTAC] > 30 wt% the K_d and α values become unrealistic

In order for Ψ_0^{cell} and Ψ_0^{expt} to be comparable, the pH electrode must measure accurate hydrogen ion activities at the cell boundary (R_c). In an oil-in-water microemulsion system it has been found that the pH values obtained with a combined electrode are consistent with this requirement, at least up to an interdroplet separation of 3.6 nm (oil $\phi = 0.6$) [9]. This suggests that reasonable hydrogen ion activities may also be obtained in the DTAC systems at least up to $\phi = 0.11$ (i.e., $2(R_c - R) = 3.6$ nm). At this stage, we cannot unequivocally state what the pH electrode measures when $\phi > 0.11$ in the DTAC system. In the $\phi > 0.11$ region the pH electrode may register a hydrogen ion activity that is averaged over radial distances between R and R_c. If this is the case, then the Ψ_0^{expt} values should be less than the actual $(\Psi_R - \Psi_{R_c})$ values. Nevertheless, postulated problems with the measurement of the hydrogen ion activity cannot account for the difference between Ψ_0^{cell} and Ψ_0^{expt} in the high ϕ region.

The lack of agreement between Ψ_0^{cell} and Ψ_0^{expt} above 7 wt% DTAC probably stems from the fact that the P–B cell model treats the solvent as a dielectric continuum and the ions as point charges. As the separation between micellar surfaces becomes smaller and the aqueous $[Cl^-]$ concomitantly becomes larger (see Table 1), the electrical double layer description should ideally take into account the ion-ion correlations and the finite size of the ions. Note that in aqueous solution the average distance between the centre of a chloride ion and the centre of the nearest water molecules is 0.312 ± 0.002 nm [27]. Between 7 wt% and 40 wt% DTAC the intermicellar spacing decreases from approximately 7 to 2 hydrated chloride ion diameters. Hence, ion-ion correlations and ion size effects become increasingly important as ϕ increases and the relationship between the time-averaged electrostatic potential and the ion density becomes less straightforward. Above 7 wt% DTAC the origin of Ψ_0^{expt} may no longer be purely electrostatic [28].

For completeness, it should be mentioned that there is an alternative, more radical way of interpreting micellar surface potentials. Recently, Healy et al. [29] presented evidence which suggests that one-component spherical micelles may behave as Nernstian colloids with the surfactant monomer acting as the potential determining ion. For example, Fig. 3 shows DTAC micellar surface potentials, in relatively dilute DTAC solutions with added NaCl, as a function of $\log_{10}$CMC. A linear least squares fit to the Ψ_0^{expt} versus $\log_{10}$CMC data in Fig. 3 gives a gradient of 56.8 ± 0.7 mV/unit change in $\log_{10}$CMC with a correlation coefficient of 0.9995. This is very close to the Nernstian slope of 59 mV. At this stage, all that can be said about the concentrated DTAC solutions is that the measured micellar surface potentials are in qualitative accord with the micelles behaving as Nernstian colloids. In other words, Ψ_0^{expt} decreases as the CMC decreases. A quantitative asessment of whether or not the DTAC micelles in concentrated micellar solution behave as Nernstian colloids cannot be made until accurate surfactant monomer activities as a function of [DTAC] are obtained.

Conclusion

Experimentally measured effective micellar surface potentials decrease as [DTAC] increases. This is consistent with dissociated counter-ions governing the electrostatic screening. Ψ_0^{expt} is in accord with the predictions of a P–B cell model, which incorporates electro-neutralization of charged surfactant headgroups by counter-ion association, up to 7 wt% DTAC. Past 7 wt %, Ψ_0^{expt} and Ψ_0^{cell} progressively diverge. More than likely, the divergence is a result of the fact that the P–B cell model ignores ion-ion correlations and the finite size of the ions. The inference is that, for aqueous systems with more than 7 wt% DTAC, the intermicellar separations and effective counter-ion concentrations are such that discrete ion effects should not be neglected.

Acknowledgements

During this study the authors had beneficial discussions with Franz Grieser, Tom Healy, Stjepan Marcelja and Barry Ninham. We thank them for their suggestions.

References

1. Gunnarsson G, Jonsson B, Wennerstrom H (1980) J Phys Chem 84:3114
2. Jonsson B, Wennerstrom H (1981) J Colloid Interface Sci 80:482
3. Jonsson B, Wennerstrom H (1987) J Phys Chem 91:338
4. Pashley RM, McGuiggan PM, Ninham BW, Brady J, Evans DF (1986) J Phys Chem 90:1637
5. Drummond CJ, Grieser F, Healy TW (1987) Chem Phys Lett 140:493
6. Kibblewhite J, Drummond CJ, Grieser F, Healy TW (1987) J Phys Chem 91:4658
7. Drummond CJ, Grieser F, Healy TW (1988) J Phys Chem 92:2604
8. Kaneko T, Miller DD, Evans DF (1990) J Solution Chem 19:457
9. Murray BS, Drummond CJ, Grieser F, White LR (1990) J Phys Chem 94:6804
10. Murray BS, Drummond CJ, Gale L, Grieser F, White LR (1991) Colloids & Surfaces 52:287
11. Balmbra RR, Clunie JS, Goodman JF (1969) Nature 222:1159
12. Malliaris A, Le Moigne J, Sturm J, Zana R (1985) J Phys Chem 89:2709
13. Malliaris A, Lang J, Zana R (1986) J Colloid Interface Sci 110:237
14. Roelants E, De Schryver FC (1987) Langmuir 3:209
15. Johansson LB-A, Soderman O (1987) J Phys Chem 91:5275
16. Drummond CJ, Grieser F, Healy TW (1986) Faraday Discuss Chem Soc 81:95
17. Maeda T, Satake I (1988) Bull Chem Soc Jpn 61:1933
18. Kale KM, Cussler EL, Evans DF (1980) J Phys Chem 84:593
19. Pashley RM, Ninham BW (1987) J Phys Chem 91:2902
20. Pashley RM, McGuiggan PM, Horn RG, Ninham BW (1988) J Colloid Interface Sci 126:569
21. Marra J, Hair ML (1989) J Colloid Interface Sci 128:511

22. Richetti P, Kekicheff P (1991) Phys Rev Lett (submitted)
23. Tanford C (1972) J Phys Chem 76:3020
24. Reiss-Husson F, Luzzati V (1964) J Phys Chem 68:3504
25. Ohshima H, Healy TW, White LR (1982) J Colloid Interface Sci 90:17
26. Bell GM, Dunning AJ (1970) Trans Faraday Soc 66:500
27. Marcus Y (1983) J Solution Chem 12:271
28. Langner M, Cafiso D, Marcelja S, McLaughlin S (1990) Biophys J 57:335
29. Healy TW, Drummond CJ, Grieser F, Murray BS (1990) Langmuir 6:506

Authors' address:

C. J. Drummond
Dept. of Applied Mathematics
The Australian National University
GPO BOX 4
Canberra, ACT, 2601
Australia

Solubilization of aluminum and calcium nitrates into W|O microemulsions and lyotropic liquid crystals stabilized by a nonionic surfactant

S.E. Friberg[†]), A. Amran[†]), and J. Sjöblom*)

[†]) Center for Advanced Material Processing, Clarkson University, Potsdam, New York, USA
*) Department of Chemistry, University of Bergen, Norway

Abstract: The phase limits were determined for W/O microemulsions and the lamellar lyotropic liquid crystal in the system of cyclohexane/aqueous solutions of aluminum and calcium nitrates/commercial tetraethyleneglycoldodecylether. Visual observation, optical microscopy, and small-angle x-ray diffraction were used.

The results showed an increase of the maximum amount of aqueous solution in the microemulsions with enhanced salt concentration as well as an enlargement of the area for the lamellar liquid crystal. The penetration of water into the surfactant layers of the liquid crystal was significantly increased after addition of salt.

Key words: Microemulsions; liquid crystals; detergents and surfactants; solubilization; salt solutions

Introduction

Recent investigations into the sol/gel process [1, 2] have shown the potential to apply W/O microemulsions in it [3–6]. It was especially observed that W/O microemulsions [6] and lyotropic liquid crystals [7] could be prepared with high concentrations of copper nitrate (2 M).

Inverse micellar solutions stabilized by a nonionic surfactant have recently also been shown to be useful to prepare silica particles of uniform size and shape [8].

We found these results to open a new area of investigation, and found an investigation into amphiphilic association structures with high salt content to be of value, considering the limited number of investigations into complete systems with salts [9, 10].

Experimental

Materials

The following chemicals were used without further purification: Nonionic surfactant, Macol LA_4, (≈tetraethylglycol dodecylether) (Mazer Chemicals), cyclohexane (100% analyzed, J.T. Baker), aluminum nitrate nonahydrate (98%, ACS reagent) Aldrich Chemical Company and calcium nitrate tetrahydrate (99%, ACS reagent). Nitric acid (70.4%, ACS certified, Baxter Healthcare Corp., Scientific Product Division) was used as acid medium to adjust the water to a pH of 1.25. The water was deionized and distilled.

Small-angle x-ray diffraction

The interlayer spacings of liquid crystalline phases were determined by small-angle x-ray diffraction (Siemens Crystalloflex 4) with a Tennelec detector system (PSD 100). Path length was 50 cm, and 0.7 mm capillaries with 0.01 mm walls were used. The measurements were done at 40 kV, 30 mA, and room temperature (22.5 °C).

Phase limits

The limits of the W/O microemulsion regions were determined by titration of the hydrocarbon/

surfactant solutions by the aqueous salt solution to turbidity. The limits were checked by visual observation of samples inside and outside the phase limit for 2 weeks. The extension of the lamellar liquid crystalline phase was determined by microscopy observation between crossed polarizers, and from the small-angle x-ray results which showed sudden changes of slope of the interlayer spacing/composition curves at the phase limit. The phase in equilibrium with the W/O microemulsions was identified by observation of two-phase samples between crossed polarizers in an optical microscope.

Results

The results will be presented in the following order. The phase diagram of the electrolyte-free system is first described, followed by a review of the modifications caused by the addition of aluminum and calcium salts to the phase regions. Finally, the small-angle x-ray results will be presented to illustrate the influence by the salt on the association structure.

Water-surfactant-cyclohexane

The system (Fig. 1) contained an isotropic solution of the hydrocarbon and the surfactant with small solubilization of water; 17.5% by weight at approximately equal amounts (by weight) of the surfactant and the hydrocarbon. The lamellar liquid crystal without hydrocarbon ranged from 43% surfactant to 71.5%. The maximum solubilization of cyclohexane in the liquid crystal was 25% in the 50–55% surfactant content range. The hydrocarbon/surfactant solution was in equilibrium with the lamellar liquid crystal; the hydrocarbon solution in the three-phase region water-lamellar liquid crystal-hydrocarbon solution contained extremely small amounts of water and surfactant. The lamellar liquid crystal of this three-phase region contained 19% cyclohexane.

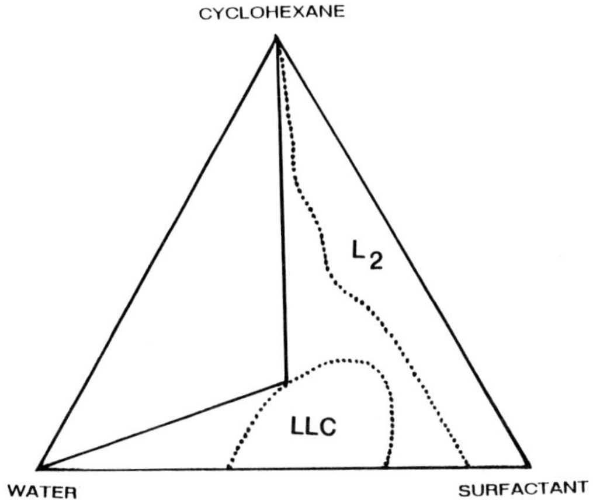

Fig. 1. The phase diagram of water (pH = 1.25, HNO_3), the nonionic surfactant (LA4), and cyclohexane. L_2 = isotropic solution; LLC = lamellar liquid crystal.

Aluminum nitrate solutions

The essential modification of the system by the addition of aluminum nitrate (Fig. 2) was the increase of water solubilization. The progression 17, 26, 32% water for solutions with, respectively, 0. 17.0, and 34.1 wt.% $Al(NO_3)_3$ in the water illustrates the almost linear increase. The difference is well illustrated by the difference in the solubilization efficiency of the surfactant at its lowest concentration. The 2/1 initial aluminum nitrate solution/surfactant ratio (Fig. 2) for the most

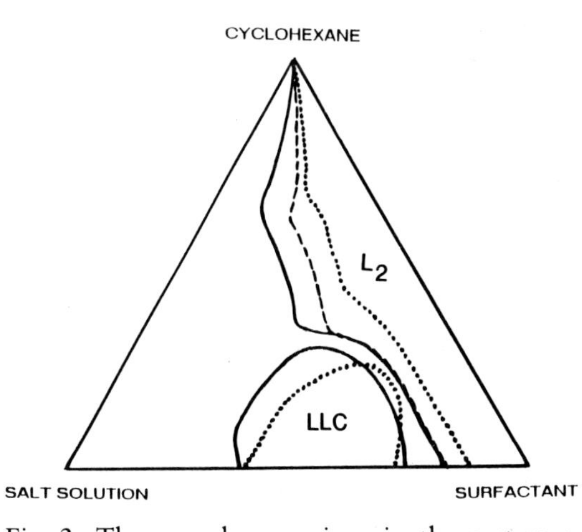

Fig. 2. The one-phase regions in the system aqueous aluminum nitrate solution, the nonionic surfactant LA4 and cyclohexane. L_2 = isotropic solution; LLC = lamellar liquid crystal.

	$Al(NO_3)_3/H_2O$ wt. ratio	wt.%
....	0	0
-----	0.205	17
———	0.515	34

concentrated aluminum nitrate solution is in contrast to the water/surfactant initial ratio of 1/2 (Fig. 1). The addition of the aluminum salt also enlarged the area of the lamellar liquid crystal. Its range in the non-hydrocarbon part was enhanced by 35%, now covering the range 38–76.5% by weight of surfactant, and the cyclohexane solubilization was increased from 25 to 29%.

Calcium nitrate solutions

The addition of calcium nitrate brought similar changes to the system (Fig. 3) with maximum water solubilization in the microemulsion as 17, 23, and 35% by weight for solutions with 0, 20.9, and 48.7% of the calcium salt in the water. The area of the liquid crystalline phase also showed a significant growth (Fig. 3).

Interlayer spacings

The interlayer spacings for the lamellar liquid crystal are shown in Figs. 4–6. The values generally increased with enhanced ratio of hydrocarbon to surfactant and with added water.

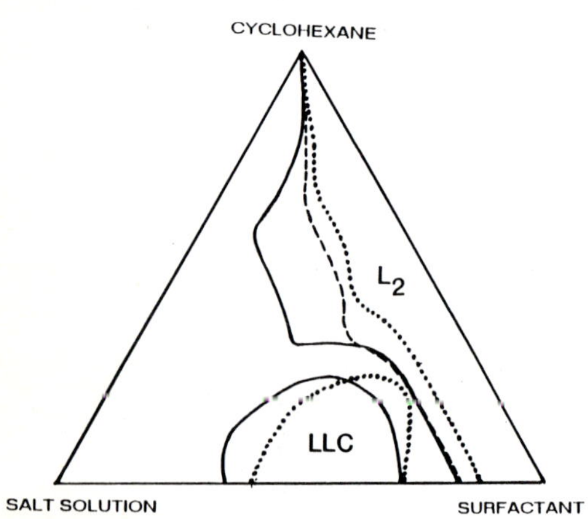

Fig. 3. The one-phase regions in the system calcium nitrate solution, the nonionic surfactant (LA4) and cyclohexane. L_2 = isotropic solution; LLC = lamellar liquid crystal.

	$Ca(NO_3)_2/H_2O$ wt. ratio	wt.%
....	0	0
------	0.264	20.9
————	0.949	48.7

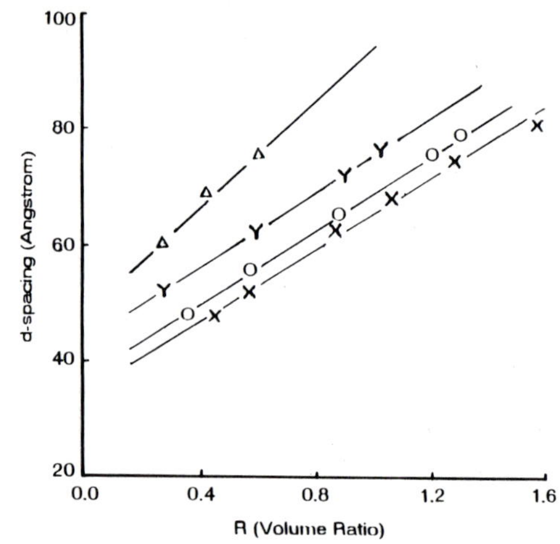

Fig. 4. Interlayer spacings in the lamellar liquid crystal in Fig. 1. Volume ratio cyclohexane/LA_4: × = 0; ○ = 0.14; Y = 0.32; △ = 0.54

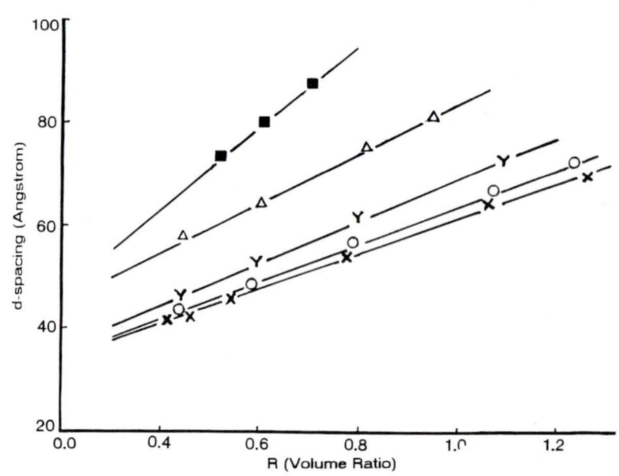

Fig. 5. Interlayer spacings in the lamellar liquid crystal in Fig. 2 with $Al(NO_3)_3/H_2O$ wt. ratio 0.515. Volume ratio cyclohexane/LA_4: × = 0; ○ = 0.14; Y = 0.32; △ = 0.54; ■ = 0.84

The values for the salt solutions were lower than those with water only, a tendency more accentuated for the calcium salt than for the aluminum salt.

The extrapolated values to zero content of aqueous salt solution were also similar, with 42.1 Å for no hydrocarbon, to 51.6 Å for the highest content of hydrocarbon (Table 1).

The deviation in the values is pronounced, as would be expected as a result of the large extrapolation. However, a statistical analysis using the

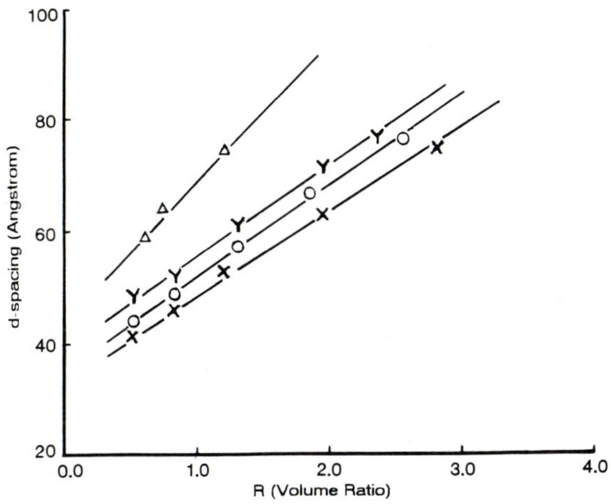

Fig. 6. Interlayer spacings in the lamellar liquid crystal in Fig. 3 with $Ca(NO_3)_2/H_2O$ wt. ratio 0.949. Volume ratio cyclohexane/LA_4: × = 0; ○ = 0.14; Y = 0.32; △ = 0.54

Table 1. Interlayer spacings in Å extrapolated to zero content of aqueous solution

Aqueous solution (wt.)	Hydrocarbon/surfactant (vol. ratio)				
	0	0.14	0.32	0.54	0.84
H_2O (100)	34 Å	37 Å	43 Å	48 Å	—
$H_2O/Al(NO_3)_3$ (65.9/34.1)	34 Å	34 Å	36 Å	45 Å	47 Å
$H_2O/Ca(NO_3)_2$ (51.3/48.7)	35 Å	35 Å	39 Å	44 Å	—

Minitab program clearly distinguished the d_0 values for water and calcium nitrate solutions as two different functions with adjusted R values of 95%.

Discussion

The general appearance of the phase diagrams is in agreement with earlier investigations on W/O microemulsions with nonionic surfactants [11–18] with a W/O microemulsion region resulting in a maximum water solubilization for a certain hydrocarbon/surfactant ratio. The fact that the isotropic solution was limited to the hydrocarbon-/surfactant axis with no solubility region emanating from the water corner show the conditions to be at temperatures well above the HLB temperature [16]. Studies of similar systems have shown the electrolyte to be dissolved in the aqueous phase, and not to exist as small polycrystalline particles [17].

The influence of the electrolytes with enhanced solubilization of the aqueous solutions for increased concentration of the salt is also in agreement with earlier results [11]. Figures 1–3 are plotted as weight percent and support the argument that volume percentages would be a more correct description of conditions has merits.

However, Table 2 demonstrates that also the volume ratios show enhanced solubilization with salt content.

It is noteworthy that this increase of solubilization with salt content has also been found for W/O microemulsion stabilized by ionic surfactants [19].

The presence of salt also caused enhanced stabilization of the lamellar liquid crystal, both with regard to the extent of the water/surfactant stability range and to maximum hydrocarbon content. This result will be compared to the water/surfactant and surfactant hydrocarbon interactions, which, viewed together, present information about the behavior.

The interaction is reflected in the extent to which water penetrates the zone B from zone A (Fig. 7) and the corresponding value from zone C. The penetration is obtained from the slope of the interlayer spacing vs. the water/(surfactant + hydrocarbon) ratio (Figs. 4–6).

The penetration fraction α is defined in the following manner. For an added solvent that is localized entirely to zone A or C (Fig. 7) the penetration $\alpha = 0$ and the variation of interlayer spacing, d, with the volume ratio of added substance to the original ones, R, is

$$d = d_0(1 + R) , \tag{1}$$

Table 2. Ratio of maximum solubilization of salt solution to water into W/O microemulsion

Aqueous solutions	$Al(NO_3)_3$/W		$Ca(NO_3)_2$/W	
wt%	17/83	34.1/65.9	20.9/79.1	48.7/51.3
Ratio wt.	1.53	1.88	1.35	2.05
Ratio volume	1.33	1.30	1.15	1.56

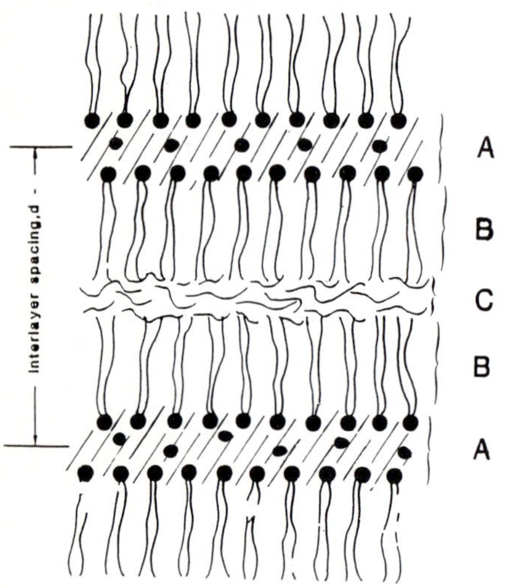

Fig. 7. The structure of a lamellar liquid crystal may be divided into three zones: A) the aqueous solution between the polar groups of the surfactant: B) the surfactant layers; and C) the space between the mirroring methyl group layers of the surfactant

in which d_0 is the interlayer spacing without solvent. If the solvent is partitioned equally into zones A, B, and C (Fig. 7) the penetration fraction $\alpha = 1$ and the interlayer spacing function becomes independent of R:

$$d = d_0 \; . \tag{2}$$

For partial penetration into zone B the penetration factor is defined from the relation

$$d = d_0(1 + (1 - \alpha)R) \; ; \tag{3}$$

this relation satisfies the conditions stated in Eqs. (1) and (2).

The α values are summarized in Table 3.

Table 3 shows two distinct trends. At first, the presence of the two metal salts enhanced the penetration of water and, secondly, its penetration was strongly reduced for highest hydrocarbon/surfactant ratio.

The latter condition is rational and expected; close to the limit of hydrocarbon content the hydrocarbon molecules are expected to influence the B zone (Fig. 7) and, hence, have an effect on the water penetration.

Information about the hydrocarbon/surfactant relationship is obtained from the curves in Figs. 4–6. The values of interlayer spacing extrapolated to zero aqueous solution content may be plotted against hydrocarbon/surfactant volume ratio and the hydrocarbon penetration from zone C to B (Fig. 7) may be calculated according to Eq. (3).

The values in Table 4 are informative.

The results illustrate two phenomena. In the first instance, the d_{os} values are equal. This means that replacing water by a salt solution did not change the conformation of the surfactant in the liquid crystalline phase. Secondly, the fact that the penetration of the hydrocarbon was different means that the presence of salt changed lateral distance between the surfactant molecules in the lamellar structure. The reason for that structural modification is evident from the values in Table 3. The salt solutions penetrated into space B (Fig. 7) from the polar groups to a higher degree than did the pure water. It appears reasonable to assume that this penetration gave an increased separation of the surfactants, allowing more hydrocarbon to penetrate into the space between the surfactant hydrocarbon chains from the methyl groups.

These results are at variance with earlier ones describing copper nitrate added to the water [7]. In that case, a strong increase of copper nitrate concentration in water did not change the water

Table 3. Penetration fraction for water

Aqueous solution (wt.)	Hydrocarbon/surfactant (vol. ratio)				
	0	0.14	0.32	0.54	0.84
H_2O (100)	0.12	0.14	0.24	0.02	—
$H_2O/Al(NO_3)_3$ (65.9/34.1)	0.50	0.44	0.43	0.46	0.13
$H_2O/Ca(NO_3)_2$ (51.3/48.7)	0.58	0.54	0.59	0.44	

Table 4. Hydrocarbon penetration, α, into surfactant and interlayer spacing, d_{os}, for the latter

Aqueous solution (wt.)	Penetration fraction, α	d_{os} Å
H_2O (100)	0.20	33.8
$H_2O/Al(NO_3)_3$ (65.9/34.1)	0.46	32.8
$H_2O/Ca(NO_3)_2$ (51.3/48.7)	0.47	33.8

penetration to a measurable degree. It must be concluded that the penetration into the nonionic surfactant layers is specific for each salt.

References

1. Brinker CJ, Scherer GW (1990) Sol-Gel Science, The Physics and Chemistry of Sol-Gel Processing. Academic Press, New York
2. Sol-Gel Technology for Thin Films, Fibers, Preforms, Electronics and Speciality Shapes. (Klein LC, ed.) (1988) Noyes Publications, New Jersey
3. Friberg SE, Yang C-C (1988) In: Doyle FM, Raghaven S, Somasundaran P, Warren GW (eds.) Innovations in Materials Processing Using Aqueous Colloid and Surface Chemistry. The Minerals, Metals and Materials Society, Warrendale, Pennsylvania, 181
4. Friberg SE, Yang C-C, Sjöblom J, Langmuir (in press)
5. Friberg SE, Amran A, Yang C-C, Sjöblom J, Colloids Surfaces (in press)
6. Friberg SE, Jones SM, Yang C-C (1992) J Disp Sci Vol 13 No 1
7. Friberg SE, Amran A, Yang C-C, Sjöblom J (1991) Colloids and Surfaces (in press)
8. Ossco-Arare K, Arriagada FJ (1990) Colloids and Surfaces 50:321
9. Ekwall P, Mandell L, Fontell K (1977) J Coll I Sci 61:3
10. Selle MH, Sjöblom J, Friberg SE, Young T, Yang C-C (1992) Progress Colloid Polym Sci (in press)
11. Saeten JO, Selle MH, Sjöblom J, Friberg SE, Gestblom B (1991) J Solution Chem 20:1149
12. Nilsson PG, Lindman B (1983) J Phys Chem 87:4756
13. Nilsson PG, Lindman B (1982) J Phys Chem 86:271
14. Stilbs P, Lindman B (1981) J Phys Chem 85:2587
15. Ravey JC, Buzier M, Picot C (1984) J Coll I Sci 97:9, 3778
16. Saito H, Shinoda K (1970) J Coll I Sci, 32:649
17. Saeten JO, Fordedal H, Skodvin T, Sjöblom J, Amran A, Friberg SE (1992) J Colloid I Sci (in press)
18. Osseo-Asare K (1990) Colloids and Surfaces 50:373
19. Friberg SE, Buraczewska I (1978) Progr Colloid PS 63:1

Authors' address:

Stig E. Friberg
Center for Advanced Materials Processing
Clarkson University
Potsdam, New York 13699-5814
USA

Partial phase behavior of ionic microemulsions stabilized by sodium dodecylsulfate and alcohol cosurfactants

S. Backlund[1]), G. Lundsten[1]), M. Ruths[1]), M. Sjöberg[2]), and T. Wärnheim[3])

[1]) Department of Physical Chemistry, Åbo Akademi, Åbo, Finland
[2]) Institute for Surface Chemistry, Stockholm, Sweden
[3]) Karlshamns Oils & Fats AB, Karlshamn, Sweden

Abstract: Ionic microemulsions, stabilized by sodium dodecylsulfate and alcohol cosurfactants, have been characterized with respect to phase behavior, particularly the composition of the different phases in Winsor type I, II, and III systems. The system containing brine, octane, sodium dodecylsulfate and 1-butanol shows a regular phase behavior, in conformity with the anticipated behavior, i.e., 2–3–2 phases. The phase transitions with increasing 1-butanol concentration are analogous to those occurring with increasing salinity, when compared with previous investigations. For the system containing brine, cyclohexane, sodium dodecylsulfate and benzyl alcohol, the behavior becomes more complex, even at low surfactant contents. The transitions are 2–3–2–3–2 phases over a wide composition range. The compositions of the phases at equilibrium are discussed in relation to structural considerations, as determined by conductivity and NMR self-diffusion measurements.

Key words: Microemulsion; partial phase equilibria; surfactant partition; cosurfactant partition; sodium dodecylsulfate; 1-butanol; benzyl alcohol; conductivity; NMR self-diffusion

Introduction

Microemulsions, systems of water, oil and amphiphile forming a single isotropic, low-viscous and thermodynamically stable solution [1] have been in the focus of surfactant research for a considerable period of time. This has also led to a more detailed and balanced description of these systems; studies of phase equilibria have elucidated many of the apparent and real complexities of these solutions [2], while a spectroscopic technique like NMR has proved to be a powerful tool when describing the structure and dynamics of microemulsions [3].

With many of the original questions resolved, the research in microemulsions has somewhat changed its direction; the interest is now more concerned with new topics such as nonaqueous microemulsion [4] or novel applications of microemulsions within material science [5] or as reaction media [6]. However, it is also clear that many basic issues have been superficially covered. Studies concerned with determining the composition of the phases and partition of the components in multiphase regions are still scarce for more complex systems.

In this paper, quintinary ionic microemulsions stabilized by an anionic surfactant, sodium dodecylsulfate (NaDDS), and an alcohol cosurfactant, either 1-butanol or benzyl alcohol (BA), have been studied. The phase behavior at constant content of surfactant has been investigated, including determination of the partition of the components in the multiphase regions. According to the Winsor nomenclature [7], studies have been made of transitions from type-I or type-III, and further to type-II systems, not induced by changes in salinity, but rather by addition of cosurfactant. Further phase transitions, illustrating the full complexity of these multicomponent systems, are shown to occur for a system containing benzyl alcohol. In part, the separated phases have also been structurally characterized, using the FT-NMR PGSE self-diffusion technique [3, 9] as well as conductivity measurements.

Materials and methods

Chemicals

Sodium dodecylsulfate (NaDDS) (BDH, specially pure) and sodium chloride (Merck, p.a.) were dried in an evacuated desiccator at 320 and 390 K, respectively. Octane (Fluka, puriss), cyclohexane (Merck, p.a.), 1-butanol (Merck, p.a.), benzyl alcohol (Merck, p.a.) were used as received. Distilled water was filtered through Millipore Q-filter before use. All samples were prepared by weight. Hyamine (R) 1622, Disulphine blue VN 150, dimidium bromide (Merck, for tenside test) and chloroform (Fluka, puriss) were used in the determinations of surfactant mass fraction.

Phase behavior

The phase behavior was determined by preparing samples at fixed composition and allowing them to equilibrate in a thermostatted bath at 298.2 K and visually determining the number of phases present. At low cosurfactant contents, coarse emulsions were observed to persist for a long time, but in most cases equilibrium was attained within a few hours. 250–350 samples were prepared for each partial phase diagram. The Krafft-point of sodium dodecylsulfate is around 300 K in 0.5 m NaCl, but is rapidly lowered upon addition of alcohol. No precipitation of NaDDS occurred in any system, even upon long time storage. The partial phase diagrams are presented as a function of α, i.e., the mass ratio (mass of hydrocarbon)/((mass of hydrocarbon) + (mass of brine)).

Chemical analysis

The partition of NaDDS was determined on separated phases by two-phase titration with Hyamine, as described by Reid et al. [8]. The estimated relative error is smaller than 2%. The partition of benzyl alcohol was determined by UV-spectrophotometry at $\lambda = 258$ nm (organic phase) or $\lambda = 255$ nm (brine phase) using a Shimadzu UV-240 instrument. The relative errors in the aqueous and middle phases are about 2%, and somewhat larger in the organic phases. Water contents were determined with a Karl Fischer Coulometer 652 (Methrohm).

Conductivity measurements

Conductivities of the separated phases were determined by means of a Wayne–Kerr bridge with automated recording fo the resistances, at 298.2 K.

Self-diffusion measurements

Self-diffusion coefficients of water and hydrocarbon were determined on a JEOL FX-100 FT NMR spectrometer at 298.2 K, with the FT-NMR PGSE method [9]. The accuracy of the determined self-diffusion coefficients is better than 5%.

Results and discussion

1-butanol system

Partial phase diagrams for the system 0.5 m NaCl (aq)-sodium dodecylsulfate-1-butanol-octane were determined at different mass fractions of surfactant, w(NaDDS) = 0.01, 0.03, and 0.05 (Fig. 1a). The area of the three-phase region, the Winsor-III system, decreases when approaching the boundary of the one-phase microemulsion, the Winsor-IV system. This is further illustrated in Fig. 1b, which shows the extension of the multi-phase regions at increasing mass fraction of surfactant and $\alpha = 0.7$. A three-phase region occurs at rather low mass fraction of surfactant, w(NaDDS) < 0.005, while the one-phase region (Winsor IV) is entered at around w(NaDDS) = 0.07.

The efficiency of the surfactant to form microemulsions is due to its enrichment at water-oil interfaces and, as stressed by Shinoda, its insolubility in the aqueous and organic phases [10]. Single alkyl chain ionic surfactants are too hydrophilic, but microemulsions are formed by adding an alcohol cosurfactant and/or by increasing the salinity.

Thus, the partition of the different components has been investigated in great detail with determining the composition of the individual phases (Fig.

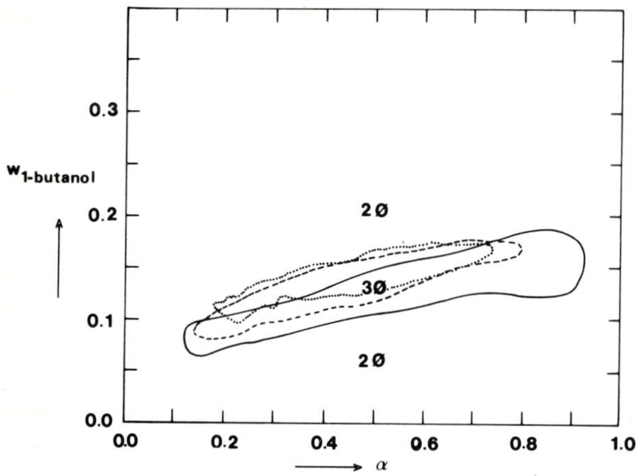

Fig. 1a. Partial phase diagram for 0.5 m NaCl(aq)-octane-1-butanol-sodium dodecylsulfate (NaDDS) at varying w(1-butanol) and three different mass fractions w(NaDDS) at 298.2 K. The mass ratio $\alpha = w(\text{octane})/(w(\text{octane}) + w(\text{brine}))$ is varied from 0 to 1. (——) w(NaDDS) = 0.01; (– – – –) w(NaDDS) = 0.03; (. . . .) w(NaDDS) = 0.05

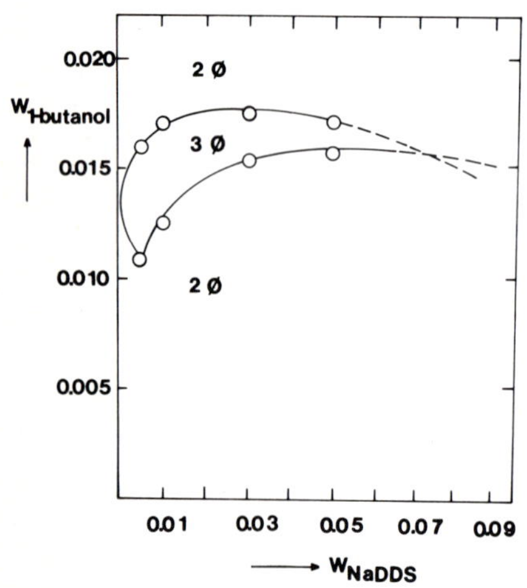

Fig. 1b. Partial phase diagram for 0.5 m NaCl(aq)-octane-1-butanol-NaDDS at $\alpha = 0.7$, varying w(1-butanol) and w(NaDDS) at 298.2 K. $n\phi$ denotes number of phases

2). In Fig. 2 the molality of the surfactant in the individual phases at increasing mass fraction of 1-butanol and at mass ratio $\alpha = 0.5$ is shown. The surfactant will, in accordance with the definition of the Winsor-I, -III, and -II systems, partition in the aqueous, middle, and organic phases, respectively. Thus, increasing the cosurfactant concentration induces the same transitions as with increasing salinity [7]. Surfactant molality in the other phases

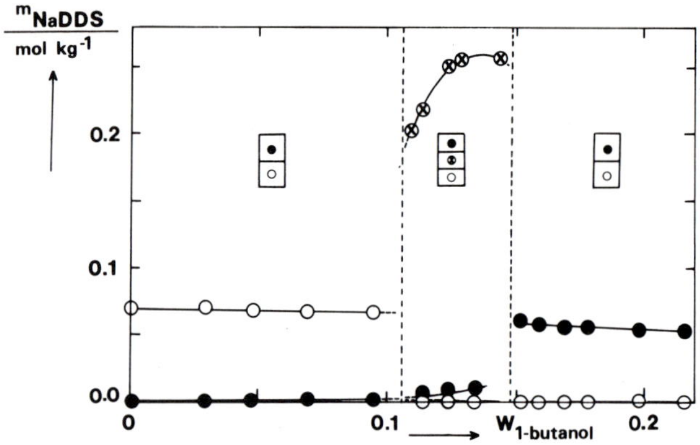

Fig. 2. The molality of NaDDS at 298.2 K in the lower (○), middle (⊗), and upper (●) phases for varying mass fraction w(1-butanol). The ratio $\alpha = w(\text{octane})/(w(\text{octane}) + w(\text{brine})) = 0.5$

is very low, but is not zero; in fact, an increase is evident upon increasing mass fraction of 1-butanol, in the organic phase in the Winsor-III region. Composition data for 1-butanol in this system has been reported previously [11]. It could be noted that the butanol molality increases in all phases with increasing total mass fraction of butanol in the system, except for the aqueous phase, which becomes saturated with butanol above w(1-butanol) = 0.05 [11].

The regular behavior of the 1-butanol system is also evident when studying the solution structure using conductivity measurements (Fig. 3). The conductivity of the lower, aqueous phase remains high (water-continuous system) and that of the upper, organic phase extremely low (oil-continuous system), throughout the progression of phase transitions. The conductivity of the middle, microemulsion phase is high throughout, showing that the system is continuous in water. Systems of similar composition have been investigated using the FT-NMR PGSE-technique, where the middle phase was shown to be bicontinuous [12], with a proposed structure of deformed, irregular surfactant-cosurfactant lamellae containing intertwined oil and water layers.

Benzyl alcohol system

The system 0.5 m NaCl (aq)-sodium dodecylsulfate-benzyl alcohol-cyclohexane has been investigated in a similar manner. The partial phase

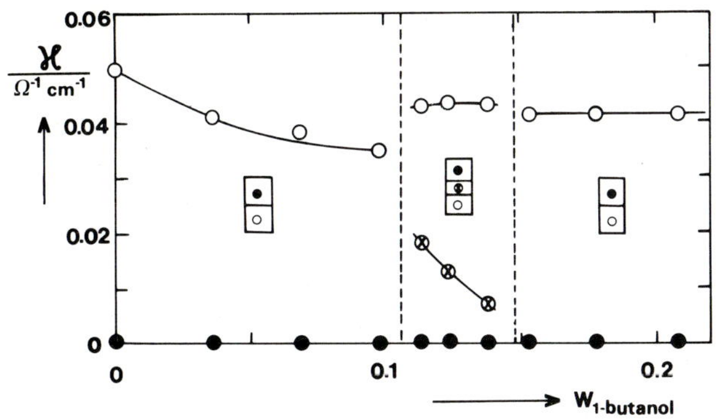

Fig. 3. The conductivity κ in the lower (○), middle (⊗), and upper (●) phases at varying mass fraction w(1-butanol) at 298.2 K and $\alpha = 0.5$

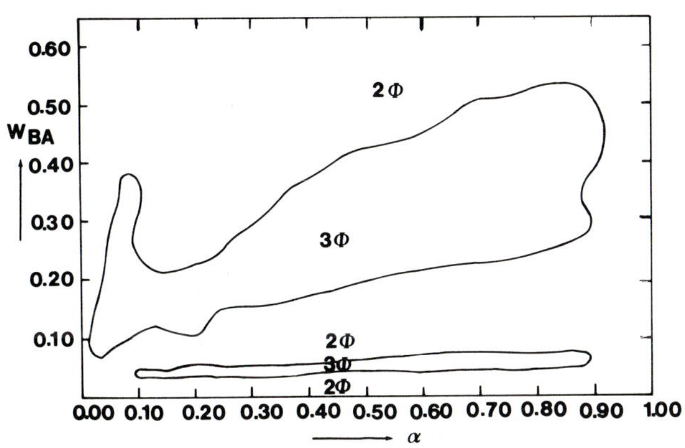

Fig. 4. Partial phase diagram for 0.5 m NaCl(aq)-cyclohexane-benzyl alcohol-sodium dodecylsulfate at varying w(BA) and mass fractions w(NaDDS) = 0.01 at 298.2 K. The mass ratio $\alpha = w(\text{cyclohexane})/(w(\text{cyclohexane}) + w(\text{brine}))$ is varied from 0 to 1

diagram of the system at w(NaDDS) = 0.01 is shown in Fig. 4, revealing the presence of two three-phase regions. The question of whether the three-phase regions are interconnected within a single three-phase body in the complete phase diagram will be addressed later on. Such a behavior can, of course, be produced in different ways in multi-component systems, but are less commonly reported in the literature.

The compositions of the separated phases are shown in Fig. 5, at $\alpha = 0.25$. The compositions follow a rather complex pattern, and only the most predominant features will be discussed in connection with the phase diagram.

At low mass fractions of benzyl alcohol, an aqueous, surfactant-rich phase is in equilibrium

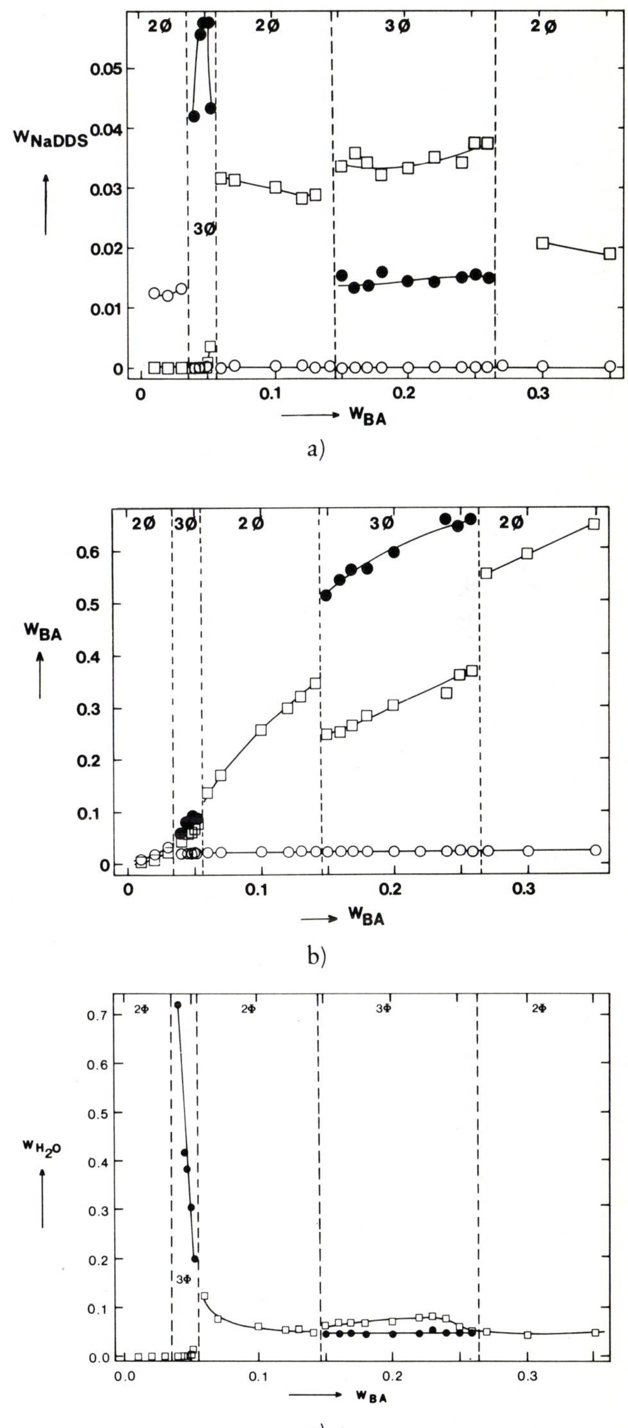

Fig. 5. The mass fraction of the different components in the system 0.5 m NaCl(aq)-cyclohexane-benzyl alcohol-sodium dodecylsulfate at varying w(BA) at 298.2 K in the lower (○), middle (●), and upper (□) phases. The mass ratio $\alpha = w(\text{cyclohexane})/(w(\text{cyclohexane}) + w(\text{brine})) = 0.25$. a) w(NaDDS); b) w(BA); c) $w(H_2O)$

with excess organic phase, a Winsor-I system. Further increasing the benzyl alcohol mass fraction leads to the separation of a third phase, a Winsor-III system, with an aqueous, a microemulsion, and an organic phase present. At even higher benzyl alcohol contents the system separates into a Winsor-II system with an excess aqueous phase in equilibrium with an organic, surfactant-rich phase. All this is analogous to the 1-butanol system.

However, the next transition is into a three-phase region, where there are appreciable amounts of surfactant present in two phases out of three. Compositional data show that the aqueous phase is saturated with benzyl alcohol and contains no surfactant. One of the other phases contains surfactant (w(NaDDS) = 0.03), a considerable amount of benzyl alcohol (w(BA) = 0.2–0.35), and some water ($w(H_2O)$ = 0.08); the second contains surfactant (w(NaDDS) = 0.015), more benzyl alcohol (w(BA) = 0.5–0.6), and some water ($w(H_2O)$ = 0.05). Finally, further increasing the benzyl alcohol content gives a two-phase system with an aqueous, surfactant-poor phase and an organic phase containing some water and the surfactant, i.e., a Winsor-II system.

The feature with two phases with reasonable high surfactant contents in the three-phase region poses an interesting fundamental problem: what type of aggregated structures, if any, are at equilibrium? To closer investigate the microstructure in the systems, determinations of water and hydrocarbon self-diffusion coefficients were performed on the separated phases in both three-phase regions; they are presented in Fig. 6. The first three-phase region is a typical Winsor-III system; the lower phase is water continuous, the middle phase has a rapid diffusion of both water and oil and should be termed bicontinuous, while the upper phase is oil-continuous. This is consistent with the conductivity measurements (Fig. 7), which show high conductivity values for the middle, microemulsion phase.

However, the characterization of the structure of the next three-phase region is less straightforward. Two phases, the middle and upper phase, both contain appreciable amounts of surfactant and water, as noted previously. The self-diffusion data are not unambigous: the hydrocarbon self-diffusion is rapid in both cases, 40–60% of pure hydrocarbon, while the water self-diffusion is comparatively slower, some 5–10% of pure water (Fig. 6).

Conductivity data give, however, further clues. The conductivity of the upper and the middle phase is about 10^{-5} cm^{-1} Ω^{-1}, clearly showing that neither of the phases is water continuous. The

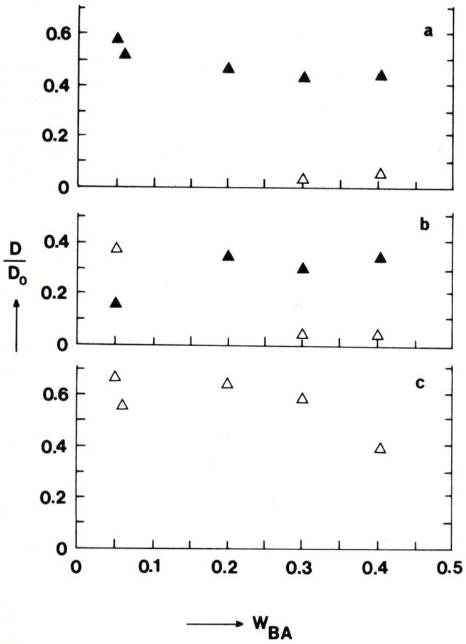

Fig. 6. Self-diffusion coefficients for (△) water and (▲) hydrocarbon in the three-phase regions of the system 0.5 m NaCl-cyclohexane-benzyl alcohol-sodium dodecylsulfate at varying w(BA) and 298.2 K. The mass ratio $\alpha = 0.5$. a) Upper phase; b) Middle phase; c) Lower phase

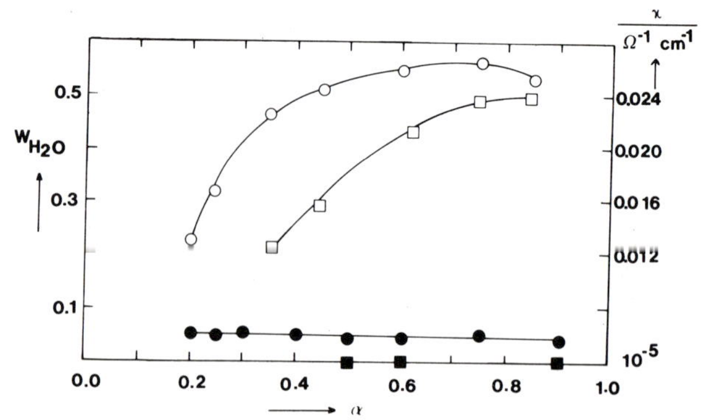

Fig. 7. Conductivity (□) and mass fraction of water (○) in the middle phases of the three-phase regions (cf. Fig. 4) at w(BA) = 0.05 (open symbols) and at w(BA) = 0.2 and 0.4 (filled symbols) as function of α at 298.2 K

conductivity of the aqueous phase is about 0.05 $cm^{-1}\Omega^{-1}$, which of course corresponds to an aqueous solution of 0.5 m NaCl. As a comparison between a typical Winsor-III system and this three-phase region, the conductivity and the water contents in both middle phases (cf. Fig. 4) are shown in Fig. 7. In a bicontinuous middle phase, the conductivity and water contents are proportional and high throughout. This contrasts to the constant conductivity and water contents of the other middle phase.

Now, consider the possible solution structure in the middle and the upper phases. Self-diffusion data and conductivity show that the phases are oil-continuous and that the water and surfactant exist in small aggregates of some sort. The large number of water molecules per surfactant molecule (40–60) in both the middle and upper phases, speaks against the possibility that the surfactant is present as monomer or as a hydrated ion-pair. The remaining alternative is thus that both phases contain small, inverse aggregates.

To summarize, two different microemulsion systems stabilized by sodium dodecylsulfate and an alcohol have been investigated in some detail. The first system, containing 1-butanol, is a common model system, and the phase diagram, composition data of separated phases and that of other measured properties also show a very regular pattern. With benzyl alcohol as cosurfactant further phase transitions occur and this has been discussed on the basis of the compositional data and structural investigations. At high concentrations of benzyl alcohol, the system enters a three-phase region where the phases contain appreciable amounts of water and surfactant. It is argued that the phases contain small, inverse aggregates.

Acknowledgements

S.B. thanks Nordiska Forskarkurser and G.L. thanks the Research Institute at Åbo Akademi for financial support.

References

1a. Danielsson I, Lindman B (1981) Coll Surf 3:391
b. Hoar TP, Schulman JH (1943) Nature 152:102
2a. Kahlweit M, Strey J (1987) J Phys Chem 91:1553 (and references therein)
b. Ekwall P, Mandell L, Fontell K (1970) J Colloid Interface Sci 33:215
c. Rance DH, Friberg S (1977) J Colloid Interface Sci 60:207
3a. Shinoda K, Lindman B (1987) Langmuir 3:135
b. Lindman B, Stilbs P, Moseley ME (1977) J Colloid Interface Sci 83:569
c. Wärnheim T, Henriksson U, Sjöblom E, Stilbs P (1984) J Phys Chem 88:5420
4. Friberg SE, Podizimec M (1984) Coll Pol Sci 262:252
5. Friberg SE (1991) paper presented at 11th Scand Symp Surf Chem, Bergen
6. Bergström K, Holmberg K (1991) paper presented at 11th Scan Symp Surf Chem, Bergen
7. Winsor PA (1954) Solvent Properties of Amphiphilic Compounds. Butterworths, London
8. Reid VW, Longman GF, Heinerth (1967) Tenside 4:292
9. Stilbs P (1987) Progr Nucl Magn Spectrosc 19:1
10. Kunieda H, Shinoda K (1982) J Disp Sci Techn 3:233
11. Backlund S, Boström G, Nordli KG, Wärnheim T (1990) Progr Coll Pol Sci 82:290
12. Ohlsson U (1988) Thesis, University of Lund, Lund

Authors' address:

S. Backlund
Dept. of Physical Chemistry
Åbo Akademi
SF-20500 Åbo
Finland

Studies of gelation of tetraethoxysilane in w/o microemulsions stabilized by nonionic surfactants NMR and phase equilibria

M. H. Selle, J. Sjöblom, S. E. Friberg[1]), T. Young[1]), and C.-C. Yang[1])

Department of Chemistry, University of Bergen, Norway
[1]) Department of Chemistry, Clarkson University, Potsdam, New York

Abstract: Acidic nonionic microemulsions have been used for gelation of tetraethyl orthosilicate (TEOS). It is shown that the release of ethanol will expand the microemulsion region after reaction. The gelation in the nonionic microemulsion (without hydrocarbon) is strongly dependent on surfactant and water. The gelation time strongly increases when surfactant is in excess of 22% by weight. The NMR spectra revealed only a minor reorganization during 4 months of the ethoxy attached to the Si atom for solutions with insufficient amount of water for gelation. This did not change with time (up to 4 months). All-in-all, the reactions with TEOS in acidic microemulsions are similar to those in organic solvents.

Key words: TEOS; microemulsion; phase equilibria: silica gel; gelation time; ^{29}Si NMR

Introduction

In the preparation of ceramics, the sol/gel technique [1] is of central importance, since it is of extreme interest in a large variety of technical applications [2] like advanced electronics and optics. The method includes hydrolysis and condensation of a metal organic compound in an alcohol solution. The reactions of tetraethoxysilane (TEOS), as a typical example, are in principle the following ones.

1) $Si(OC_2H_5)_4 + 4H_2O = Si(OH)_4 + 4C_2H_5OH$
2) $Si(OH)_4 = SiO_2 + 2H_2O$

These reactions are pH-dependent. At high pH values distinct silica particles are formed. At low pH the exchange of OH or alkoxy groups is extremely slow and few rearrangements take place in the structure [1]. Hence, the reorganization to a regular high-density polymer to form particles does not take place. Instead, a highly branched and irregular polymer is formed and its network includes the entire ethanol solution. The basic properties of this gel have been studied [3, 4], as well as those of the dry xerogel [5, 6].

The process has decisive advantages because of the high purity of the final product and, in addition, the molecular mixing of oxides is easily achieved. However, the method also has distinct drawbacks. The precursors are metal organic compounds that are expensive and a pronounced shrinkage is experienced, thus preventing the formation of large three-dimensional structures.

The first disadvantage was recently alleviated, when the microemulsion/gel method was introduced [7]. The microemulsion allows solubilization of inorganic salts to huge concentrations into a hydrocarbon and the gelation as well as the xerogel formation [7] are similar to the corresponding processes in the sol/gel ethanol solutions.

The studies of the hydrolysis and condensation reaction have so far been limited to w/o microemulsions stabilized by an ionic surfactant/pentanol combination [8]. However, we realized the need for gels of this kind stabilized with a nonionic surfactant, and our preliminary studies of these [9–11] revealed the potential for systems with extremely high salt concentrations.

In the present article, we introduce the hydrolysis and condensation in such systems. The result demonstrates the huge composition regions for gelation and also the influence of the water content on the reactions.

Experimental

Materials

The nonionic surfactants Macol LA4 (LA4) (Mazer Chemicals) and Triton N42 (Rohm and Haas Nordiska AB), hexane (Fisher Co.) and dodecane (Merck), silicon tetraethoxide ($Si(OEt)_4$) (Aldrich 99 + %) and (Hüls AG, 98%) were all used without purification. Hydrochloric acid (HCl) from Baker Chemicals was used to make the 4 M HCl solution and as the acid medium to adjust the H_2O to pH2.

Determination of phase diagrams

The w/o microemulsion region in the system (LA4/hexane/H_2O(pH2)) at room temperature ('25°C) was determined by preparing various weight ratios of LA4/hexane which were titrated with water under mixing. In the system (Triton N42/dodecane/4 M HCl) the components were weighed into glass vials and mixed. The boundary line of the phase regions was checked by long-time storage of the samples.

After equilibrium, 20 and 45% by weight silicon tetraethoxide was added to the Triton N42 system while 66.7 and 73.4% were added to the LA4 system.

Gel preparation

Silicon tetraethoxide was added to a glass vial of the amphiphilic association systems and vortexed to clear solution with closed cap and left to stand. Gelation was estimated by visual observation of the flow properties.

NMR

A JEOL-FX 90 AMHz NMR was used to obtain the ^{29}Si spectrum of the system (LA4/hexane/H_2O(pH2)) at room temperature. Neither TMS nor relaxation agent was added to the system, and each signal in the spectrum was recorded relative to $Si(OEt)_4$ signal.

Results

The results are presented in the following order. First the w/o microemulsion regions for both systems are shown in order to demonstrate the similarity between the behavior of the surfactants. The regions for isotropic solution after the initial reaction of the TEOS with water follow next, followed by the areas in which gelation took place. The variation of gelation time with composition is shown and, finally, NMR results show the influence on the reaction path by the initial level of water.

Microemulsion regions of the two systems

Both systems show a characteristic increase of water solubilization for surfactant/hydrocarbon ratios in the range 0.15–0.25 (Fig. 1). After the first increase the maximum water solubilization (W) approximately could be linearized as a function of the weight percent of surfactant (S) in the intervals $15 < S < 35$ (LA4 system) and $15 < S < 25$ (Triton N42 system):

$$W = 0.72S + 9.5 \quad \text{(LA4 system)} \tag{1}$$

and $$W = 3.6S - 23.1 \quad \text{(Triton system).} \tag{2}$$

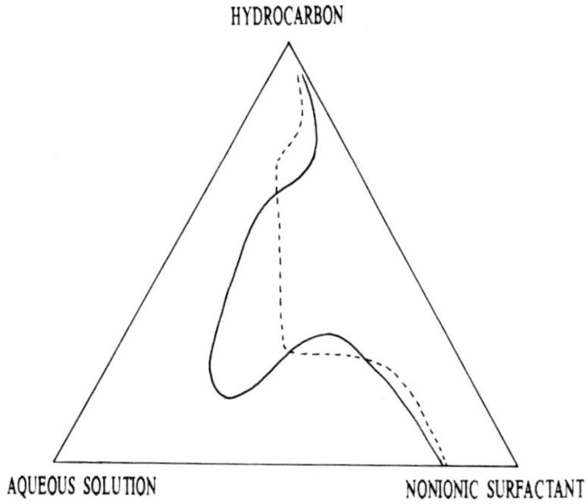

Fig. 1. Microemulsion regions in the system H_2O (4 M HCl)/ Triton N42/ dodecane (———) and H_2O (0.01 M HCl)/LA4/hexane (– – –)

Triton N42 is a more efficient solubilizing agent for water than is LA4, but the general shape of the two curves is similar.

The strong reduction of the isotropic solubilization region at high surfactant/hydrocarbon ratios was in both cases due to the presence of a lamellar liquid crystal. Its phase area was not determined.

Isotropic solution regions after addition of TEOS

Addition of silicon tetraethoxide gave solutions in huge regions not limited by the original microemulsion areas (Fig. 2). The two surfactant/hydrocarbon combinations gave similar regions for the isotropic solutions with one difference: the dodecane combination did not produce isotropic solutions without addition of surfactant. For the hexane/LA4 combination the following limitations were found along the water/hexane axis. 41.5% hexane was the maximum for 73.4% of silicon tetraethoxide counted on a total weight, and 51.5% hexane for 66.7% of the silicon compound. The silicon tetraethoxide was soluble in the hexane/surfactant solution, thus allowing some water to be dissolved with greater solubility for lower content of silicon tetraethoxide for both systems (Fig. 2).

Gel formation

The gels were formed only at higher water content than 48.5% by weight, as marked in Fig. 1 for

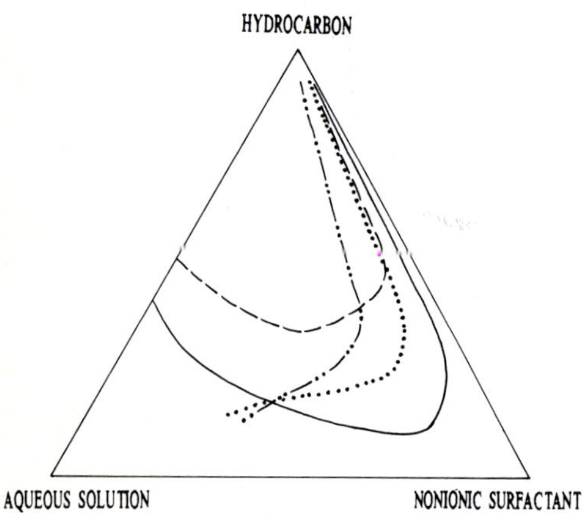

Fig. 2. The solution region after addition of silicon tetraethoxide. (—··) 20%, (····) 45%, (– – –) 66.7 and (———) 73.4% of TEOS, respectively

73.4% of silicon tetraethoxide counted on total weight. The solutions with water content less than 48.5% remained as transparent liquids during the time of the experiments (approximately 4 months for the LA4/hexane system).

In the system with Triton N42 a different approach was chosen. The surfactant and water were now combined directly with TEOS and gelation was observed for varying amounts of TEOS. The gelation was found for TEOS content in excess of a minimum value that depended on the water/surfactant ratio. The minimum TEOS concentration in wt.% was approximately

$$[\text{TEOS}] = -0.3S + 30 \tag{3}$$

with surfactant concentration also in wt.%. This result means that there is a maximum number of 27 water molecules per TEOS molecule to form a gel, and that the surfactant has no influence on the gelation.

On the other hand, the gelation time in the Triton N42 system depends strongly on the concentration of water and surfactant. The gelation time increased drastically with increasing amounts of Triton N42 and reduced amounts of water in the samples.

Three different weight ratios between water and silicon tetraethoxide were investigated, i.e., 70/30, 50/50, and 40/60. The geleation time was determined as a function of surfactant concentration (% by weight) (Fig. 3). This figure shows that with a small amount of Triton N42 present, the gelation time was short and constant with surfactant content. Addition of 22% by weight Triton N42 or more gave a linear increase of the gelation time.

The gelation time was plotted as a function of the mole ratio of water and surfactant (w/s), (Fig. 4). This figure shows that a very fast gelation takes place for water concentration (moles) in excess of 50 times the surfactant concentration.

NMR spectra

The NMR spectra for a sample with low water content and no hydrocarbon present (H_2O/Si mole ratio = 0.62) are presented in Fig. 5 for different times after preparation. The spectra show two features of interest [12]. At first, the unreacted silicon tetraethoxide is a significant component, also after 4 months. Secondly the strongest growing high resolution component is the Si–O–Si$(OEt)_3$ or

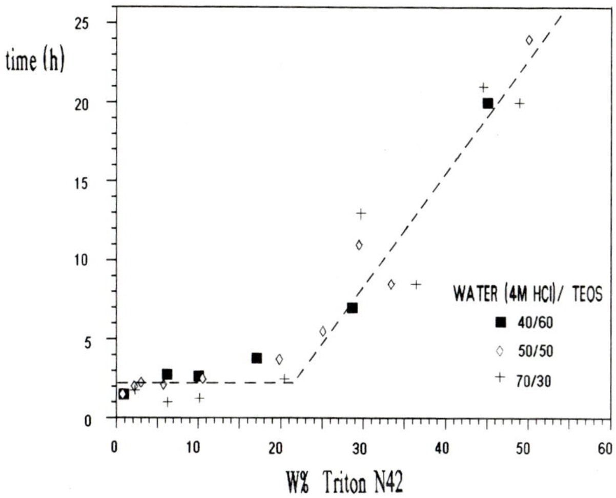

Fig. 3. The gelation time as a function of Triton N42 concentration (% by weight)

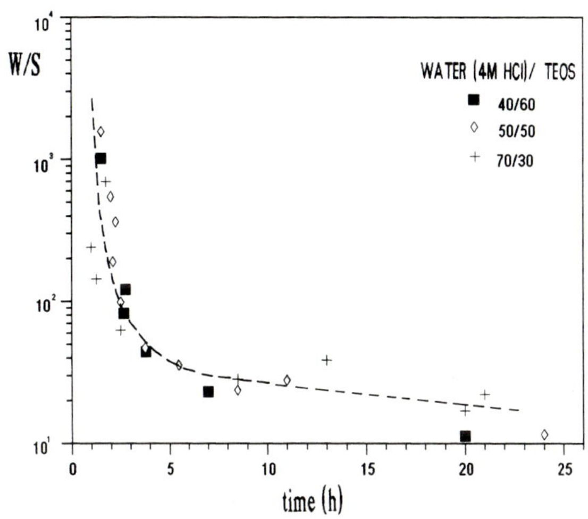

Fig. 4. The gelation time as a function of the mole ratio of water and Triton N42 (w/s)

$Si–O–Si(OH)(OEt)_2$. All other components including $(–Si–O–)_2Si(OEt)_2$ are reduced after 4 months.

An increase of the water content to an equal molar ratio of the silicon tetraethoxide gave different result (Fig. 6). Now, the signal for silicon tetraethoxide completely disappears after 8 days. The $Si–O–Si(OEt)_3/Si–O–Si(OH)(OEt)_2$ was still a significant component, but the $(SiO)_2Si(OEt)_2/(SiO)_2(OH)(OEt)_2$ were more prevalent. In addition, the triple connected $(SiO)_3Si(OEt)/(SiO)_3SiOH$ was also significant.

An increase of the water/silicon ratio to 1.65 gave an enhanced trend toward more condensation

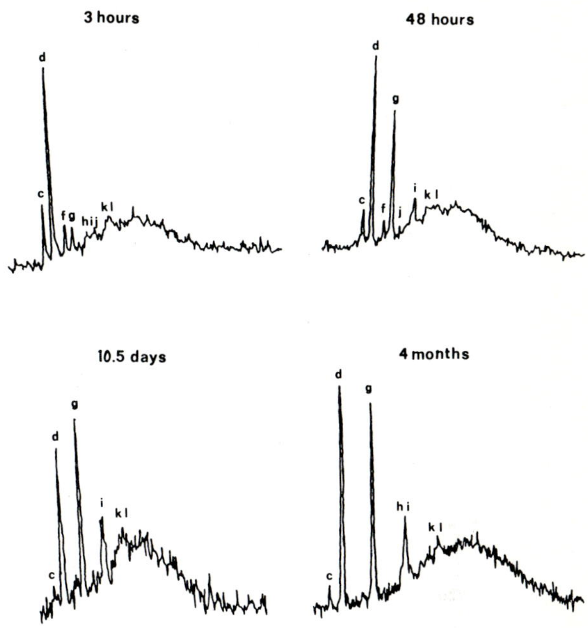

Fig. 5. NMR spectra for a sample with H_2O/Si mole ratio = 0.62 at different times after preparation. (No hydrocarbon present).
Figures 5, 6, and 7: a = $Si(OH)_3(OEt)$, b = $Si(OH)_2(OEt)_2$, c = $Si(OH)(OEt)_3$, d = $Si(OEt)_4$, e = $(SiO)Si(OH)_2(OEt)$, f = $(SiO)Si(OH)(OEt)_2$, g = $(SiO)Si(OEt)_3$, h = $(SiO)_2Si(OH)OEt)$, i = $(SiO)_2Si(OEt)_2$, j = $(SiO)_2Si(OH)_2$, k = $(SiO)_3Si(OH)$ and l = $(SiO)_3Si(OEt)$

products. The silicon tetraethoxide signal now was insignificant already after 5 hours (Fig. 7). Even the single connections $Si–O–Si(OH)_2(OEt)$, $Si–O–(OH)(OEt)_2$ and $Si–O–Si(OEt)_3$ were still significant, but not as dominant as the double connections $(SiO)_2–Si(OH)(OEt)$, $(SiO)_2Si(OEt)_2$, or the triple ones $(SiO)_3Si(OH)$ and $(SiO)_3Si(OEt)$. After 8 days the latter were completely dominant, and after 4 months the single connections were nonexistent.

Discussion

It is obvious that a colloidal system like a microemulsion system can have both advantages and disadvantages with regard to xerogel formation. The key to an understanding of the benefits of such a system is usually in the proper understanding of the microstructure of the microemulsion system. We will now summarize previous information about the structure of the acidic nonionic microemulsion utilized in this study.

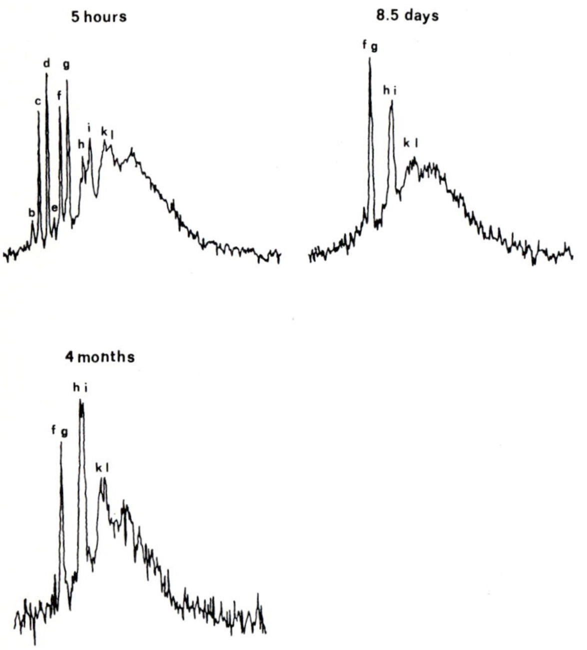

Fig. 6. NMR spectra for a sample with equal molar ratio of H_2O and Si at different times after preparation (see Fig. 5 for notation)

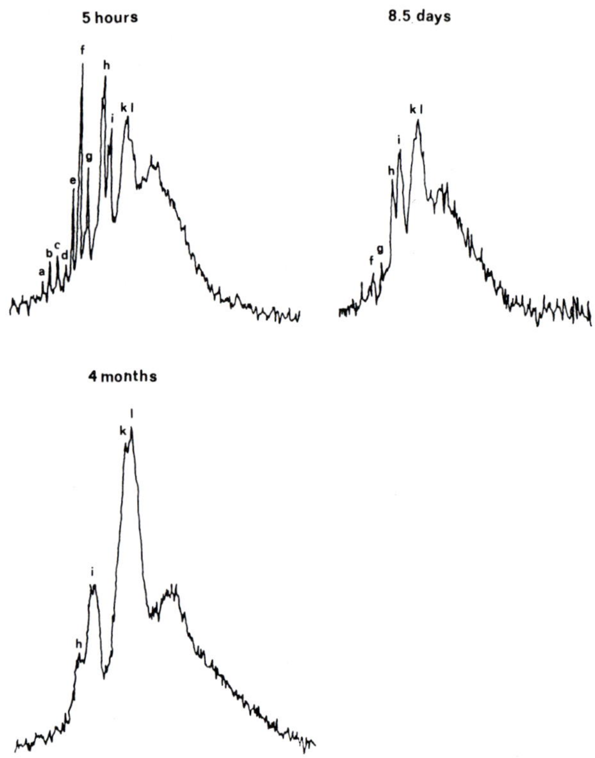

Fig. 7. NMR spectra for a sample with H_2O/Si mole ratio = 1.65 at different times after preparation (see Fig. 5 for notation)

Structure of nonionic acidic microemulsion

For microemulsion formation of nonionic surfactant the most central property is to balance the hydrophilic and lipophilic properties at the HLB temperature [13–16]. Basically, an acid will change the HLB range of the polyoxyethylene-based nonionic surfactant. This is due to the interaction between the acidic protons and the ether-oxygen in the oxyethylene chain [17]. As a consequence, the nonionic surfactant will become more polar and this will raise the HLB temperature of the system. Maximal solubilization of oil and water can hence be expected to take place at somewhat higher temperatures in the acidic solutions as compared to the neutral aqueous system.

The system Triton N42/water(H^+)/dodecane has been a good model system when previously utilizing acidic colloidal systems [18, 19]. A reason for this is that the system can hold more than 50% by weight of 4 M HCl at room temperature. Current spectroscopic investigations based on NMR self-diffusion and dielectric spectroscopy have focused on the structure of these microemulsions.

A reliable experimental method for structural interpretation of microemulsions is NMR self-diffusion measurements [20–25]. From the level of the self-diffusion of the components, qualitative information can be derived about the microstructure of the system.

The water self-diffusion coefficients for the system water or 4 M HCl/dodecane/Triton N42 have been determined [18]. The water self-diffusion coefficient first increases with increasing water (4 M HCl) content and then, around 30–40%, starts to decrease. However, the self-diffusion coefficient always stays at a low level. (The same situation is valid for the acid-free system.) The hydrocarbon diffusion is rapid and comparable to the diffusion of pure dodecane on the time-scale of the diffusion measurement, i.e., 100 ms. The information from these experiments indicates that the structure of the nonionic microemulsion system is based on closed aggregates, i.e., droplets in dodecane.

Further, the same microemulsion system (in the presence of electrolyte) show a large dielectric dispersion in the MHz region [19]. The origin of this dispersion is an interfacial polarization or a so-called Maxwell–Wagner–Sillars dispersion [26–28]. Such polarizations occur in dispersed systems where one phase is conducting. The level of the dispersion depends on the conductivity and aggregate shape. The result from these measurements supports the NMR data in explaining the aggregate shape of the nonionic microemulsion system [19, 29].

Figures 3 and 4 reveal two interesting details about the gelation in the oil-free microemulsion system. With a large excess of water to surfactant the reaction proceeds within 2–3 h, but below 20 molecules of water to Triton N42 there is a conspicuous prolongation of the gelation time. It is also interesting to notice the surfactant dependence. There is a clear retardation of the gelation when the amount of surfactant exceeds 22% by weight.

For the microemulsion systems consisting of closed aggregates the surfactant membrane will most likely constitute a barrier, preventing the reaction to take place immediately. This obstruction will be most efficient at low water contents. However, when ethanol starts to form there will be fundamental changes with regard to solution structures. It is well known from previous investigations that closed aggregate structures will break down when a low chain alcohol is added. This means that the surfactant barrier will be less efficient and the gel formation will accelerate.

The microemulsion structure did not influence the quantitative aspects of the reaction. As pointed out in the results section, gels were not formed at lower water percentages than 48.5% by weight for 73.4% TEOS. These values show a minimum water/TEOS molecular ratio of 2.0 in accordance with Eqs. (1) and (2).

The NMR spectra revealed the fact that little or no reorganization takes place at low pH values of the groups attached to the silicon atom [1]. There was only insignificant changes in the spectra between 10 days and 4 months (Fig. 4). There was no tendency for groups such as $SiOSi(OEt)_3$ to reorganize to form SiO_2 and $Si(OEt)_4$. In the same manner, the partially hydrolyzed/condensed species at higher water/TEOS ratios remained constant with time (Figs. 6, 7).

Conclusion

The reactions of TEOS in water-in-oil microemulsions stabilized by nonionic surfactants were found to be similar to those in organic solvents.

Acknowledgements

S.E.F. thanks the New York State Commission for Science and Technology for support through the CAMP program at Clarkson University. J.S. acknowledges the Royal Norwegian Council for Scientific and Industrial Research (NTNF) for a travel and subsistence grant to Clarkson University, USA.

References

1. Brinker CJ, Scherer GW (1990) Sol-Gel Science, The Physics and Chemistry of Sol-Gel Processing. Academic Press, New York
2. Sol-Gel Technology for Thin Films, Fibers, Performs, Electronics and Speciality Shapes. LC Klein (ed.) (1988) Noyes Publications, New Jersey
3. Sakka S, Kozuka H, Adachi T (1988) J Non-Cryst Solids 102:263
4. Scherer GW, Pardanek SA, Swiatek RM (1988) J Non-Cryst Solids 107:14
5. Cheng JJ, Wang D-W (1988) J Non-Cryst Solids 100:288
6. Shukla BS, Johri GP (1988) J Non-Cryst Solids 101:263
7. Friberg SE, Yang C-C (1988) In: Doyle FM, Raghaven S, Somasundaran P, Warren GW (eds.) Innovations in Materials Processing Using Aqueous, Colloid and Surface Chemistry. The Minerals, Metals and Materials Society, 181
8. Friberg SE, Yang C-C, Sjöblom J (1992) Langmuir 8:372
9. Friberg SE, Amran A, Yang C-C, Sjöblom J, Colloids Surfaces (in press)
10. Saeten, JO, Førdedal H, Amran A, Friberg SE, J Colloid Interface Sci (in press)
11. Saeten JO, Selle MH, Sjöblom J, Friberg SE, Gestblom B (1991) J Solution Chem 20:12
12. Irwin AD, Holmberg JS, Jonas J (1988) J Non-Cryst Solids 101:249
13. Staito H, Shinoda K (1970) J Colloid Interface Sci 32:649
14. Shinoda K (1972) J Colloid Interface Sci 34:278
15. Shinoda K, Kunieda H (1973) J Colloid Interface Sci 42:381
16. Shinoda K, Friberg SE (1975) Adv Colloid Interface Sci 4:281
17. Schott H, Royce AE (1984) J Pharm Sci 73:793
18. Sjöblom J, Söderlund H, Wärnheim T (1986) In: Setton R (ed.) Chemical Reactions in Organic and Inorganic Constrained Systems. D. Reidel Publishing Company, 305

19. Gestblom B, Sjöblom J (1988) Langmuir 4:360
20. Lindman B, Stilbs P, Moseley ME (1981) J Colloid Interface Sci 83:569
21. Stilbs P, Moseley ME, Lindman B (1980) J Magn Reson 40:401
22. Stilbs P, Rapacki K, Lindman B (1983) J Colloid Interface Sci 95:583
23. Wärnheim T, Sjöblom E, Henriksson U, Stilbs P (1984) J Phys Chem 88:5420
24. Nilsson PG, Lindman B (1982) J Phys Chem 86:271
25. Nilsson PG (1984) Thesis, University of Lund, Lund
26. Maxwell JC (1982) A Treatise on Electricity and Magnetism. Clarendon, Oxford
27. Wagner KW (1914) Arch Electrotech 3:100
28. Sillars RW (1937) Proc Inst Elec Engs London 80:378
29. Sjöblom J, Gestblom B, In: Lindman B, Friberg S (eds.) Organized Solutions. Marcel Dekker (in press)

Authors' address:

M. H. Selle
Dept. of Chemistry
University of Bergen
N-5007 Bergen
Norway

Adsorption of surfactants at interfaces – studies by small-angle neutron scattering*)

R. H. Ottewill

School of Chemistry, University of Bristol, U.K.

Abstract: Small-angle neutron scattering provides a valuable technique for the study of adsorption at interfaces. It is particularly valuable for particles in nonaqueous media where the surfactant is present during particle formation and, hence, is an essential factor to ensure colloid stability of the dispersion. The method of contrast variation, using mixtures of hydrogenated and deuterated solvents can be used and from the data obtained an analysis of the amount adsorbed estimated. This technique is also applicable to micelles and work is described on 6-(eicosyldimethylammonio) hexanoate. In addition, neutron reflectivity studies on the latter material at the air-water interface are described; these studies allow a comparison to be made of the area occupied at a micelle surface and at a planar interface.

Key words: Neutron scattering; calcium carbonate; polystyrene; adsorption; surfactant

Introduction

The stability of particles in the colloidal sense frequently depends on materials adsorbed to the surface [1, 2]. The types of materials employed include surface active agents, natural or synthetic macromolecules, chemically grafted chains etc. Consequently, characterisation of adsorption at interfaces has always been a topic of major interest [3, 4]. Classically, adsorption isotherms have usually been obtained by starting with the basic materials, e.g. a powder or dispersion, and determining the uptake on the surface; a variety of techniques is available for this procedure. However, in a number of cases, to ensure ultimate colloid stability, dispersions are formed in the presence of a stabilising agent so that the material is strongly adsorbed or grafted and thus becomes an integral part of the particle [5–7]. For examination of the latter situation a technique is required which allows discrimination to be made between the adsorbed layer and the core particle. Over the past 15 years the technique of small-angle neutron scattering has been developing rapidly [8], both experimentally and theoretically for the examination of adsorbed states [9, 10]. A major advantage of the method is based on the significant difference between the scattering cross-section of hydrogen and deuterium. This can be exploited with both particulate systems and in the study of adsorption at planar interfaces using the technique of neutron reflectivity [11, 12]. The present work gives a brief outline of the application of small-angle neutron scattering to the study of adsorption.

Small-angle neutron scattering

The basis of a neutron-scattering experiment is illustrated schematically in Fig. 1. When the neutrons are generated in a reactor the primary beam passes through a velocity selector which determines the wavelength of the beam and the spread of wavelengths [13]. Neutrons can also be generated by a pulsed source [14]. After collimation with the neutron guides the beam passes through the sample, often contained in a quartz cuvette, and

*) Invited Lecture, 11th Scandinavian Symposium on Surface Chemistry, June 1991

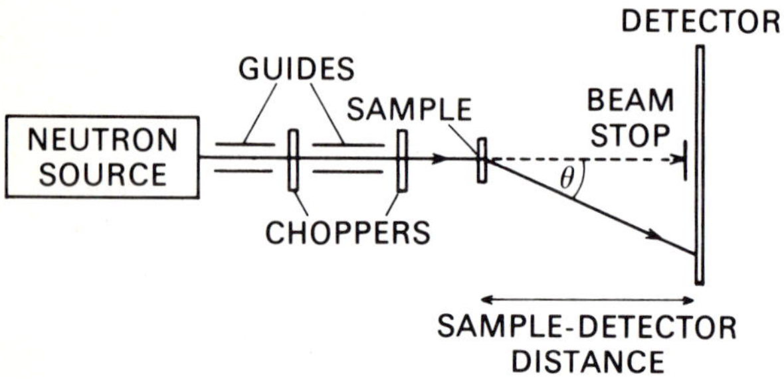

Fig. 1. Schematic representation of small-angle neutron scattering equipment

reaches the beam stop. The neutrons scattered by the sample at various angles reach the multi-detector. The latter is an extremely important part of the instrument, and usually consists of a 64×64 array of detector cells, i.e. in total, 4096 cells. The neutrons captured by the cells are converted into electrical pulses which can be registered on a computer for subsequent analysis. For isotropic samples the results can be radially averaged, with each radius on the detector corresponding to a value of the scattering angle θ. Any anisotropy of scattering by the sample is readily detected in the two-dimensional array of data. The range of θ values can be adjusted by changing the sample-detector distance. Frequently, a material such as water which has considerable incoherent scattering from the H-atoms and gives a constant intensity profile as a function of θ is used as a standard substance to convert the scattered intensity into absolute units, the number of neutrons scattered per unit of time per steradian. For elastic scattering the magnitude of the scattering vector is defined by the quantity

$$Q = 4\pi \sin(\theta/2)/\lambda \, , \tag{1}$$

where λ is the wavelength of the incident beam.

Theory

Dispersions of homogeneous spherical particles

For a dispersion of spherical particles, each of volume V_p, containing N_p particles per unit volume, the intensity of scattering $I(Q)$ for non-interacting particles at a scattering vector of magnitude Q is given by [15]

$$I(Q) = A(\rho_p - \rho_m)^2 \, N_p \, V_p^2 \, P(Q) \, , \tag{2}$$

where ρ_p is the coherent neutron scattering-length density of the particles and ρ_m is the coherent neutron scattering-length density of the medium. A is a constant which can be obtained by calibration in order to convert intensities into absolute units. $P(Q)$ is the single-particle scattering function given by [16] as

$$P(Q) = [3(\sin QR - QR \cos QR)/(QR)^3]^2 \, , \tag{3}$$

with R being the particle radius. We note also that $V_p = 4\pi R^3/3$, and that the volume fraction of the system $\phi = N_p V_p$.

However, few colloidal systems are really monodisperse and usually some polydispersity has to be introduced in order to compare calculated curves with experimental results. A convenient distribution for computational purposes is the zeroth-order log-normal distribution [17], namely,

$$p(R) = \frac{\exp[-(\ln R/R_m)^2/2\sigma_0^2]}{(2\pi)^{1/2} \, \sigma_0 \, R_m \exp(\sigma_0^2/2)} \, , \tag{4}$$

where R_m is the modal radius of the particles and σ_0 is a parameter which gives a measure of the width of the distribution. It is related to the standard deviation σ_d by

$$\sigma_d = R_m[\exp(4\sigma_0^2) - \exp(3\sigma_0^2)]^{1/2} \tag{5}$$

For a narrow distribution the shape is Gaussian and given by

$$\sigma_d = \sigma_0 R_m$$

In order to allow for polydispersity, Eq. (2) can be summed over the distribution of R values given by Eq. (4).

Concentric sphere particles

The concept of a spherical core particle surrounded by an outer concentric shell of another material has many applications in the study of colloidal systems. The outer shell could, for example, be an adsorbed layer of surfactant, a layer of polymer, a macromolecule, a grafted layer, etc. For the present purpose it will be assumed that the core particle is homogeneous with a radius R_1 and a coherent scattering length ρ_c and that the shell is homogeneous also with a thickness $R_2 - R_1$ and a coherent neutron scattering length of ρ_s. The composite particle can then be dispersed in a medium of coherent scattering length density ρ_m giving for the scattered intensity [7, 18].

$$I(Q) = AN_p[(\rho_s - \rho_m)(A_2 - A_1) + (\rho_c - \rho_m)A_1]^2 \quad (6)$$

with

$$A_2 = 3V_T[(\sin QR_2 - QR_2\cos QR_2)/(QR_2)^3]^2 \quad (6a)$$

and

$$A_1 = 3V_c[(\sin QR_1 - QR_1\cos QR_1)/(QR_1)^3]^2 . \quad (6b)$$

Also,

$$V_T = 4\pi R_2^3/3 \quad \text{and} \quad V_c = 4\pi R_1^3/3 .$$

For the condition that $R_2 = R_1$, i.e. no shell present, then

$$I(Q) = AN_p[(\rho_c - \rho_m)A_1]^2 , \quad (7)$$

which is identical with Eq. (2). Moreover, if the shell is matched by the dispersion medium, which is equivalent to putting $\rho_s = \rho_m$ then Eq. (2) is again obtained. Alternatively, the core particle can be matched by putting $\rho_m = \rho_c$, which gives,

$$I(Q) = AN_p[(\rho_s - \rho_m)(A_2 - A_1)]^2 \quad (8)$$

an equation for scattering by a concentric shell.

Coherent scattering length densities

In Table 1 a list is given of the scattering length densities of various dispersion media, surfactants and particulate materials. The values listed indicate the substantial changes which can be made in the coherent scattering length density by replacing hydrogen atoms with deuterium atoms. This enables the study of hydrogenated molecules to be made, for example, on deuterated particles and vice versa since excellent contrast can then be achieved between the core and shell. Moreover, by using either pure liquids or mixtures of hydrogenated and deuterated solvents a wide variation in ρ_m can be achieved.

Applications

Adsorption on to a bare particle

One of the conditions which is unique to neutron scattering is that ρ_m can be made zero. This occurs, for example, for a mixture of 8% D_2O and 92% H_2O by volume (see Table 1). From Eq. (6), we then obtain

$$I(Q) = AN_p[(\rho_c - \rho_s)A_1 + \rho_s A_2]^2 . \quad (9)$$

Thus, starting with a bare particle of radius R_1 and coherent scattering length ρ_c, e.g. on *h*-polystyrene particles by adsorption of a surfactant molecule, e.g. d_{23}-dodecanoic acid, a concentric shell can be built up of thickness $R_2 - R_1$ and scattering length density ρ_s. This provides a useful way of determining an adsorption isotherm [10, 19].

Table 1. Coherent scattering length densities

Material	Formula	$\rho/10^{10}\ cm^{-2}$
Solvents		
Water	H_2O	− 0.56
Deuterium oxide	D_2O	6.40
h_{26}-Dodecane	$C_{12}H_{26}$	− 0.46
d_{26}-Dodecane	$C_{12}D_{26}$	6.71
h_8-Toluene	C_8H_8	0.94
d_8-Toluene	C_8D_8	5.63
Surfactants		
d_{23}-Dodecanoic acid*)	$C_{11}D_{23}COOH$	6.43
6-(Eicosyldimethylammonio)**) hexanoate	$C_{20}H_{41}N(CH_3)_2(CH_2)_5COO^-$	− 0.08
$C_{10}E_5$†)	$C_{10}H_{21}(CH_2CH_2O)_5OH$	0.17
Alkyl aryl sulphonate†)	$C_{30}H_{53}SO_3^-$	0.35
Substrates		
h-Polystyrene	$(C_8H_8)_n$	1.41
Calcium carbonate	$CaCO_3$	4.72

Densities: *) 0.97 g cm^{-3}; **) 0.933 g cm^{-3}; †) assumed 1.0 g cm^{-3}

By extrapolating the experimental data to $Q \to 0$ using a plot of $\ln I(Q)$ against Q^2, we obtain for $\rho_m = 0$ and $Q \to 0$,

$$I(Q \to 0) = AN_p[\rho_s V_s + \rho_c V_c]^2 , \quad (10)$$

where V_s is the volume of the shell of adsorbed material. Thus the adsorbed layer material parameter is, $\rho_s V_s$. Moreover, V_s can be written as

$$(R_2 - R_1) S_c ,$$

where S_c is the surface area of the core particle. Also,

$$\rho_s = b_s / v_s ,$$

where b_s is the isotopic scattering length of the adsorbate and v_s its molecular volume, giving

$$\rho_s V_s = (R_2 - R_1) S_c b_s / v_s$$

and the number of molecules in the concentric shell as,

$$n_s = (R_2 - R_1) S_c / v_s$$

and

$$\rho_s V_s = n_s b_s .$$

Hence, for the particle with an adsorbed layer

$$I(0)_{p+s} = AN_p[\rho_p V_p + n_s b_s]^2$$

and for the base particle,

$$I(0)_p = AN_p[\rho_p V_p]^2 ,$$

giving

$$n_s = \frac{V_p \rho_p}{b_s} \left[\frac{[I(0)_{p+s}]^{1/2}}{[I(0)_p]^{1/2}} - 1 \right] . \quad (11)$$

Alternatively expressing the adsorption in mols of material per unit area, we obtain

$$\Gamma_s = n_s / S_c N_{AV} ,$$

where N_{AV} is Avogadro's number.

In Fig. 2 an adsorption isotherm is given for the adsorption of d_{23}-dodecanoic acid on a hydrogenated polystyrene latex of mean diameter 410 Å. Measurements were made at 22 °C and pH 8.1. The plateau region of the isotherm corresponded to an area per molecule of 28 ± 5 Å^2. All the results were obtained in a medium composed of 8% D_2O giving $\rho_m = 0$.

An isotherm is also shown in Fig. 3 for the adsorption of decyl pentaoxyethylene glycol monoether ($C_{10}E_5$) on a hydrogenated polystyrene latex of diameter, 312 Å. The results were obtained using a medium composed of 30% H_2O and 70% D_2O, $\rho_m = 4.28 \times 10^{10}$ cm^{-2}, and analysed using Eq. (6). The temperature was maintained at 25 °C. The plateau corresponded to an area per molecule of 71 ± 5 Å^2.

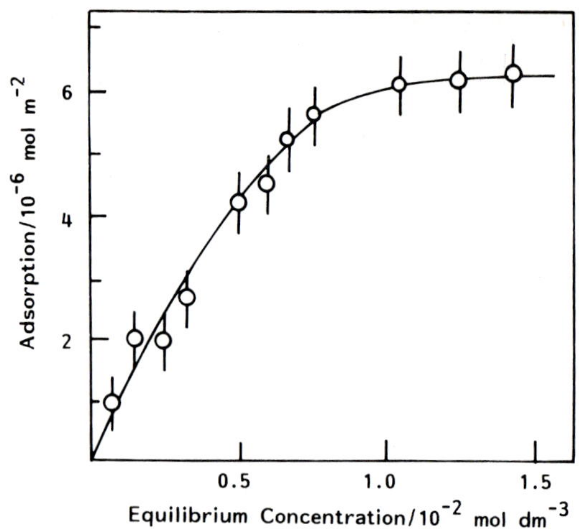

Fig. 2. Adsorption isotherm for d_{23}-dodecanoic acid on h-polystyrene at pH 8.1 and 22 °C; electrolyte concentration 2.2×10^{-2} mol dm^{-3}. ○, experimental points obtained at $\rho_m = 0$, Eq. (11)

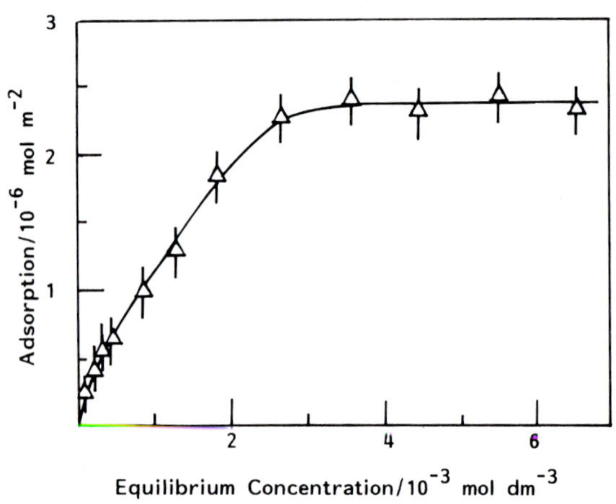

Fig. 3. Adsorption isotherm for $C_{10}E_5$ on h-polystyrene at 25 °C. △, experimental points obtained at $\rho_m = 4.28 \times 10^{10}$ cm^{-2}; data analysed by fitting intensity curves using Eq. (6)

Particles with an integral adsorbed layer

An example of a colloid system where the adsorbed layer is an integral part of the particle and vital for the colloid stability of the system is that of calcium carbonate in a non-aqueous environment. For the preparation a microemulsion is prepared in a non-aqueous environment with, as the aqueous core, a solution of calcium hydroxide stabilised by an alkyl aryl sulphonic acid [7, 18]. In order to form the carbonate carbon dioxide is passed into the system whilst water is removed by distillation. The final product is a dispersion of an alkaline earth metal carbonate in a hydrocarbon environment stabilised by a layer of surfactant.

Equation (6) can be applied to examine systems of this type, but instead of working at $\rho_m = 0$, which gives insufficient information, a variation of ρ_m can be used utilising mixtures of hydrogenated and deuterated solvents. The effect of changes of ρ_m can also be simulated and Fig. 4 shows curves of $P(Q)$ against Q calculated for a particle with $R_1 = 28$ Å, $\sigma_0 = 0.10$, $R_2 = 46$ Å and ρ_c and ρ_s equal to 4.72×10^{10} cm^{-2} ($CaCO_3$) and 0.35×10^{10} cm^{-2} (surfactant) respectively. The match point between this composite particle and the medium is given by

$$\bar{\rho} = (\rho_s V_s + \rho_c V_c)/V_T ,$$

and for the conditions considered was 1.45×10^{10} cm^{-2}. The values of ρ_m are given in the legend to Fig. 4. As ρ_m approaches $\bar{\rho}$ the curve of $P(Q)$ against Q shows progressively a more pronounced peak which rises above $P(Q) = 1$. This behaviour results, as shown previously [7, 18], in negative values for the radius of gyration of the particle just below the contrast match point. When there is a large difference between the value of ρ_m and $\bar{\rho}$ then the curve has essentially the form expected for an isotropic sphere.

Some experimental observations on calcium carbonate particles stabilised by an alkyl aryl sulphonate in a non-aqueous environment are shown in Fig. 5. The results were obtained on the same dispersion (1.0% w/w) but using different media to obtain a variation of ρ_m. Figure 5a shows the results obtained in a mixture of 11% d_{26}-dodecane and 89% h_{26}-dodecane by volume. This corresponded to $\rho_m = 0.35 \times 10^{10}$ cm^{-2} so that the shell was matched by the medium giving predominantly scattering from the core; the curve extends to Q values of ca. 0.20. The results shown in Fig. 5b were obtained in h_8-toluene only. The ρ_m value of 0.94×10^{10} cm^{-2} is just below the value of the contrast match point of the composite particle hence the scattering was of low intensity. The experimental results closely resemble the curve

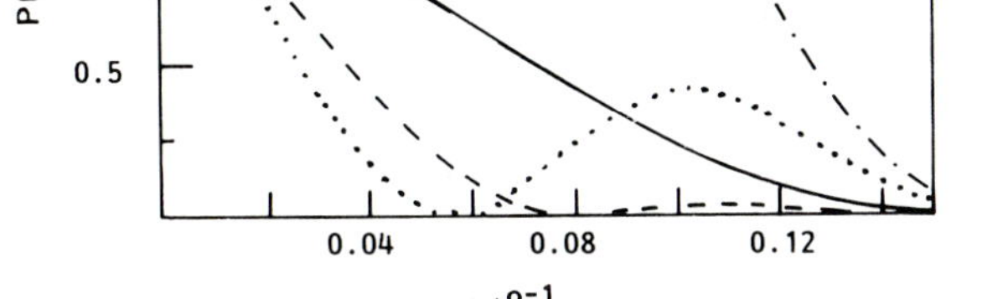

Fig. 4. Calculated curves of $P(Q)$ against Q using Eq. (6) with $R_1 = 28$ Å, $R_2 = 46$ Å, $\rho_c = 4.72 \times 10^{10}$ cm^{-2}, $\rho_s = 0.35 \times 10^{10}$ cm^{-2} and various values of ρ_m:- ———, 0.5×10^{10} cm^{-2}; —·—·, 1.0×10^{10} cm^{-2}; ....., 2.0×10^{10} cm^{-2}; – – –, 4.72×10^{10} cm^{-2}

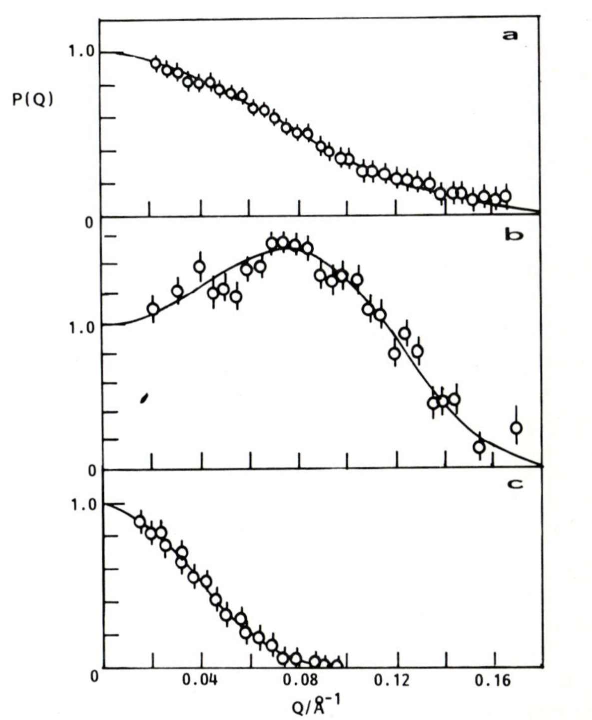

Fig. 5. Experimental results plotted as $P(Q)$ against Q for 1% calcium carbonate dispersions in dispersion media with various values of ρ_m:- a) 0.35×10^{10} cm^{-2}; b) 0.94×10^{10} cm^{-2}; c) 4.72×10^{10} cm^{-2} (Table 2)

Table 2. Small-angle neutron scattering – calcium carbonate dispersions

Dispersion medium	$\rho_m/10^{10}$ cm^{-2}	R_1/Å	R_2/Å	$(R_2 - R_1)$/Å
11% d_{26}-dodecane, 89% h_{26}-dodecane	0.35	22	40	18 ± 1
h_8-toluene	0.94	22	40	18 ± 1
72% d_{26}-dodecane, 28% h_{26}-dodecane	4.72	22	40	18 ± 1

corresponding to this situation in the simulated plots (Fig. 4). In Fig. 5c the dispersion was examined in a medium composed of 72% d_{26}-dodecane and 28% h_{26}-dodecane. This gave $\rho_m = 4.72 \times 10^{10}$ cm^{-2} corresponding to the value of the calcium carbonate core. In this situation scattering comes predominantly from the outer shell. It can be seen from a comparison between Figs. 5a and 5c that $P(Q)$, in the latter, decays rapidly towards a Q value of ca. 0.08 Å^{-1} as expected for a spherical particle with larger dimensions than in Fig. 5a. The values of R_1 and R_2 calculated by fitting curves to the experimental data are listed in Table 2. The results give a remarkable consistency with $R_1 = 22 \pm 1$ Å and $R_2 = 40 \pm 1$ Å and illustrate the sensitivity of scattering, with contrast variation of the medium, to the core-shell nature of the particle.

The value of R_1 obtained for the core particle immediately enables the surface area of the average particle to be obtained and, since the thickness of the layer is known $(R_2 - R_1)$, then the total volume of the adsorbed shell can be obtained. Assuming that the layer is homogeneous and dividing by the molecular volume of the adsorbate gives the area per adsorbed molecule as being ca. 40 Å^2. This value, although approximate, appears to be consistent with the molecular geometry of the adsorbate material.

Micellar systems with a core-shell structure

The zwitterionic surfactant 6-(eicosyldimethylammonio) hexanoate (see Table 1) is highly soluble in water and has a low critical micelle concentration (ca. 10^{-5} mol dm^{-3}) at ambient temperatures [20, 21]. Small-angle neutron scattering studies of the fully hydrogenated material indicated that the micelles formed were spherical with a radius of 30.3 ± 0.5 Å. Further studies with a deuterated alkyl chain ($C_{20}D_{40}H$–) indicated that the alkyl chains formed a spherical core with a radius of 24.6 ± 0.5 Å. However, for this unique type of molecule it was of considerable interest to examine the location of the head group and its orientation at the outer surface of the micelle. For this purpose, studies were carried out with the molecule, $C_{20}H_{41}\overset{+}{N}(CD_3)_2(CH_2)_5COO^-$, i.e. with the two methyl groups on the N-atom deuterated. A series of experiments was carried out in which ρ_m for the medium was varied from just below to just above the value of $\bar{\rho}$ for the micelle (0.72×10^{10} cm^{-2}). The results are shown in Fig. 6; as anticipated, the scattered intensity was rather low, as indicated by the error bars. As demonstrated in the earlier simulations (Fig. 4) the shapes of the curves as a function of ρ_m gave clear evidence that the micelle had a core-shell structure with the deuterated groups in the outer shell. The parameters that gave the best fit to the experimental data are listed in Table 3.

These measurements indicated that the $-CD_3$ groups formed a shell of thickness, 3.3 ± 0.4 Å on the outside of the micelle. This suggests that there is little fluctuation with time of the charged nitrogen centres and that the head group arrangement is rather compact. As the total head group shell thickness was found to be ca. 5 Å [20], it also suggested that the $-COO^-$ groups were spaced slightly further out in the shell than the $-\overset{+}{N}(CD_3)_2$-groups. Hence, it seems reasonable to conclude that the $-\overset{+}{N}-$ and the $-COO^-$ charges are close to the surface of the micelle and that the vector joining the charged centres has a direction almost

Table 3. Contrast variation – $C_{20}H_{41}\overset{+}{N}(CD_3)_2COO^-$

Medium	$\rho_m/10^{10}$ cm^{-2}	$(R_2 - R_1)$/Å	R_1/Å
10% D_2O	0.14	2.90 ± 1.0	25
15% D_2O	0.48	2.92 ± 1.0	25
20% D_2O	0.83	3.40 ± 1.0	25
25% D_2O	1.18	3.93 ± 1.0	25

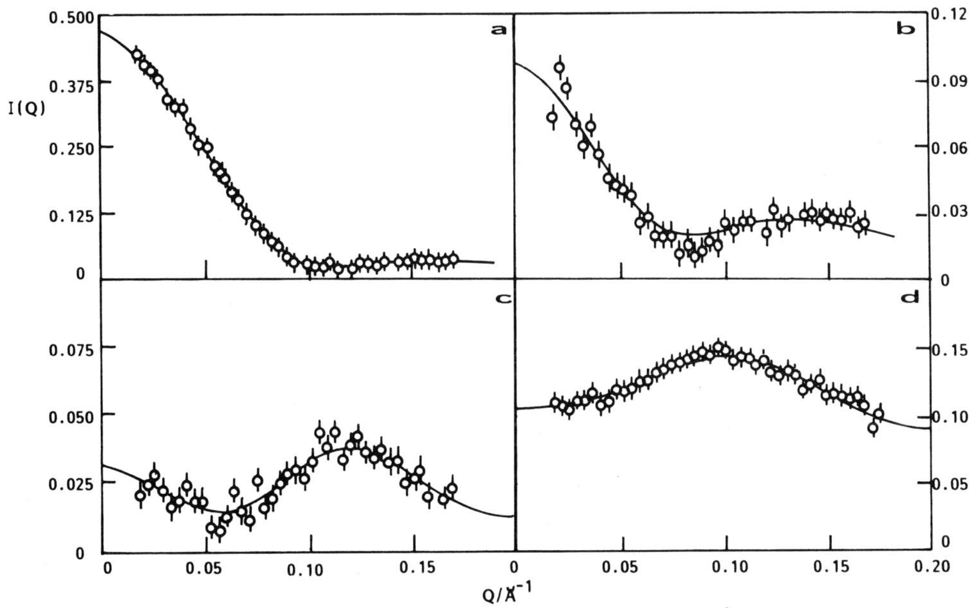

Fig. 6. Experimental results plotted as $I(Q)$ against Q obtained with solutions of $C_{20}H_{41}\overset{+}{N}(CD_3)_2(CH_2)_5COO^-$ at various values of ρ_m:- a) 0.14×10^{10} cm^{-2}; b) 0.48×10^{10} cm^{-2}; c) 0.83×10^{10} cm^{-2}; d) 1.18×10^{10} cm^{-2} (Table 3)

parallel to the micelle surface. From the association number and the total radius of the micelle the area per head group was found to be of the order of 100 Å^2.

Adsorption at the air-water interface

A classical problem in surface chemistry is the determination of the area occupied by a molecule of a surfactant when adsorbed at the air-solution interface [22, 23]. The usual approach has been to apply the Gibbs' adsorption equation with its inherent assumptions to determine the surface excess and hence to calculate the area per molecule from this. With simple nonionic systems this approach gives results which compare favourably with x-ray data obtained on crystals and determinations using a Langmuir–Adam trough and insoluble monolayers. However, for ionic surfactants on various salt concentrations and for mixed surfactant systems analysis is more complex and a direct determination is required. Over the last decade the instrumentation used for small-angle neutron scattering has been adapted for reflectivity measurements and this provides a direct approach to examining molecules at interfaces.

In consideration of the core-shell model, we considered the adsorbed layer to be a homogeneous shell of thickness $R_2 - R_1$ and coherent scattering length density ρ_s. In the case of a planar interface, we can consider the adsorbed material to form a homogeneous layer of thickness d and coherent scattering length density ρ_s between a liquid substrate and air, as illustrated in Fig. 7. This situation may be investigated using an incident neutron beam intensity I_0, reflecting it from the layer at an angle θ and measuring the reflected intensity I_R. The reflectivity R can be expressed in terms of refractive

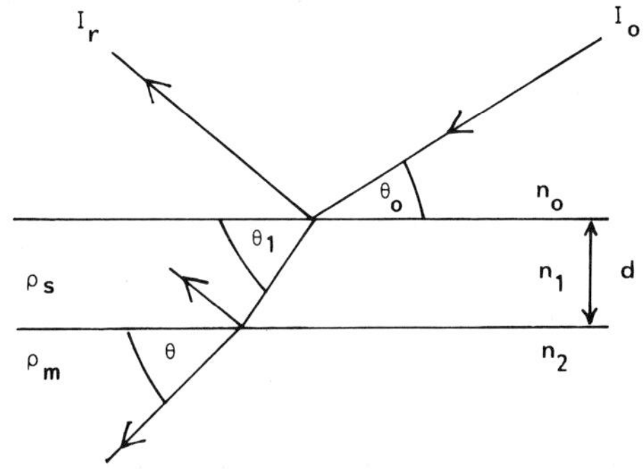

Fig. 7. Schematic illustration of neutron reflectivity from a film covered planar interface

index in the form [24, 25]

$$\frac{I_r}{I_0} = R = \left[\frac{r_{01} + r_{12}\exp(2i\beta)}{1 + r_{01}r_{12}\exp(2i\beta)}\right]^2 ,$$

where r_{ij} is the Fresnel coefficient at the ij interface such that $r_{ij} = (p_i - p_j)/(p_j + p_j)$ with $p_i = n_i \sin\theta_i$ and $\beta = 2\pi n_1 d \sin\theta_1/\lambda$. In terms of a neutron beam the refractive index of the material for specular reflection is given by

$$n_i = 1 - \rho_i \lambda^2/2\pi ,$$

with ρ_i being the coherent scattering length density. The refractive index for neutrons for most materials is slightly less than one. For air $n = 1$ and hence $\rho = 0$. Moreover, as stated earlier, it is possible with a mixture of 8% D_2O and 92% H_2O to obtain $\rho_m = 0$ and hence for this situation $n = 1$ also, and the air $-$ $H_2O:D_2O$ interface is null reflecting. This gives a situation in which with deuterated materials in the layer reflected intensities can be readily measured. Indeed, measurements can be made both with variation of contrast in the medium and by substitution of various parts of the adsorbed molecule, with deuterated groups; thus, variation of contrast can be achieved with the layer as well. For neutron reflectivity the angle θ_0 is small and although to obtain a variation of Q, measurements could be made at various angles, it is usually simpler experimentally to fix θ_0, typically at ca. 1.5°, and vary the wavelength of the beam. In the case of pulsed sources a spread of wavelengths is obtained which makes them very suitable for reflectance measurements.

Figure 8 gives an example of neutron reflectivity measurements carried out on the ultra-long chain surfactant, 6-(eicosyl dimethylammonio) hexanoate. The material used for this experiment had an alkyl chain of composition $C_{20}D_{40}H$– and was the same material as used for small-angle neutron scattering measurements on micelles [20]. The substrate used was composed of 10^{-4} mol solution of the surfactant in 8% D_2O and 92% H_2O ($\rho_m = 0$) at 25 °C. Analysis of the reflection results shown in Fig. 8 gave a d value of 23.9 ± Å [26] which compares favourably with the length of the eicosyl chain found in the micelle, 24.6 ± 0.6 Å. The head group of the molecule $-\overset{+}{N}(CH_3)_2(CH_2)_5COO^-$ has a coherent scattering length density of 0.497×10^{10} cm^{-2} and hence is essentially close to its match point in null reflecting water. The results suggest that the head group is buried in the water

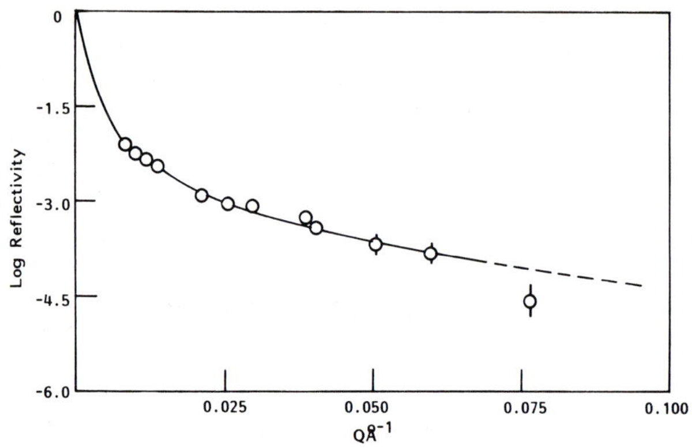

Fig. 8. Experimental results plotted as log reflectivity against Q for $C_{20}D_{40}H\overset{+}{N}(CH_3)_2(CH_2)_5COO^-$ on null reflecting water, $\rho_m = 0$

with the alkyl chains above the surface. The area per molecule obtained was 73 Å^2, which indicates that the surfactant chains are much more closely packed at the air-water interface than in the micelle [20].

Acknowledgements

The work described has involved many collaborators whose enthusiastic support is warmly acknowledged. These include Drs. N. M. Harris, D. J. Hodge, R. G. Laughlin, I. Marković and A. R. Rennie. Thanks are also due to SERC for support and the Institut Laue Langevin for neutron facilities.

References

1. Ottewill RH (1967) In: Schick MJ (ed) Nonionic Surfactants. Marcel Dekker, New York, Chap. 19, p 627
2. Hough DB, Thompson L (1987) In: Schick MJ (ed) Nonionic Surfactants. Marcel Dekker, New York, Chap 11, p 601
3. Parfitt GD, Rochester CH (1983) Adsorption from Solution at the Solid/Liquid Interface. Academic Press, London
4. Ottewill RH, Rochester CH, Smith AL (1983) Adsorption from Solution. Academic Press, London
5. Barrett KEJ (1975) Dispersion Polymerisation in Organic Media. Wiley, London
6. Antl L, Goodwin JW, Hill RD, Ottewill RH, Owens SM, Papworth S (1986) Colloids and Surfaces 17:67
7. Marković I, Ottewill RH, Cebula DJ, Field I, Marsh JF (1984) Colloid Polym Sci 262:648
8. Kostorz G (1979) In: Kostorz G (ed) Treatise on Materials Science and Technology. Academic Press, London, p 227

9. Cebula DJ, Thomas RK, Harris NM, Tabony J, White JW (1978) Faraday Disc Chem Soc 65:76
10. Harris NM, Ottewill RH, White JW (1983) In: Ottewill RH, Rochester CH, Smith AL, Adsorption from Solution. Academic Press, London, p 139
11. Penfold J, Thomas RK (1990) J Phys Condens Matter 2:1369
12. Lee EM, Thomas RK (1989) Physica B156 & 157:525
13. ILL (1988) Guide to Neutron Research Facilities at the Institut Laue-Langevin. Grenoble: Institut Laue-Langevin
14. Heenan RK, Osborn R, Stanley HB, King SM, Mildner DFR, Furusaka M (1992) J Appl Cryst (in press)
15. Jacrot B (1976) Rep Prog Phys 39:911
16. Guinier A, Fournet G (1955) Small-Angle Scattering of X-rays. Chapman and Hall, London
17. Espenscheid WF, Kerker M, Matijević E (1964) J Phys Chem 68:3093
18. Marković I, Ottewill RH (1986) Colloid Polym Sci 264:65
19. Harris NM (1980) PhD thesis, University of Oxford
20. Hodge DJ, Laughlin RG, Ottewill RH, Rennie AR (1991) Langmuir 7:878
21. Hodge DJ (1989) PhD thesis, University of Bristol
22. Adam NK (1941) The Physics and Chemistry of Surfaces. Oxford University Press, London
23. Adamson AW (1967) Physical Chemistry of Surfaces. Interscience, New York
24. Jenkins FA, White HE (1951) Fundamentals of Optics. McGraw Hill, London
25. Penfold J (1991) In: Lindner P, Zemb T (ed) Neutron, X-ray and Light Scattering. North Holland, Amsterdam, p 261
26. Ottewill RH, Rennie AR, Simister E (1992) (to be published)

Author's address:

Professor Dr. R. H. Ottewill
School of Chemistry
University of Bristol
Cantock's Close
Bristol BS8 1TS
England

Surfactant and cosurfactants in lamellar liquid crystals and adsorbed on solid surfaces. IV. The model system sodium *p*-octylbenzenesulfonate/benzyl alcohol and α-alumina

A. M. Blokhus[1]), M. Ruths[2]), J. Sjöblom[1]), Ø. Moksheim Andersen[1]), Z. Ma[3]), and S. E. Friberg[3])

[1]) Department of Chemistry, University of Bergen, Norway
[2]) On leave from the Department of Physical Chemistry, Åbo Akademi, Finland
[3]) Center for Advanced Materials Processing, Clarkson University, Potsdam, New York, USA

Abstract: The competitive adsorption of sodium *p*-octylbenzenesulfonate (SOBS) and benzyl alcohol (BzOH) onto solid α-alumina at 35 °C with and without 0.1 M NaCl has been studied using high-performance liquid chromatography (HPLC). Calculations of the individual area requirements of SOBS and BzOH in the adsorbed film were undertaken at the saturation level (plateau) of the SOBS adsorption isotherm. The lamellar liquid crystalline D phase of the system SOBS/BzOH/water was investigated by means of x-ray measurements in order to determine the area requirements. The discrepancies found between the area requirements in the D phase and on the α-Al_2O_3 reveal that the electrostatical headgroup repulsion will, on the solid surface, prevent the SOBS molecules to pack as in the D phase.

Key words: Ionic surfactant; co-surfactant; α-alumina; adsorption; lamellar liquid crystal; molecular packing

Introduction

The adsorption of surfactants at the solid/liquid interface is a topic of great interest due to its importance in many applications. In recent years considerable effort has been made to study adsorption of anionic surfactants onto oppositely charged mineral oxide surfaces, including determination of adsorption isotherms, zeta-potentials, heats of adsorption, etc. [1–5]. Much attention has been devoted to elucidating the mechanisms of adsorption and the molecular organization of the surfactants at the solid surface. Generally, the adsorption isotherms obtained for these systems are divided into different regions. At low surface coverage the surfactant molecules adsorb individually through electrostatic interaction, whereafter a sudden increase in the adsorption occurs that marks the onset of surfactant aggregation through lateral hydrophobic interaction before a leveling off at the plateau of the adsorption isotherm takes place.

However, studies of the incorporation or adsolubilization of organic molecules into the adsorbate is a topic that has attracted less attention, although the solubilization of organic water-insoluble compounds in micellar systems is extensively studied. Recently, some work has been published on surfactant-based separation techniques that utilize adsolubilization [6–10]. We have in previous contributions [11–14] studied the competitive adsorption of anionic surfactants (sodium dodecylsulfate, sodium dodecanoate) and medium-chain alcohols (1-butanol, benzyl alcohol) on solid α-alumina. In these studies the focus was on the adsolubilization of alcohols, when the surfactant has reached its saturation level, i.e., at the plateau of the adsorption isotherm of the surfactant. Parallels were drawn between the packing conditions in the mixed adsorbed surfactant/alcohol layer at the solid surface and in the lamellar liquid crystalline D phase of the same surfactant/alcohol/water system. In this phase the surfactant/co-surfactant molecules have a well-defined packing in a bilayer structure.

The present work extends our models to involve an alkyl-aryl surfactant in the system sodium *p*-octylbenzenesulfonate (SOBS)/benzyl alcohol (BzOH)/water. In order to carry out these investigations we have used high-performance liquid chromatography (HPLC) for separation and quantification of SOBS and BzOH in mixture. All the investigations were carried out at 35 °C, well above the Krafft point of SOBS, also in the presence of electrolyte.

Experimental

Materials

Sodium *p*-octylbenzenesulfonate (SOBS) was obtained from Aldrich (> 97%). Sodium chloride and benzyl alcohol were both supplied from Merck as puriss grade; the compounds were used without further purification. The solid adsorbent was α-alumina, Al_2O_3, obtained from Ventron. The particle size was specified to be 40 μm, and the surface area was 103 $m^2 g^{-1}$, according to BET measurements. Both NaCl and Al_2O_3 were dried at 115 °C before use.

Methods

The region of the lamellar liquid crystalline D phase was determined by visual inspection of mixed samples equilibrated at 35 °C, and the samples were also inspected through two crossed polaroids. The low-angle x-ray diffraction patterns were obtained from a Kiessig low-angle camera (R. Seifert) with a Tenndec position-sensitive gas ionization detector system using Ni-filtered K_α copper line. The samples were carefully mixed; air bubbles were removed by centrifugation and transferred to capillaries by suction.

The adsorption experiments were performed at 35 °C by mixing 0.5 g α-Al_2O_3 with 5-ml volumes of aqueous solutions (with or without 0.1 M NaCl) containing various concentrations of SOBS and benzyl alcohol. The test tubes were placed in a water bath and vigorously shaken for 24 h. Thereafter, the mixture was centrifuged and the supernatant liquid was used for further analysis. The adsorption density, Γ_i (mol/m^2) of species i was calculated according to the equation

$$\Gamma_i = \frac{\Delta c_i \times V^0}{m \times a^s}, \qquad (1)$$

where Δc_i(mol/l) is the concentration difference due to adsorption of species i, V^0 (l) is the volume of the liquid phase, m (g) is the mass of alumina and a^s ($m^2 g^{-1}$) its specific surface area. For the determination of the adsorption isotherm of SOBS (without any benzyl alcohol added) the equilibrium concentration of SOBS was determined by UV spectroscopy at 261 nm. For the determination of equilibrium concentrations of both SOBS and benzyl alcohol in mixture, however, high-performance liquid chromatography (HPLC) was used. HPLC was performed with a solvent delivery system (SP-8700, Spectra-Physics), an automated sampling system (ISS-100, Perkin-Elmer), and a photodiode array detector (HP-1040A, Hewlett-Packard) interfaced to a lab-data system (VG Multichrome). The column (15 × 0.5 cm, 3 μm ODS-Hypersil) was slurry-packed and provided with a precolumn (2.5 × 0.5 cm, 40 μm LC-18 Pelliguard). Two solvents were used for elution: A) $CH_3OH:H_2O$ (1:4) and B) $CH_3OH:H_2O$ (9:1). The eluation profile was: 0–5 min, 40–100% B in A (linear gradient); 5–8 min, 100% B (isocratic); 8–10 min, 100–40% B in A (linear gradient). The flow rate was 0.8 ml min^{-1}, and the chromatographic monitoring wavelengths were every 2 nm between 220 and 280 nm. Prior to injection (6 μl) all samples were filtered through a 0.45-μm Millipore membrane filter.

Results

In Fig. 1 an HPLC-chromatogram for a mixture of SOBS and BzOH is given. The figure shows that the separation of the two peaks is very good. The retention times, t_R, of the peaks are 2.30 and 4.40 min, respectively. From UV-spectra, BzOH and SOBS could be identified at t_R 2.30 min and 4.40 min, respectively.

Figure 2 gives the adsorption isotherm of SOBS on α-Al_2O_3 with and without 0.1 M NaCl at 35 °C. In both cases the isotherms show the characteristic features of anionic surfactant adsorption onto a positively charged oxide: relatively low adsorption at low equilibrium concentration followed by a marked increase in the adsorption density due to the onset of hydrophobic interactions, and, there-

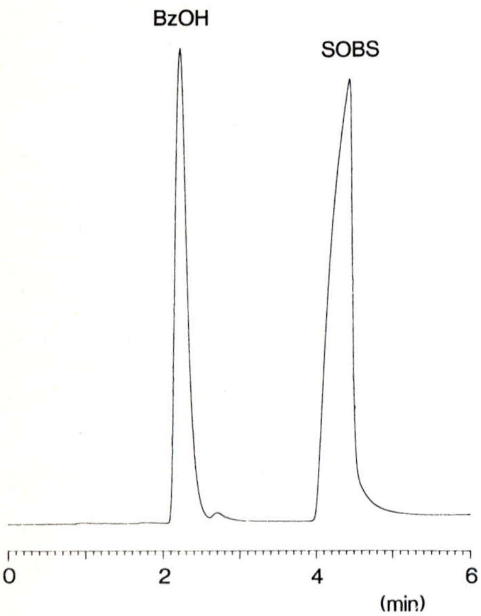

Fig. 1. An HPLC-chromatogram for SOBS (t_R = 4.40 min) and BzOH (t_R = 2.30 min) in mixture

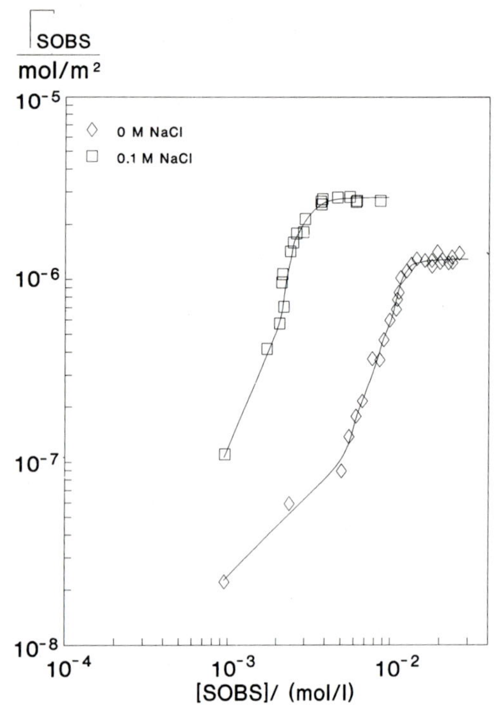

Fig. 2. Adsorption isotherm for SOBS from aqueous solutions with and without NaCl on solid α-alumina at 35 °C

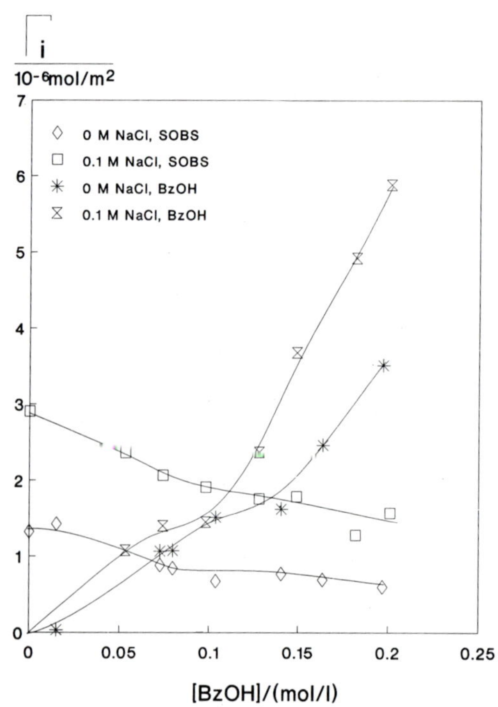

Fig. 3. The adsorption density of SOBS and benzyl alcohol at saturation level of SOBS as a function of the benzyl alcohol concentration

after, a leveling off which indicates a saturation in packing at the solid surface. The adsorption density for the same equilibrium concentrations is always higher in the presence of NaCl. In the absence of NaCl the isotherm levels off at a SOBS equilibrium concentration around 1.3×10^{-2} M which corresponds to the critical micelle concentration (CMC) for SOBS at 35 °C, according to our conductivity measurements.

Figure 3 presents the shift in the adsorption saturation value Γ_i^{sat} for the surfactant on α-Al_2O_3 upon addition of benzyl alcohol, with and without NaCl present. The figure also includes the corresponding adsorption density of the benzyl alcohol. The readings of Γ_i^{sat} refer to the saturation level on the isotherm of SOBS at a concentration equal to 3.3×10^{-2} M SOBS. The adsorption of SOBS is found to decrease with increasing alcohol concentrations, while the BzOH adsorbs to a greater extent on the solid surface. This indicates that the SOBS is replaced by the benzyl alcohol on the solid surface as the concentration of benzyl alcohol in bulk increases.

The lamellar liquid crystalline phase D in the system water/SOBS/benzylalcohol was found to be located between 25 and 43 wt% of water, 20 and 35 wt% of benzyl alcohol, and 30 and 50 wt% of SOBS, as presented in Fig. 4. The location of the D phase is similar to that in other anionic surfactant/

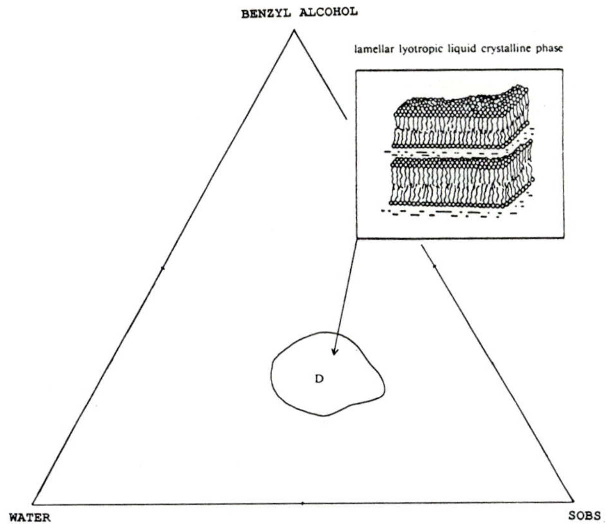

Fig. 4. Partial phase diagram of SOBS/BzOH/H_2O showing the D phase region

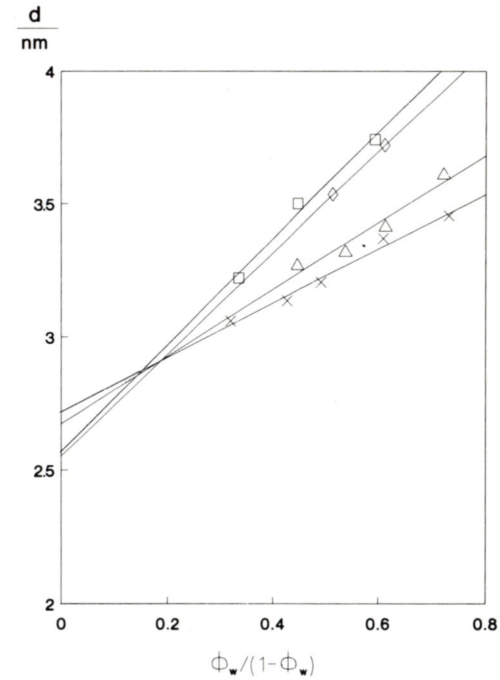

Fig. 5. Repetition distances (d) for the sample in the D phase in the model system SOBS/benzyl alcohol/water as a function of the water content

alcohol systems [15]. The x-ray characterization of the D phase for the system water/SOBS/benzyl alcohol is given in Fig. 5. Extrapolating to zero content of water gives an alkyl chain bilayer length of 2.6 nm. The maximal swelling at the water-rich borderline is about 3.5 nm, which gives the water layer an extension of ca 1.0 nm.

Discussion

In order to describe the conditions on the solid surface of the surfactant and cosurfactant at the saturation level of the adsorption isotherm, parallels with the packing conditions in the lamellar liquid crystalline phase D are drawn. This phase has a microstructure consisting of planar hydrocarbon bilayers with intervening water layers, as illustrated in Fig. 4 (the inserted picture). The surface charge density of the aggregates in the lamellar D phase is dependent on the total planar area A^{tot} ascribed to the ionic and the hydroxylic groups. By denoting the x-ray repetition distance d, the geometries give

$$1/\rho = d \times A^{tot} \times N_A \times w_s/2 \times M_s \,, \qquad (2)$$

where ρ is the density, N_A is Avogadro's constant, w_s is the weight fraction, and M_s is the molar mass of the surfactant. As reported earlier [11, 16], the total area A^{tot} can be linearized in a plot versus the molar ratio of alcohol to surfactant, and the individual area of the surfactant A_s^0 and cosurfactant A_{ROH}^0 can be evaluated according to

$$A^{tot} = A_s^0 + (n_{ROH}/n_s)\,A_{ROH}^0 \,, \qquad (3)$$

At the adsorption plateau the surfactant and cosurfactant will form a mixed surfactant layer. Spreading out of one monolayer from the D phase, it has been shown [13] that the total area A^{tot} to be covered at the solid surface should obey

$$A^{tot} = A_s^0 + (\Gamma_{ROH}/\Gamma_s)\,A_{ROH}^0 \,, \qquad (4)$$

where Γ_{ROH} and Γ_s is the adsorption density of alcohol and surfactant, respectively.

The total area ascribed to the surfactant adsorbed at the solid surface is obtained from

$$A^{tot} = 1/N_A \times \Gamma_s \,, \qquad (5)$$

where Γ_s is the adsorption density of the surfactant.

In Fig. 6 plots based on these equations are presented. A good linearity is obtained between the total area A^{tot} and the adsorption densities (molar ratio) of BzOH and SOBS. From the plot the individual areas ascribed to the surfactant and co-

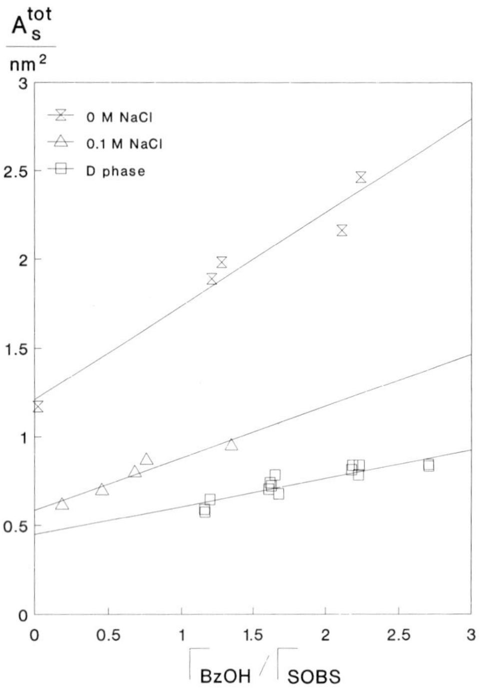

Fig. 6. The total area A^{tot} ascribed to the surfactant on alumina and in the D phase as a function of the adsorption densities (molar ratio) of BzOH and SOBS

surfactant adsorbed at the solid surface or in a D phase are obtained. For the surfactant at the solid surface the area ascribed is 1.21 nm^2 and 0.59 nm^2 without and with NaCl added, respectively. For the benzyl alcohol adsorbed in the surfactant layer at the solid surface the analogous areas are 0.53 and 0.29 nm^2. The high values obtained in the electrolyte-free systems illustrate that, in this case, the adsorbed layer is not reminiscent of a condensed film. However, by screening the headgroup interactions of the ionic SOBS in the adsorbed film with NaCl a more closely-packed film arrangement is obtained at the solid surface. This is consistent with our earlier findings on the electrolyte effect on the competitive adsorption of sodium dodecylsulfate and benzylalcohol on solid alumina [13]. The area requirements of the species in the adsorbed film showed pronounced dependence on the salinity level in the sense that an increase in electrolyte content decreased the area requirements of the constitutents dramatically until limiting area requirements in the adsorbed film were obtained. The adsorbed surfactant-cosurfactant film at the solid surface is in a condensed state at the highest electrolyte content, thus revealing a very good agreement with the packing conditions in the lamellar liquid crystalline D phase of the same system.

From the plot in Fig. 6 the individual area ascribed to SOBS and benzyl alcohol in the lamellar liquid crystalline D phase is found to be 0.45 nm^2 and 0.16 nm^2, respectively. These values are significantly lower than those found in the adsorbed film, 0.59 and 0.29 nm^2, respectively. This indicates that the close packing arrangements in the D phase are not obtained in the corresponding adsorbed layer at the solid alumina surface. In our study of the system sodium dodecylsulfate/benzyl alcohol a very good agreement for the individual areas in the adsorbed film on the solid alumina surface (at electrolyte level above 0.2 M NaCl) and in the D phase was found. The area requirements obtained in the D phase were then 0.37 nm^2 and 0.20 nm^2 for SDS and benzyl alcohol, respectively, and 0.40 nm^2 and 0.24 nm^2 in the adsorbed layer on α-Al_2O_3. A comparison of the area requirements of SDS and SOBS in the D phases shows that, for SOBS, a somewhat higher value is obtained, while for the benzylalcohol the area requirements indicate a more efficient packing when combining BzOH with SOBS rather than SDS. Obviously, a stronger molecular interaction is obtained in the D phase when both the surfactant and cosurfactant molecules contain a benzene ring, which seems quite reasonable for geometric reasons. In the adsorbed layer at the solid surface, however, an equivalent close-packing is not achieved, as seen from the discrepancies in the areas. Obviously, salinities higher than 0.1 M NaCl are required in order to screen the lateral repulsion between the sulfonate headgroups and attain interaction forms as in the corresponding D phase. The experimental conditions in our study (35 °C) did not allow an increase of salinity to 0.2 M, since the SOBS started to precipitate.

Acknowledgments

M. Ruths thanks the Academy of Finland and the Nordic Council of Ministers for financial support. J. Sjöblom acknowledges the Royal Norwegian Council for Scientific and Industrial Research (NTNF) for a travel and subsistence grant to Clarkson University.

References

1. Hough DB, Rendall HM (1983) In: Parfitt GD, Rochester CH (eds) Adsorption from Solution at the

Solid/Liquid Interface. Academic Press, New York, p 247
2. Chander S, Fuerstenau DW, Stigter D (1983) In: Ottewill RH, Rochester CH, Smith AL (eds) Adsorption from Solution. Academic Press, New York, p 187
3. Fuerstenau DW (1971) In: Hair ML (ed) The Chemistry of Biosurfaces. Marcel Dekker, New York, Vol 1, p 143
4. Somasundaran P, Goddard ED (1979) Mod Aspects Electrochem 13:207
5. Partyka S, Rudzinski W, Brun B, Clint JH (1989) Langmuir 5:297
6. Harwell JH, O'Rear EA (1989) In: Scamehorn JF, Harwell JH (eds) Surfactant-Based Separation Processes. Marcel Dekker, New York, p 155
7. Scamehorn JF, Harwell JH (1988) In: Surfactants in Chemical/Process Engineering. Marcel Dekker, New York, p 77
8. Lee C, Yeskie MA, Harwell JH, O'Rear EA (1990) Langmuir 6:1758
9. Esumi K, Shibayama M, Meguro K (1990) Langmuir 6:826
10. Zhu B-Y, Zhao X, Gu T (1988) J Chem Soc Faraday Trans I 84:3951
11. Sjöblom J, Blokhus AM, Høiland H (1990) J Colloid Interface Sci 136:584
12. Sjöblom J, Blokhus AM, Sun WM, Friberg S (1990) J Colloid Interface Sci 140:481
13. Blokhus AM, Sjöblom J (1991) J Colloid Interface Sci 141:395
14. Ruths M, Sjöblom J, Blokhus AM (1991) J Colloid Interface Sci 145:108
15. Ekwall P (1975) In: Brown GH (ed) Advances in Liquid Crystals. Academic Press, New York, Vol 1, p 1
16. Danielsson I, Friman R, Sjöblom J (1982) J Colloid Interface Sci 86:442

Author's address:

Anne Marit Blokhus
Department of Chemistry
University of Bergen
N-5007 Bergen
Norway

Amine functionalized surfactants – pH effects on adsorption and interaction

P. M. Claesson[1]), P. C. Herder[2]), M. W. Rutland[3]), Å. Waltermo[1]), and B. Anhede[4])

[1]) The Surface Force Group, Department of Physical Chemistry, The Royal Institute of Technology, Stockholm, and The Institute for Surface Chemistry, Stockholm, Sweden
[2]) Casco Nobel AB, Stockholm, Sweden
[3]) Department of Applied Mathematics, Research School of Physical Sciences and Engineering, Institute of Advanced Studies, The Australian National University, Canberra, Australia
[4]) Berol Nobel AB, Stenungsund, Sweden

Abstract: The interactions between layers of two hydrolyzable surfactants, dodecylamine (DA) and tetraoxyethylene dodecylamine (TEDA), adsorbed on negatively charged muscovite mica have been investigated. It was found that the variation of the interaction and adsorption as a function of pH was similar, but not identical, for the two kinds of surfactants. At a concentration of 10^{-4} M at low pH (below 8) both types of surfactants adsorb electrostatically to form a monolayer. The resulting surfaces have a hydrophobic character, particularly in the case of dodecylamine. A series of events takes place as the pH is increased. For DA the first event is that uncharged molecules adsorb within the monolayer. This does not occur to any significant degree for TEDA due to steric hindrance within the adsorbed layer. In the next step a bilayer builds up on the surface. The binding of the outer layer is stronger for DA than for TEDA. The DA bilayer becomes uncharged at pH 10.3, and liquid dodecylamine droplets, which phase separate from the bulk solution at this pH-value, precipitate on the surface. At pH-values above 11–12 the affinity between both types of surfactants and the surface decreases significantly, resulting in a less ordered first layer. In addition, a strong surface charge develops that prevents multilayer build-up.

Key words: Adsorption; amine surfactants; interaction; pH effects; surface forces; dodecylamine; tetraoxyethylene dodecylamine

Introduction

The adsorption of surfactants, polymers, and polymer/surfactant combinations onto solid surfaces are topics of great technical and fundamental interest. Some fundamental studies are primarily concerned with the kinetics of adsorption, others with the equilibrium structures formed on the surfaces, and still others with understanding of some particular process like flotation. A range of methods like ellipsometry [1], calorimetry [2] or neutron scattering [3] can be used for obtaining valuable information in this area. Another method that may be employed is the surface force technique [4]. It primarily focuses on the intermolecular forces acting between adsorbed layers. The information obtained with this method is the total force as a function of surface separation. The results are then interpreted in terms of various force contributions (e.g., double-layer forces, van der Waals forces, hydration forces or steric forces) and in terms of the structure of the adsorbed layer (e.g., thickness and preferred orientation of adsorbed molecules).

Hydrolyzable amine surfactants are of great importance as collectors in flotation. It is often observed that the efficiency of a surfactant as a collector is optimal over a rather narrow pH-range [5]. To get a fundamental understanding of this behavior, one has to consider the pH dependence of the properties of the surfactant bulk solution, the surfactant solution/air and surfactant solution/solid interfaces. The bulk solution properties of dodecylamine (DA) have been thoroughly studied. The pK_b of the dissociation of a dodecylammonium ion (DAH^+) to a DA molecule is 10.63 [6], and the logarithm of the association constant of a DAH^+ and a DA molecule is reported as -3.12. The knowledge of the different equilibria involved in hydrolyzable surfactant systems can be used for mapping the concentration of the various species as a function of pH. Pugh [5] and Laskowski et al. [7] analyzed the dodecylamine/water system. As the pH increases, the concentration of neutral DA and singly charged dimers increase at the expense of that of DAH^+ ions. At a total concentration of about 10^{-4} M the dodecylamine/water system phase separates into one water-rich phase and a liquid DA phase at pH 10.3 [5]. The properties of the phase separated dodecylamine droplets have been investigated, and it was found that, as expected, the droplets are positively charged at pH-values below about 11. However, at higher pH-values the droplets were negatively charged. This negative charge was interpreted in terms of adsorption of OH^- ions.

A change in pH also affects the intermolecular forces acting between the surfactant molecules. In the case of alkyl amines, an increase in pH will result in a decreasing number of charged species. Consequently, the predominance of electrostatic forces over other force contributions will be reduced at higher pH-values, which has several consequences. The surface tension of dodecylamine surfactant solutions decreases significantly between pH 9 and 10 [7]. The reason is that a larger number of molecules can adsorb at the interface without a resulting increase in free energy associated with the creation of a strongly charged electrical double-layer. The increase in surface tension observed at even higher pH-values is related to the phase-separation of the bulk solution. Studies of surface rheological properties of dodecylamine solutions, which are markedly pH-dependent, confirm the greatly increased adsorption at the solution/air interface above pH 9 [8].

The change in intermolecular forces with pH also affects the association of the surfactants in other ways. For instance, as a result of the decreasing electrostatic repulsion, the critical micelle concentration (CMC) and the Krafft point decrease with increasing pH for DAHCl [9]. Also, the concentration range over which micelles can exist decreases due to a decreasing solubility [9]. The types, extent, and properties of liquid crystalline phases are, of course, also markedly affected by the degree of ionization of the hydrolyzable surfactant [10, 11]. For instance, Wärnheim et al. have shown that the area per molecule in the lamellar phase of octylamine/octylammonium chloride increases, and the alkyl chain order decreases significantly with the number of charged groups in the lamellar [10].

In this study, we directly measure the forces acting between adsorbed layers of DA, and between layers of an ethoxylated amine surfactant, as a function of surface separation and pH. We are thus able to follow how the forces operating in the system vary with pH. Further, the data obtained from surface force measurements are used for elucidating how the adsorption to a strongly negatively charged surface and the structure of the adsorbed layer depend on the pH-value. The work presented here is a summary of some results obtained in our group in recent years. Interaction forces between mica surfaces in octylammonium chloride and dodecylammonium chloride solution at low pH-values have been studied by Herder [12]. Some aspects of the pH-dependent interactions in amine surfactant solutions are discussed in [13, 14].

Materials and methods

Chemicals

Dodecylammonium chloride was obtained from Eastman Kodak and used as received. Tetraoxyethylene dodecylamine or N-[2-[2-[2-(2-hydroxyethoxy)ethoxy]ethoxy]ethyl]-N-dodecylamine was synthesized by Berol Nobel AB, Stenungsund, Sweden. The synthesis will be described elsewhere [14]. Sodium hydroxide and hydrochloric acid were from Merck. The water used was purified by a modified Millipore unit as described elsewhere [12, 13]. The substrate surfaces in these experiments were muscovite mica, a layered aluminosilicate mineral [15]. Each layer is strongly negatively charged (0.48 nm^2/charge) due to isomorphous substitution of aluminum for silicon. This lattice charge (or total dissociation charge) should not be confused with the surface charge which, of course, is influenced by the type and concentration of adsorbing cations in solution. In the crystal this charge is compensated mainly by potassium ions located between the aluminosilicate sheets. When mica is immersed in water, the potassium ions on the surface are easily exchanged for other types of cations, e.g., protons [16]. The mica surface does not contain any hydroxyl groups.

Surface force measurements

Surface forces were determined with either a Mark II [4] or a Mark IV [17] surface force apparatus (SFA). Two molecularly smooth muscovite mica surfaces were first glued with an epoxy resin (Epon 1004) onto optically polished half-cylindrical silica discs. The surfaces were then mounted in the surface force apparatus in a crossed cylinder configuration. One of the surfaces was mounted on a piezo electric tube. The surface separation was varied by changing the voltage applied to this tube. The surface separation, D, was determined interferometrically to within 2 Å using fringes of equal chromatic order. The force (F_c) was determined from the deflection of a double cantilever spring supporting the other surface. The force between crossed cylinders normalized by the local geometric mean radius (R), which was about 2 cm, is related to the free energy of interaction per unit area between flat surfaces (G_f) via the Derjaguin approximation [18]:

$$F_c(D)/R = 2\pi G_f(D) \ . \qquad (1)$$

This relation is valid provided that $D \ll R$.

Whenever the gradient of the force with respect to the surface separation (dF_c/dD) exceeds the spring constant, the mechanical system is in an unstable region and the surfaces will jump to the next stable region [19].

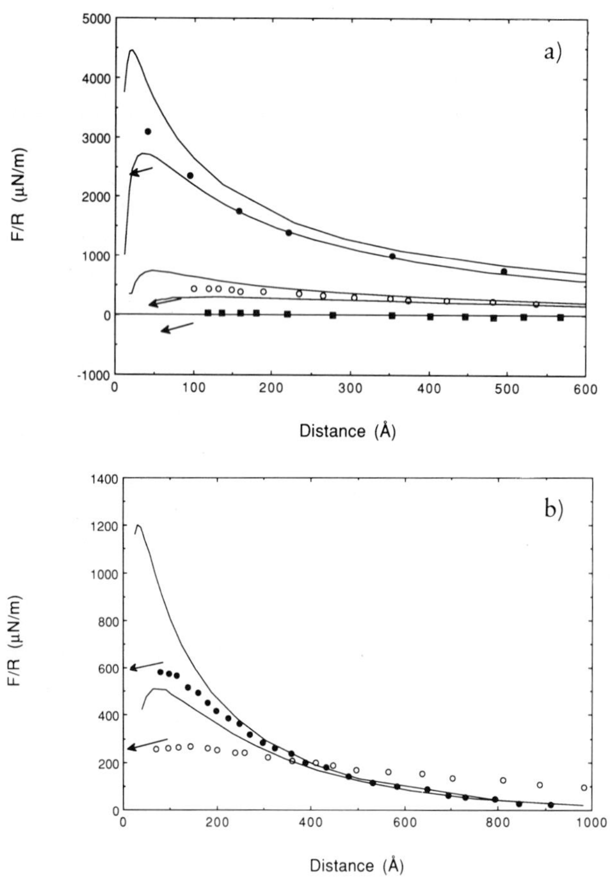

Fig. 1. Force normalized by radius as a function of surface separation between mica surfaces immersed in a range of dodecylammonium chloride solutions at pH 5.5–6. The solid lines represent calculated DLVO forces, assuming interaction at constant charge (upper lines) or interaction at constant potential (lower lines). Arrows represent inward jumps caused by the presence of an attractive force. a) Forces in water (filled circles), 3×10^{-6} M (unfilled circles), and 9×10^{-6} M (filled squares) DAHCl solutions. The parameters used for calculating the double-layer force are for water: surface potential -175 mV, Debye-length 960 Å, and for 3×10^{-6} M DAHCl: surface potential -70 mV and Debye-length 700 Å. b) Forces in 4×10^{-5} M (unfilled circles) and 1×10^{-4} M (filled circles) DAHCl solutions. The solid lines were calculated by using a surface potential of $+60$ mV and a Debye-length of 250 Å

Force analysis

The measured forces were analyzed using DLVO theory. The van der Waals force was always approximated with the non-retarded van der Waals interaction between mica in water (Hamaker constant 2.2×10^{-20} (J)). This slightly overestimates the van der Waals interaction since adsorption of surfactants, screening by electrolyte, and retardation effects all weaken this force contribution. The double-layer force was calculated in the non-linear Poisson Boltzmann (PB) approximation using a numerical procedure [20]. In the analyses it has been assumed that the origin of the van der Waals force and the plane of charge are located at the surface of any adsorbed surfactant layer.

Results

Forces in dodecylamine solutions

The force between mica surfaces immersed in pure water (pH 5.6) is dominated by a repulsive double-layer force at large separations (Fig. 1a). This repulsion is overcome by an attractive van der

Waals force at a separation below about 3–4 nm. By fitting double-layer forces calculated in the PB-approximation to the measured forces, one can obtain an estimate of the apparent surface charge density. In this case, we obtained an area/charge of 27 nm^2, which is much larger than the area per charge in the mica lattice (0.48 nm^2). Hence, the apparent surface charge is less than 2% of the lattice charge (total dissociation charge), which is due to adsorption of protons. It should, however, be noted that the values obtained for the surface charge density by fitting forces calculated in the PB-approximation to measured forces are apparent (too low) values due to the neglect of ion-ion correlation effects in that theory [21]. The normalized adhesion force ($F(0)/R$) between the mica surfaces in contact (at $D = 0$) is 30–50 mN/m.

Addition of a small amount of DAHCl to the solution has a strong effect on the surface forces. At a DAHCl concentration of 3×10^{-6} M (pH 5.5–6.0) the height of the repulsive force maximum is only about 15% of that observed in pure water (Fig. 1a). The apparent area per charge at this concentration is 160 nm^2. The repulsive force maximum is located at a slightly larger separation than expected from DLVO-theory, indicating the presence of a long-range hydrophobic attraction [12, 13, 22] or possibly a collapse of the electrostatic double-layer as discussed by Podgornic and Parsegian [23]. The normalized adhesion force, 120 mN/m, is larger than in pure water, showing that the surfaces have a somewhat hydrophobic character.

On increasing the DAHCl concentration to 9×10^{-6} M, the surfaces become uncharged (Fig. 1a). A further increase in surfactant concentration results in a recharging of the surfaces. This is illustrated for DAHCl concentrations of 4×10^{-5} M and 1×10^{-4} M (pH 5.5–6.0) in Fig. 1b. Clearly, the adsorption of cationic surfactants and of protons is sufficiently strong to slightly overcompensate the mica lattice charge. The repulsive force maximum is located at a larger separation than expected from DLVO-theory. The normalized adhesion force at DAHCl concentration of 1×10^{-4} M has a value between 200 and 300 mN/m. The adhesion force is related to the interfacial tension (γ_{SL}) according to [24]:

$$F(0)/R = \alpha\gamma_{SL} , \quad (2)$$

where the value of the constant α varies between 3π (for soft surfaces) and 4π (for hard surfaces). The surfaces used by us deform considerably when in adhesive contact (due to flattening of the glue) so by using the factor 3π, we deduce an interfacial tension of 27 ± 5 mN/m. In the concentration range 9×10^{-6} M–1×10^{-4} M, the thickness of the surfactant layer adsorbed on mica varies in different experiments between 5 and 9 Å.

At pH 8.3 (1×10^{-4} M DAHCl) the long-range double-layer force, the attractive force, and the adhesion force are of a similar strength as those at pH 5.5 (Fig. 2). However, the layer thickness has increased significantly to about 16–20 Å on each surface.

The long-range double-layer force observed at pH 9.5 at a DAHCl concentration of 1×10^{-4} M (Fig. 2) is similar to those observed at lower pH-values. However, at pH 9.5 a bilayer is present on each surface. No long-range attractive force and no strong adhesion force are observed; instead, a non-DLVO repulsion with a range of about 10 Å is observed at short separations. This repulsion is due to a combination of dehydration of polar groups [25] and steric confinement [26]. It is not clear which of these contributions is most important. As a comparison, the water layer thickness, determined by Wärnheim et al. in the lamellar octylamine/octylammonium chloride system is 7–9 Å at a water content of 80 mol% [10] (this was the only water content studied by Wärnheim et al.). In the surface force experiments a weak minimum is observed outside the range of the steric/hydration

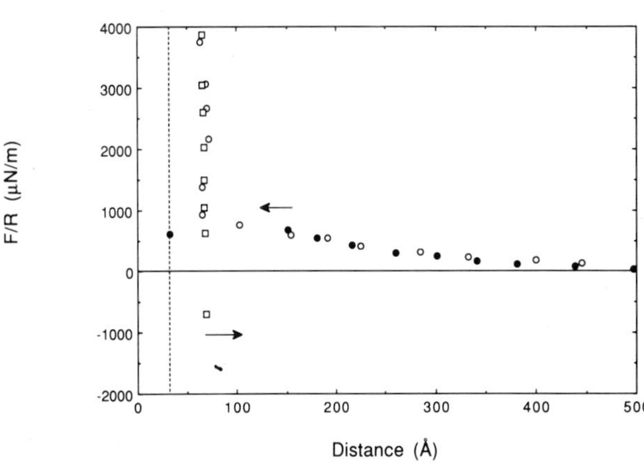

Fig. 2. Force normalized by radius as a function of surface separation between mica surfaces immersed in a 10^{-4} M DAHCl solution at pH 8.3 (filled circle) and at pH 9.5 measured on approach (unfilled circles) and on separation (unfilled squares). Arrows represent jumps. The dashed line represents the layer thickness at pH 8.3

repulsion. The depth of this minimum is similar to those observed between layers of other types of nonionic hydrophilic surfactant headgroups [27].

A further increase in pH results in a decrease in the strength of the double-layer force. At pH 10.3 (DAHCl concentration 1×10^{-4} M) the bilayers, with a thickness of 35 Å, are uncharged. The forces observed are initially consistent with a van der Waals attraction with a weak attractive minimum and a very steep hydration/steric force at smaller separations (Fig. 3). A bilayer can also be formed at low pH, but then only at a much higher surfactant concentration. This is, of course, due to the large headgroup repulsion between charged alkylammonium groups, and the high free energy associated with a strongly charged electrical double-layer. At a pH-value of 5.5 a bilayer is formed at a DAHCl concentration of 6×10^{-3} M. The forces observed in this case are displayed in Fig. 3. A strong repulsive double-layer force, corresponding to an apparent area per charge of 0.4–0.5 nm^2, dominates the long-range interaction.

When the surfaces are kept in contact with the 10^{-4} M surfactant solution at pH 10.3–10.5 a further adsorption occurs with time. Once the adsorption has become considerably larger than that corresponding to a bilayer only the general features of the force curve are reproducible. A typical force curve, measured at pH 10.5, is displayed in Fig. 4. A weak double-layer force dominates the long-range interaction followed at shorter separations by a steric repulsion. On separation a very weak attractive minimum is sometimes observed. When the pH is decreased again the multilayer disappears.

When the pH is increased directly from about 10 or below to above 12 no multilayer build-up is observed. Instead, a strongly repulsive double-layer force dominates the long-range interaction (Fig. 4). A strong short-range repulsion is observed at distances below 90 Å. The range and poor reproducibility of the distance dependence of this short-range force indicate that the structure of the adsorbed layer can be thought of as a poorly ordered bilayer. Note that all changes in surface forces with pH are reversible.

Bilayer formation and CMC

At pH 5.5 a bilayer is formed on each surface at or above a DAHCl concentration of 6×10^{-3} M. As a comparison, the CMC is around 1×10^{-2} M. However, if the surfactant concentration is kept constant at 10^{-4} M and the pH is increased, a

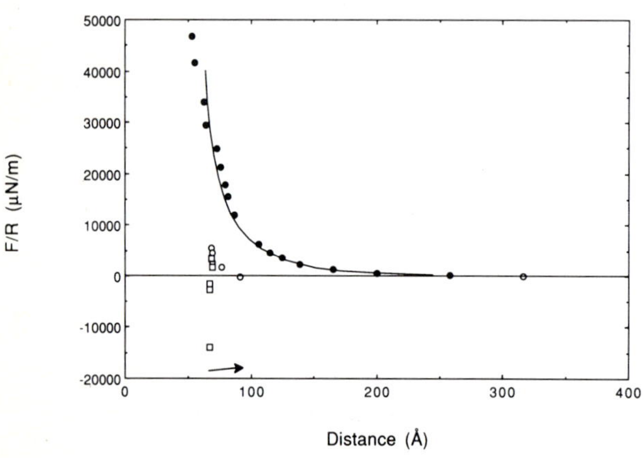

Fig. 3. Force normalized by radius as a function of surface separation between mica surfaces coated with a bilayer of DAHCl. Filled circles represent the forces measured in a 6.5×10^{-3} M surfactant solution at pH 5.6, and unfilled symbols represent forces measured in a 10^{-4} M surfactant solution at pH 10.3. The circles represent forces measured on approach, and squares are forces measured on separation. The arrow indicates a jump. The solid line is a calculated force using the following parameters for the double-layer force: surface potential + 220 mV, Debye-length 39 Å

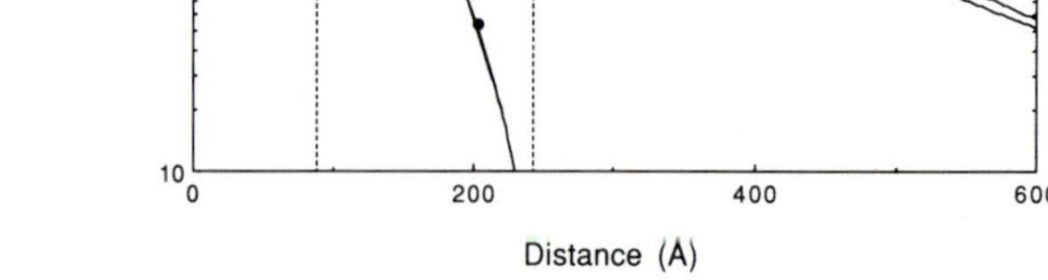

Fig. 4. Force normalized by radius as a function of surface separation between mica surfaces immersed in a 10^{-4} M DAHCl solution at pH 10.5 (unfilled circles) and at pH 12.3 (filled circles). Dashed lines represent the layer thicknesses. Solid lines are calculated forces. The parameters used for calculating the double-layer force are for pH 10.5: surface potential ± 35 mV, Debye-length 145 Å, for pH 12.3: surface potential − 47 mV, Debye-length 26 Å

bilayer forms at around pH 9–9.5. The CMC at this pH-value is about 2×10^{-4} M. Hence, it appears that a bilayer structure develops on the surface when the surfactant concentration exceeds about half the CMC. However, it is likely that the packing density in the bilayer continues to increase until the CMC is reached. One observation that supports this notion is the steeper steric/hydration repulsion at pH 10.3 compared to the one at pH 9.5 [13].

Forces in tetraoxyethylene dodecylamine (TEDA) solutions

The forces between mica surfaces immersed in a 3×10^{-5} M KBr solution without and with 1×10^{-4} M tetraoxyethylene dodecylamine (TEDA) solution have also been measured. Without any TEDA the interaction is well desribed by DLVO theory (Fig. 5). The forces observed below pH 8 with 10^{-4} M TEDA are also displayed in Fig. 5. In this case, a very weak double-layer force is present at separations above 100 Å. The apparent area per charge, obtained from a PB analyses is less than 36 nm^2. Hence, the adsorption of the surfactant in cationic form together with adsorption of protons is, just as in the case of DAHCl, sufficiently strong to neutralize the mica lattice charge. The adsorbed layer of TEDAH$^+$ differs from that of the corresponding DAH$^+$ layer in two important respects: it is about twice as thick, and its hydrophobic character is less, as demonstrated by the lower adhesion force.

An increase in pH to between 9.4 and 10 results in a dramatic change in the measured force (Fig. 6). The weak double-layer force and the strong attraction observed at lower-pH-values are absent and the surfactant coated surfaces now have a hydrophilic character. A weak minimum is located at a separation of 70–80 Å, and a short-range steeply rising repulsive force is observed at shorter separations. The surfactant layer can, under a high compressive load, be squeezed together to a thickness of 25–30 Å per surface. The observed repulsion is a result of entropic confinement [26] and dehydration [25] of polar groups. It was noted that, in some cases, the outer layer could be squeezed out from between the surfaces. The surfaces then reached monolayer contact at the same position as at lower pH values. The force needed in order to reach monolayer contact varied considerably, indicating an inhomogeneous adsorption in the outer layer.

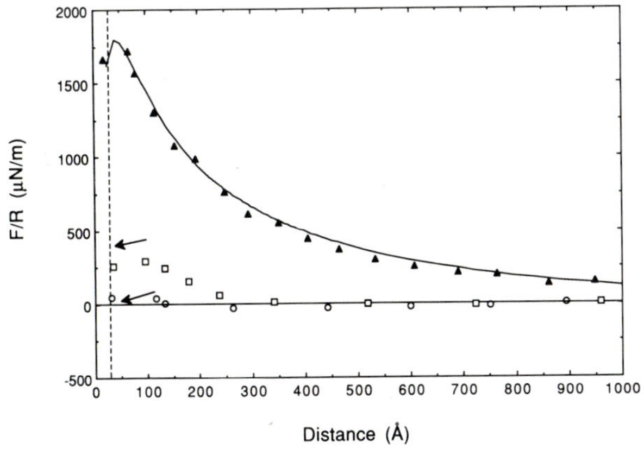

Fig. 5. Force normalized by radius as a function of surface separation between mica surfaces immersed in 3×10^{-5} M KBr without TEDA (filled triangles), and with 10^{-4} M TEDA at pH 7.9 (unfilled squares) and pH 5.9 (unfilled circles). The dashed line represents the layer thickness and the arrows are jumps. The solid lines represent calculated DLVO forces, assuming interaction at constant charge. The Debye-length is 450 Å and the surface potential 105 mV

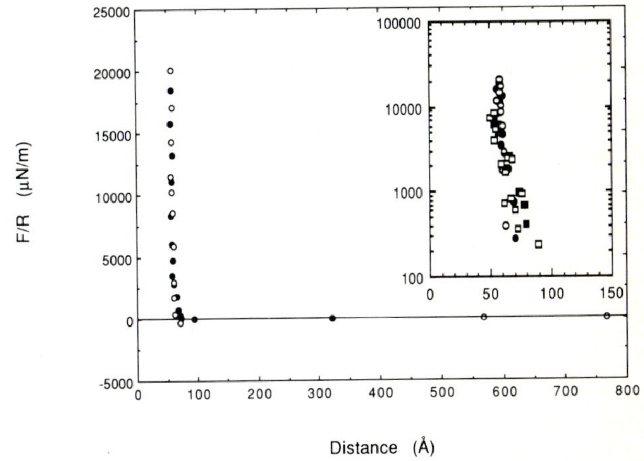

Fig. 6. Force normalized by radius as a function of surface separation between mica surfaces immersed in a 10^{-4} M TEDA solution at pH 9.4. Filled symbols represent forces measured on approach and unfilled symbols are forces measured on separation. The insert shows the short-range repulsion on a logarithmic scale

Discussion

Driving force for adsorption

The surfactants studied here are hydrolyzable, and they exist in cationic form at low pH-values and in uncharged form at high pH-values. This change in the state of the surfactant obviously affects the affinity for the negatively charged mica surface and the intermolecular forces within and between the surfactant layers.

In general terms, the adsorption process can be described as follows: At low pH-values there will be a strong electrostatic attraction between the negatively charged mica surface and the cationic surfactant. The driving force for adsorption is expected to be primarily electrostatic in origin. Once the surface charge has been neutralized, further adsorption is opposed by electrostatic forces given the free energy cost of creating an electrical double-layer [28]. Additional adsorption is also opposed by steric hindrance within the layer, whereas removal of hydrocarbon tails from contact with water favors additional adsorption. As the pH-value increases, it will become increasingly possible to remove hydrocarbon moieties from the aqueous phase without paying the price of building up a strongly charged electrical double-layer. Finally, at even higher pH-values, when almost all surfactants are uncharged, the electrostatic affinity between the surface and the surfactants will decrease. This will have consequences for the structure of the inner layer and, hence, also for the outer surfactant layer.

Whether these driving forces for adsorption will promote an adsorption in monolayers/bilayers or in smaller aggregates will depend largely on the number of charged sites on the surface and the size of the surfactant ion.

The structure of the adsorbed surfactant layers at low pH-values

At low pH-values the surfactants adsorb electrostatically to the negatively charged mica surface. When one surfactant is adsorbed for each negative lattice site the layer thickness should be about 7–8 Å for DAH^+ and about twice this value for $TEDAH^+$. The measured layer thickness is in excellent agreement with this over a range of pH-values (Fig. 7) and concentrations. The resulting surface is hydrophobic due to a preferential orientation of the charged groups toward the surface

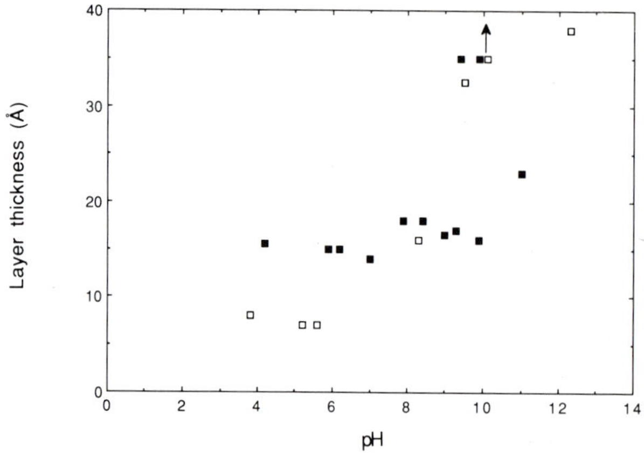

Fig. 7. The thickness of the adsorbed surfactant layer on mica as a function of pH for DA (unfilled squares) and for TEDA (filled squares). The surfactant concentration was 10^{-4} M. The layer thickness was measured at a value of $F/R > 10.000$ μN/m. For such high forces, the layer thickness is insensitive to the exact value of the compressive force. The arrow indicates the very large layer thickness observed at pH 10.3–10.5.

and the hydrocarbon tails toward the solution. This orientation of the molecules gives rise to a relatively strong adhesion between the monolayer coated surfaces. An adhesion force of 200–300 mN/m is observed for DAH^+ (Fig. 8). As a comparison, the adhesion force between Langmuir Blodgett (LB) layers having CH_3 and CH_2 groups toward the aqueous phase is considerably higher (350–500 mN/m) [29]. Hence, it is clear that the adsorbed layer is not perfect, but contains surfactant adsorbed with a reversed orientation in, or above, the inner layer (Fig. 9a). As expected, the ethylene oxide chain in $TEDAH^+$ makes the adsorbed surfactant layer less hydrophobic than that formed by DAH^+ (Fig. 8). It appears that $TEDAH^+$ adsorbs preferentially with the charged ammonium group toward the surface and the hydrocarbon and ethylene oxide chains directed toward the solution. The adhesion force between charged LB layers having a 1:1 mixture of CH_3 and COOH groups at the surface is about 40 mN/m [30], i.e., considerably lower than the value of 90–130 mN/m obtained for uncharged $TEDAH^+$ monolayers.

The structure of the DA layer at pH-values above 8

As the pH-value is increased to slightly above 8, the thickness of the adsorbed DAH^+/DA layer

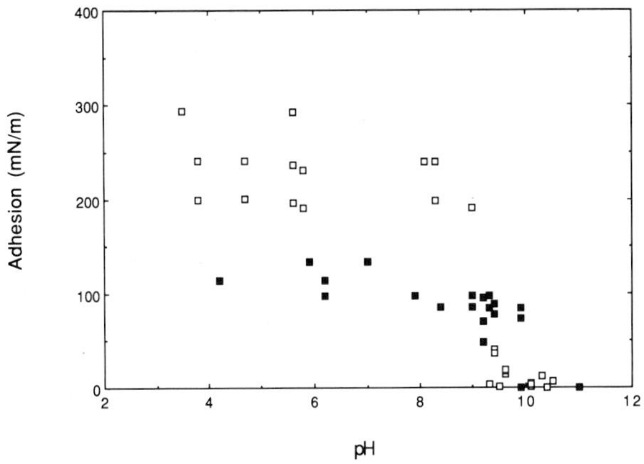

Fig. 8. The pull-off force between two mica surfaces immersed in a 10^{-4} M surfactant solution as a function of pH. The surfactant was DA (unfilled squares) or TEDA (filled squares)

increases (Fig. 7a), but the surface remains essentially hydrophobic (Fig. 8) with a surface charge density very similar to that observed at lower pH-values. Clearly, an additional adsorption of uncharged species driven by a hydrophobic interaction takes place within the monolayer, as illustrated in Fig. 9a. The fact that uncharged species adsorb, despite comprising only 0.5% of the molecules in bulk solution, reflects the high free energy cost of creating a highly charged surface. The fact that no increase in layer thickness with concentration is observed at low pH-values, where almost all surfactants are charged (Fig. 7b) also demonstrates that electrostatic forces oppose additional adsorption of charged surfactants. The high adhesion force observed at pH 8.3 indicates that most of the uncharged molecules that have adsorbed in the monolayer are oriented with the polar group toward the surface, despite the resulting high interfacial tension against water. This suggests that there is either a favorable interaction between uncharged amine-groups and mica, or between ammonium- and amine-groups in the layer.

At pH-values between 9.5 and 10.3, the layer thickness and adhesion force reveal that a bilayer has formed on each surface. This additional adsorption, due to a hydrophobic interaction occurs without any increase in surface charge density. Hence, neutral molecules predominate in the outer layer (Fig. 9a). A charged bilayer can be obtained at pH 5–6, but only at a considerably higher surfactant concentration. The thickness of the bilayer obtained at pH 5–6 is considerably smaller than that of the bilayer obtained at high pH. The structure of the bilayer obtained at low pH is illustrated in Fig. 9c.

At pH 10.3 hydrated dodecylamine droplets phase separate from the bulk surfactant solution. At this pH-value no repulsive double-layer force between the adsorbed layers exists and no force barrier prevents the dodecylamine droplets from precipitating onto the surface. This results in a poorly defined structure of the adsorbed layer. The large layer thickness and the low adhesion force

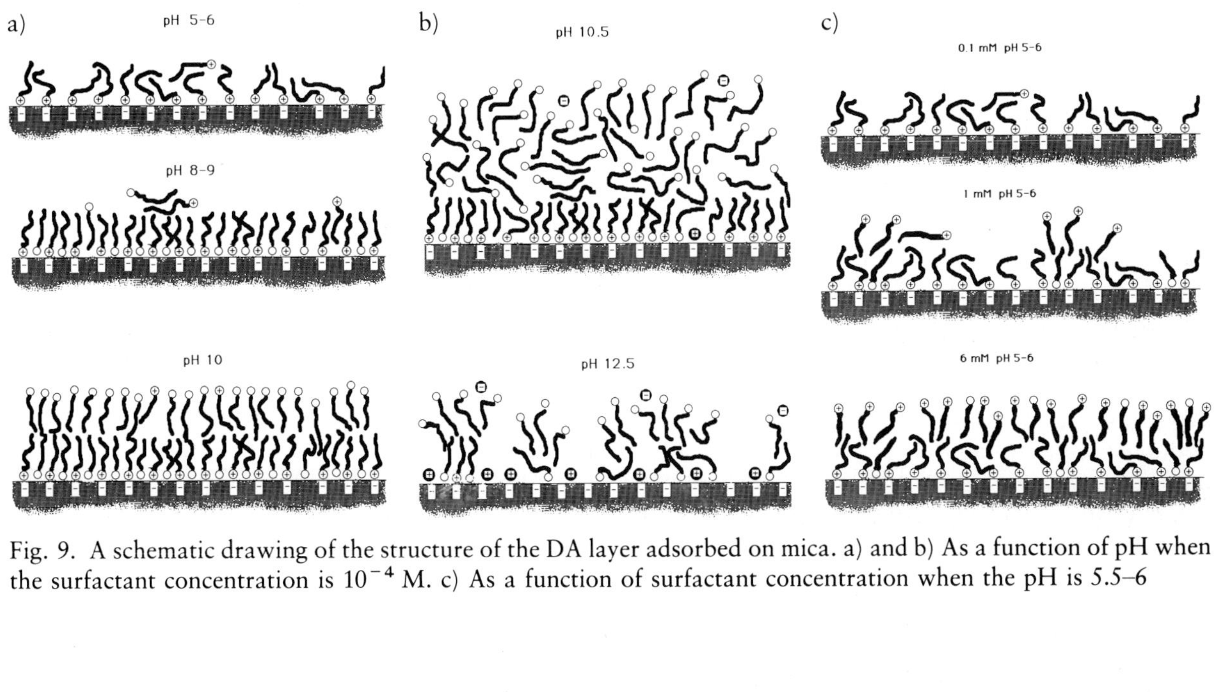

Fig. 9. A schematic drawing of the structure of the DA layer adsorbed on mica. a) and b) As a function of pH when the surfactant concentration is 10^{-4} M. c) As a function of surfactant concentration when the pH is 5.5–6

show that considerably more than a bilayer is present on the surface, and that the interface is hydrophilic, as illustrated in Fig. 9b.

At pH 12.3 only a small fraction of the molecules are charged. Hence, adsorption of the surfactants fails to neutralize the mica surface charge. Consequently, the surfaces are strongly negatively charged. The phase separated dodecylamine droplets are also negatively charged at this pH-value due to adsorption of hydroxide ions [7]. Hence, no precipitation of such droplets is observed at this pH-value. It seems likely that, not only are surfactants present in the interfacial region, but also other cations (sodium in this case) are adsorbed on the surface and hydroxide ions are associated with the adsorbed surfactant layer. A tentative structure of the adsorbed layer is given in Fig. 9b.

The structure of the TEDA layer at pH-values above 8

The only difference between TEDA ($C_{12}H_{25}NHE_4$) and DA (C_{12} $H_{25}NH_2$) is that the former surfactant is ethoxylated. As a result, the polar part of the molecule is significantly larger. At low pH the charged nitrogen atom is adsorbed to the surface with both the hydrocarbon and ethylene oxide chains directed toward the solution. In this configuration the area occupied by the molecule is comparable to the mica lattice charge (0.48 nm^2). Hence, one would expect that the steric hindrance in this layer is significant already at the point where the adsorbed amount is such that the mica lattice charge is neutralized. This is a consequence of the fact that the cross-sectional area of the surfactant is similar to the area per negative mica lattice charge. Hence, unlike the situation for DA, hardly any further adsorption takes place in the monolayer when the pH is increased. We also note that the interfacial tension between the TEDAH$^+$ monolayer and water is considerably lower than between the DAH$^+$/DA monolayer and water. Consequently, the hydrophobic driving force for an additional adsorption is lower in the TEDAH$^+$ case. Nevertheless, the surfactant adsorbs in a bilayer structure at pH-values between 9.4 and about 10. The orientation of the molecules in the outer layer is clearly such that most molecules have the hydrophilic ethylene oxide chain directed towards the solution. The comparatively weak hydrophobic character of the monolayer and the large size of the polar group prevent a tight packing in the outer layer. This explains the observation that the outer layer in many cases can be removed from between the surfaces by applying an external force. It appears likely that the adsorption in the outer layer is not homogeneous, but that surface aggregates are formed at these high pH-values.

At even higher pH-values (above 11), where most of the surfactant is uncharged, the electrostatic driving force for adsorption decreases. Just as was the case for DA, this results in a break-up of the clear monolayer/bilayer structure observed at lower pH-values [14].

Adsorption to mica compared with other surfaces

Finally, we would like to stress that the forces governing adsorption of charged surfactants on hydrophilic, oppositely charged surfaces are expected to be the same in most cases. However, the structures formed on the surfaces will depend on the particular system under study. For the case of highly charged muscovite mica and a small surfactant like DAHCl imperfect monolayers and bilayers are easily formed, in line with the predictions by Yeskie and Harwell [31]. As the size of the polar group increases, the layers formed will be even less ordered, as demonstrated by the mica/TEDA system. For other less highly charged surfaces, like silica at low pH, the same driving forces for adsorption as observed in this study are expected to give rise to small surface aggregates, as discussed by Rupprecht and Gu [32]. Hence, the number of charged sites on the surface is of prime importance for the surface association of surfactants.

Conclusions

The adsorption of both DA and TEDA on negatively charged mica surfaces can be rationalized by the following scheme. Initially, the adsorption is electrostatically driven. The surfactants thus adsorb with the cationic nitrogen atoms on the surface. The hydrocarbon (and in the case of TEDA, the ethylene oxide) chains are directed toward the solution. A further adsorption onto the essentially hydrophobic surfaces is driven by a hydrophobic interaction, but counteracted by electrostatic and steric forces. Bilayer coated surfaces can be obtained by either increasing the surfactant concentration or the pH. As the pH is increased, it is the uncharged molecules that preferentially adsorb on the surfaces. The effects of introducing an ethylene oxide chain attached to the nitrogen atom are i) to cause the hydrophobic interaction between the monolayer and surfactants in solution to decrease, and ii) to increase the size of the polar group to such an extent that steric hindrance within the layers prevent a tight packing. At pH-values below about 10 the properties of the adsorbed layer are consistent with those expected from an essentially monolayer/bilayer structure. At pH-values above 11–12 most surfactants are uncharged, and the adsorbed layers are less ordered.

References

1. Arnebrant T, Nylander T, Cuypers PA, Hegg PO, Larsson K (1984) In: Mittal KL, Lindman B (eds.) Surfactants in Solutions. Plenum Press, New York, 1984, vol 2, p 1291, Cuypers PA, Corsel JW, Janssen MP, Kop JMM, Hemker HC, Hermens WTh, ibid vol 2, p 1301
2. Rouquerol J, Partyka S (1982) J Chem Tech Biotechnol 31:584
3. Rennie AR, Lee EM, Simister EA, Thomas AK (1990) Langmuir 6:1031
4. Israelachvili JN, Adams GE (1978) J Chem Soc Faraday Trans 1 74:975
5. Pugh RJ (1986) Colloids Surfaces 18:19
6. Somasundaran P, Ananthapadmanabhan KP (1979) In: Mittal KL (ed.) Solution Chemistry of Surfactants, Vol 2. Plenum Press, New York
7. Castro SH, Vurdela RM, Laskowski JS (1986) Colloids Surfaces 21:87
8. Kizling J (private communication)
9. Dai Q, Laskowski JS (1991) Langmuir 7:1361
10. Wärnheim T, Bergenståhl B, Henriksson U, Malmvik AC, Nilsson P (1987) J Colloid Interface Sci 118:233
11. Wärnheim T, Henriksson U, Malmvik AC, Bergenståhl B (1988) Liquid Crystals 3:791
12. Herder PC (1990) J Colloid Interface Sci 134:346
13. Rutland MW, Waltermo Å, Claesson PM, Langmuir 8:176
14. Waltermo Å, Sjöberg M, Anhede B, Claesson PM, submitted
15. Güven N (1971) Zeitschrift Krystallographie 134:196
16. Pashley RM (1981) J Colloid Interface Sci 83:531
17. Parker JL, Christenson HK, Ninham BW (1989) Rev Sci Instruments 60:3135
18. Derjaguin BV (1934) Kolloid Z 69:155
19. Horn RG, Israelachvili JN (1981) J Chem Phys 75:1400
20. Chan DY, Pashley RM, White LR (1980) J Colloid Interface Sci 77:283
21. Guldbrand L, Jönsson B, Wennerström H, Linse P (1984) J Chem Phys 80:2221; Kjellander RK, Marcelja S (1986) J Phys Chem 90:1230; Attard P, Mitchell J, Ninham BW (1988) J Chem Phys 89:4358
22. Israelachvili JN, Pashley RM (1984) J Colloid Interface Sci 98:500, Kékicheff P, Christenson HK, Ninham BW; (1989) Colloids Surfaces 40:31
23. Podgornic R, Parsegian VA (1991) Chem Phys 154:477
24. Johnson KL, Kendall K, Roberts AD (1971) Proc R Soc London A324:301; Muller VM, Yushchenko VS, Derjaguin BV (1983) J Colloid Interface Sci 92:92; Christenson HK, Claesson PM (1990) ibid 139:589; Neuman AW, Moy E (1990) ibid 139:591
25. LeNeuve DM, Rand RP, Parsegian A (1976) Nature (London) 259:601
26. Israelachvili JN, Wennerström H (1990) Langmuir 6:873
27. Claesson PM, Eriksson JC, Herder CE, Pezron E, Pezron I, Bergenståhl B, Stenius P (1990) Faraday Discussions Chem Soc 90:129
28. Gunnarsson G, Jönsson B, Wennerström H (1980) J Phys Chem 84:3114
29. Claesson PM, Blom CE, Herder PC, Ninham BW (1986) J Colloid Interface Sci 114:234; Herder PC (1990) J Colloid Interface Sci 134:336
30. Berg JM, Claesson PM (1989) Thin Solid Films 178:261
31. Yeskie MA, Harwell H (1988) J Phys Chem 92:2346
32. Rupprecht H, Gu T (1991) Colloid & Polymer Sci 269:506

Author's address:

P. M. Claesson
The Surface Force Group
Dept. of Physical Chemistry
The Royal Institute of Technology
S-100.44 Stockholm
Sweden

Progress in Colloid & Polymer Science Progr Colloid Polym Sci 88:74–85 (1992)

Surfactant adsorption and surface micellization

T. Gu,[1]) B.-Y. Zhu[2]), and H. Rupprecht[3])

[1]) Department of Chemistry, University of Bergen, Norway
[2]) Department of Chemistry, Peking University, China
[3]) Institut für Pharmazie, Universität Regensburg, FRG

Abstract: This paper briefly reviews a new adsorption theory particularly applicable to surfactant adsorption. A one-step model of surface micellization is first introduced. By using the mass-action law treatment, an S-type isotherm equation is derived which explains the adsorption of nonionic surfactants from aqueous solution onto polar adsorbents. However, the adsorption of ionic surfactant generally shows two-step character. First the surface-active species are adsorbed through electrostatic attraction and/or van der Waals interaction. The second step involves hydrophobic interaction if the adsorption is from aqueous solution, and hydrogen bonding or other polar interaction if from non-polar solvent. Under these conditions a general adsorption isotherm equation is derived. The model indicates that the size of surface micelles is usually much smaller than that of corresponding bulk micelles. From thermodynamics of adsorption, it is evident that when dealing with adsorption from aqueous solution, the second step, i.e. surface micellization, is an entropy-driven process similar to the bulk micellization.

Key words: Surfactant; adsorption; surface; micellization

Introduction

The adsorption of surfactants at solid/liquid interfaces is of considerable interest in a number of industrial processes. When adsorbed onto a solid surface, a surfactant can modify the surface from hydrophilic to hydrophobic or vice versa. It can cause a finely divided solid to be easily dispersed in a liquid or to make it flocculate heavily. All these functions are important in a multitude of processes like ore flotation, oil recovery, detergency, cosmetics, pharmaceuticals, etc. Much work has been done in this area, and a great number of papers has been published on this topic [1, 2].

The interpretation of surfactant adsorption isotherms has been intensely pursued by surface and colloid chemists for many years. The surfactant adsorption isotherms can generally be divided into three types, i.e., Langmuir or L-type, S-type, and two plateaus or LS-type [3–6]. For L-type isotherms, represented by the Langmuir equation, the following assumptions [7] are made: 1) homogeneous interface; 2) adsorption completed at monolayer coverage; 3) insignificant lateral interactions, and 4) the solvent and solute have equal molecular cross-sectional area. Although many systems appear to show this type of isotherm, the assumed conditions are rarely, if ever, encountered in practice. In particular, the latter two assumptions are clearly not fulfilled. Consequently, the use of this equation, even for L-type isotherms, may conceal the mechanism of adsorption processes. Based on different models for the surfactant adsorption, a number of other isotherms has been suggested [8–12]. However, a universal theory to interpret the various types of surfactant adsorption isotherms is obviously not obtained.

Structures of the adsorbed layer of surfactants have also been suggested. Gaudin and Fuerstenau [14] introduced the hemimicelle concept (close-packed local monolayer) and Harwell et al. [15], the admicelle structure (close-packed local bilayer) for adsorption and aggregation of ionic surfactants on polar solids. In addition, the authors assumed the adsorbates to establish more or less homogeneous flat layers on the solid, where the surfactant ions are in mutual contact by hydrophobic interactions, i.e., aggregated and oriented perpendicular to the surface.

We have investigated surfactant adsorption during recent years [17, 18]. Recently, a two-step adsorption mechanism, i.e., an adsorption through direct interaction between the surface-active species and the solid surface, and later a hydrophobic interaction between the adsorbed species, were adopted to qualitatively explain the LS-type adsorption isotherms [16]. An empirical approach which allows to estimate the aggregation number of surface micelle (n)* from the first and second adsorption plateaus ($n = \Gamma_2/\Gamma_1$) was proposed [16]. The surface solubilization of Chrome Azurol S (an anionic dye without adsorption onto negatively charged silica), provided evidence for surface hydrophobic aggregation or surface micellization [19]. Based on the mass-action law treatment of surface micellization, a semi-empirical approach was developed which also allows to estimate the aggregation number of surface micelle, as well as the equilibrium constant (K) and the standard free energy of surface micellization (ΔG^0_{sm}) [20, 21]. The results show that the aggregation values obtained by the mass-action law treatment are in good agreement with those obtained by the previous empirical approach. Thereafter, based on the mass-action law treatment of one-step formation of surface micelles, an equation was derived that explains the S-type adsorption isotherm and allows to calculate n, K, and ΔG^0_{sm}. This treatment fits the adsorption data of nonionic surfactants on silica [3]. Very recently, based on the two-step adsorption model and the mass-action treatment, a general adsorption isotherm equation was derived which can describe the various types (including L-, S-, and LS-types) of isotherms [4, 5]. Based on this general equation and the thermodynamics of adsorption, a more detailed picture of surfactant adsorption and surface micellization has been developed [3–6, 16–21].

*) In this paper the terms of surface micelle and surface micellization are used instead of the ambiguous terms of hemimicelle and hemimicellization, respectively.

One-step model [3]

It is reasonable to apply the mass-action law to adsorption equilibria and, as a first crude approximation, to only involve the surface micellization. Consider now the aggregation of n monomers of nonionic surfactant on a surface site (s) to form a surface micelle (sm). At equilibrium

$$S + n \text{ Monomer} \rightarrow \text{Surface Micelle} \tag{1}$$

and

$$K = a_{sm}/a_s a^n \ , \tag{2}$$

where n is the average aggregation number of the surface micelle, a is the activity of monomers in solution, and for dilute solutions $a = C$, the concentration of nonionics; if $C < CMC$, a_{sm} and a_s are activities of surface micelles and unoccupied surface sites, respectively, and K is the equilibrium constant. Approximately, a_{sm} and a_s can be given by

$$a_{sm} = \Gamma/n \tag{3}$$

and

$$a_s = (\Gamma_\infty - \Gamma)/n \ , \tag{4}$$

respectively, where Γ is the amount of surfactant adsorbed at C and Γ_∞ is the limiting adsorption at high concentrations. Thus, Eq. (2) becomes

$$K = \Gamma/(\Gamma_\infty - \Gamma)C^n \ . \tag{5}$$

Equation (5) can be put in the form

$$\Gamma = \Gamma_\infty KC^n/(1 + KC^n) \ . \tag{6}$$

Equation (6) can be considered a combination of the Langmuir and Freundlich adsorption isotherms; it has the correct limiting values at high concentrations, and formally reduces to the Freundlich equation at low concentrations. Equation (5) can be expressed as

$$\log[\Gamma/(\Gamma_\infty - \Gamma)] = \log K + n \log C \ . \tag{7}$$

If the data obey this equation, the plot of $\log[\Gamma/(\Gamma_\infty - \Gamma)]$ vs $\log C$ gives a straight line. The slope of the straight line gives n and the intercept gives $\log K$. If surface micellization occurs, the value of n should be > 1. With multi-site adsorption (i.e., each adsorbed molecule occupies more than one site) n should be < 1, but > 0. When $n = 1$, Eq. (6) is the Langmuir equation. Actually, in the latter two cases no surface micelles form. It should be emphasized that, according to Eq. (1), the n value is never less than zero. This treatment is particularly suitable for the adsorption of nonionic surfactants on polar adsorbents. Figure 1 shows the adsorption of octylmethylsulfoxide (OMS), decylmethylsulfoxide (DEMS), and Triton X-100 (TX100) on two silica gels (i.e., narrow pore ($N\text{–}SiO_2$) and wide pore ($W\text{–}SiO_2$) silica gels), plotted according to Eq. (7). The values of n and K obtained from the slopes and intercepts of the straight lines in Fig. 1 are given in Table 1. These values of parameters define the isotherms represented by the solid curves in Figs. 2 and 3, which fit the experimental data.

Two-step model and the general adsorption isotherm equation [4, 5, 18]

Generally, it is reasonable to consider that the adsorption of surfactants should include two steps. First, according to the nature of adsorption, there must be an attraction between surface-active species and the solid surface. The second step originates for the character of surfactants. The surfactant

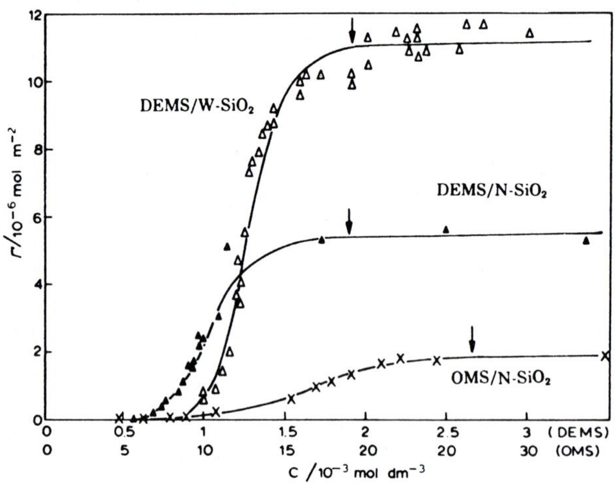

Fig. 2. Adsorption isotherms of OMS and DEMS on silica gels at 25 °C [3]. Points are experimental, curves are calculated from Eq. (6). Vertical arrows mark the CMC for surfactants

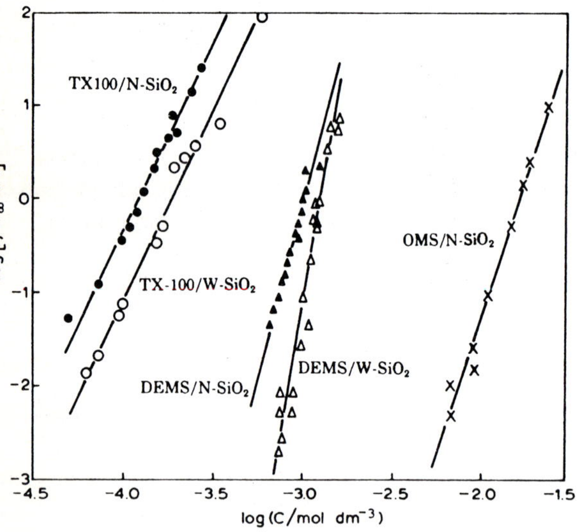

Fig. 1. Adsorption of OMS, DEMS, and TX100 on silica gels at 25 °C, plotted according to Eq. (7) [3]

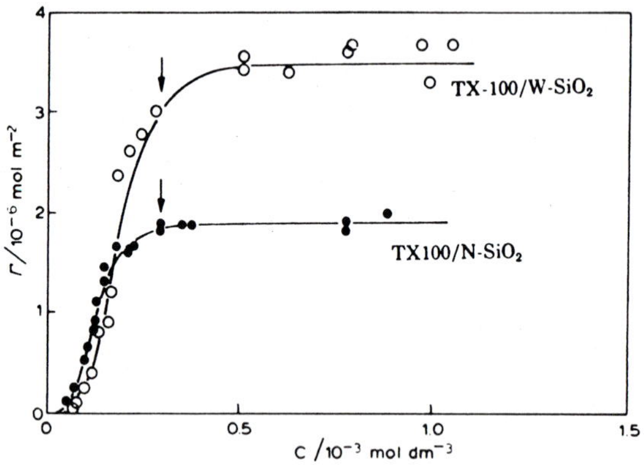

Fig. 3. Adsorption isotherms of TX100 on silica gels at 25 °C [3]. Vertical arrows mark the CMC for surfactants

Table 1. Values of CMC, SMC, n, k, ΔG^0_{sm}, and ΔG^0_m for nonionic surfactants on silica gels from aqueous solutions at 25 °C [3]

Surfactant	CMC (10^{-3} mol·dm^{-3})	Adsorbent	SMC (10^{-3} mol·dm^{-3})	n	K	$-\Delta G^0_{sm}$ (kJ·mol^{-1})	$-\Delta G^0_m$ (kJ·mol^{-1})
OMS	26.5	$N\text{–}SiO_2$	12	5.7	$1.54 \cdot 10^{10}$	10.2	9.0
DEMS	1.88	$N\text{–}SiO_2$	0.8	7.4	$1.50 \cdot 10^{22}$	17.1	15.6
DEMS	1.88	$W\text{–}SiO_2$	1.0	11.8	$1.33 \cdot 10^{34}$	16.5	15.6
TX100	0.29	$N\text{–}SiO_2$	0.07	3.9	$1.74 \cdot 10^{15}$	22.3	20.2
TX100	0.29	$W\text{–}SiO_2$	0.1	4.0	$7.25 \cdot 10^{14}$	21.2	20.2

is a molecule consisting of a hydrophilic head and a hydrophobic tail. The hydrocarbon chain is responsible for the hydrophobic effects and results from a formation of micelles.

Similarly, one may expect that the adsorbed surface-active species should also exhibit similar hydrophobic effects and form surface micelles. These surface micelles may be of different sizes and shapes, including small and more or less spherical micelles, closepacked local monolayers, and closepacked local bilayer, etc.

Thus, the basic assumption in the general theory is that the adsorption of surfactants occurs in two steps. In the first step the surface-active species are adsorbed as individual ions or molecules (depending on the type of surfactant involved) on the surface through electrostatic attraction (only present in the case of ionic surfactant, where the surface active ions are adsorbed on the oppositely charged solid surface) and/or specific (i.e., van der Waals) attraction. At equilibrium

$$\text{Site} + \text{Monomer} \rightarrow \text{Adsorbed Monomer}\ , \tag{8}$$

with the corresponding equilibrium constant

$$k_1 = a_1/a_s a\ , \tag{9}$$

where a_1 is the activity of adsorbed monomers. In the second step the adsorption increases dramatically as surface micelles form on the adsorbent through hydrophobic interaction between tails of the surface active species. At equilibrium

$$(n-1)\ \text{Monomers} + \text{Adsorbed Monomer} \rightarrow \text{Surface Micelle} \tag{10}$$

and

$$k_2 = a_{sm}/a_1 a^{n-1}\ , \tag{11}$$

where a_{sm} is the activity of surface micelles and n is the aggregation number, as previously.

The surface concentrations of adsorbed monomer (Γ_1), surface micelles (Γ_{sm}), and unoccupied sites (Γ_s) can be used instead of a_1, a_{sm}, and a_s, respectively. Thus, Eqs. (9) and (11) become

$$k_1 = \Gamma_1/\Gamma_s C \tag{12}$$

and

$$k_2 = \Gamma_{sm}/\Gamma_1 C^{n-1}\ , \tag{13}$$

respectively.

The general adsorption isotherm equation can be derived from the combination of Eqs. (12), (13) and the following two additional conditions

$$\Gamma = \Gamma_1 + n\Gamma_{sm} \tag{14}$$

and

$$\Gamma_\infty = n(\Gamma_s + \Gamma_1 + \Gamma_{sm})\ ; \tag{15}$$

both Γ and Γ_∞ can be measured from experiments. The general isotherm equation is

$$\Gamma = \frac{\Gamma_\infty k_1 C\left(\frac{1}{n} + k_2 C^{n-1}\right)}{1 + k_1 C(1 + k_2 C^{n-1})}\ . \tag{16}$$

Equation (16) has two important limiting cases. If $k_2 \rightarrow 0$ and $n \rightarrow 1$, it reduces to the Langmuir equation

$$\Gamma = \frac{\Gamma_\infty k_1 C}{1 + k_1 C}\ . \tag{17}$$

If $n > 1$ and $k_1 C \ll 1$, Eq. (16) reduces to an S-type equation (6):

$$\Gamma = \frac{\Gamma_\infty k_1 k_2 C^n}{1 + k_1 k_2 c^n} = \frac{\Gamma_\infty K C^n}{1 + K C^n}\ , \tag{18}$$

where $K = k_1 k_2$. In view of this result, the one-step model is only a limiting case of the more general two-step model. For the adsorption of nonionic surfactants on silica gel from aqueous solution, neither electrostatic nor other strong interactions between the adsorbate and adsorbent are present, thus, k_1 is small and $k_1 \ll 1$. This explains why the first plateau is usually absent in the adsorption isotherms of these systems and that the data can fit the *S*-type equation.

The various surfactant adsorption isotherms, including L-, S-, and LS-types, can be obtained from the general Eq. (16) if the appropriate values of parameters k_1, k_2, and n are taken [4]. If k_1 is small enough, Eq. (16) reduces to S-type Eq. (6). When k_1 is large enough, then LS-type isotherms are obtained (see Fig. 4). On the other hand, the larger the k_2, the steeper the second step on the isotherm and the lower the critical surface micelle concentration (SMC) (see Fig. 5). These general conclusions are in agreement with the requirements of the model. (More sample computations can be found in [4, 18].)

The constants k_1, k_2, and n in Eq. (16) can be determined from adsorption data. If the concentrations are high enough i.e., $k_1 C \gg 1$,. Eq. (16) be-

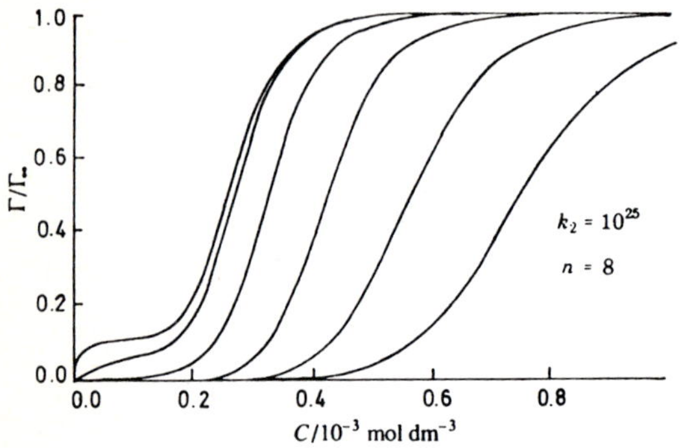

Fig. 4. Calculated isotherms according to general Eq. (16) with $k_2 = 10^{25}$, $n = 8$, for different values of k_1. Curves from left to right: $k_1 = 10^5$, 10^4, 10^3, 10^2, 10 and 1, respectively [4]

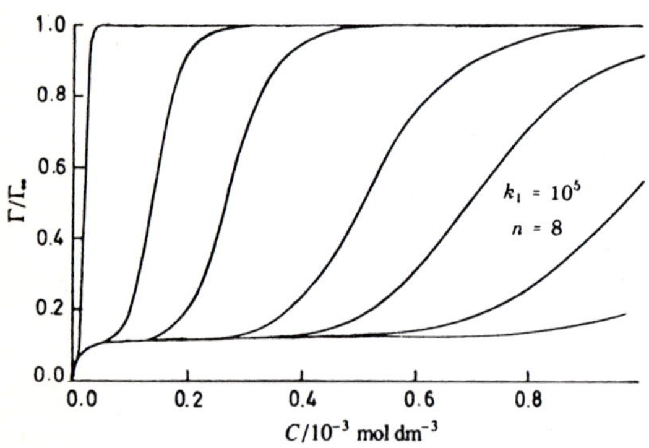

Fig. 5. Calculated isotherms according to general Eq. (16) with $k_1 = 10^5$, $n = 8$, for different values of k_2. Curves from left to right: $k_2 = 10^{33}$, 10^{27}, 10^{25}, 10^{23}, 10^{22}, 10^{21}, and 10^{20}, respectively [4]

comes

$$\Gamma = \frac{\Gamma_\infty\left(\frac{1}{n} + k_2 C^{n-1}\right)}{1 + k_2 C^{n-1}} . \tag{19}$$

Since Γ_∞ can be obtained from adsorption data at high concentrations, it is not difficult to obtain the optimum n and k_2 from Eq. (19) by trial and error with computer procedures. The value of k_1 can be evaluated from the adsorption data at low concentrations via a limiting Langmuir-type equation. Then by using Eq. (16) with these values of constants, one can calculate the theoretical adsorption isotherm. Figure 6 shows the typical LS-type adsorption isotherms of the cationic surfactants, tetradecylpyridinium bromide (TPB) [16] and cetyltrimethylammonium bromide (CTAB) [21] on silica gel. The agreement between the calculated

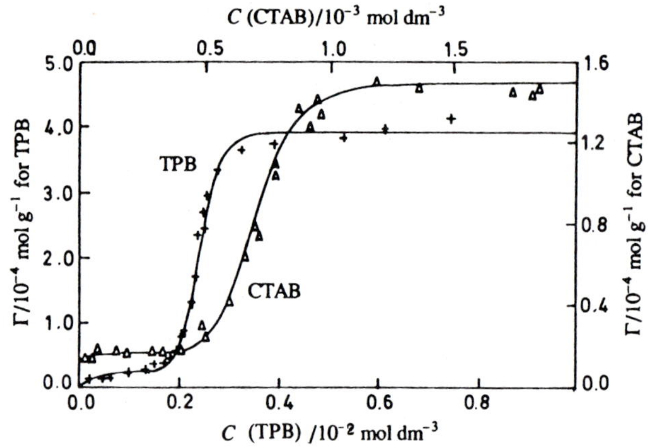

Fig. 6. Adsorption isotherms of TPB and CTAB on silica gel at 25 °C. Experimental points: +, TPB [16], Δ, CTAB [21]. Curves; calculated from Eq. (16). TPB: $\Gamma_\infty = 0.39$ m mol·g^{-1}, $k_1 = 5 \times 10^3$, $k_2 = 4.7 \times 10^{30}$, $n = 12.7$; CTAB: $\Gamma_\infty = 0.15$ m mol·g^{-1}, $k_1 = 5 \times 10^5$, $k_2 = 2.4 \times 10^{25}$, $n = 9.0$

curves and the experimental points is quite satisfactory. (More examples can be found in [5, 18].)

Thermodynamics of surface micellization [3, 5, 18]

From k_2 the standard free energy of surface micellization (ΔG^0_{sm}) can be calculated by using the equation

$$\Delta G^0_{sm} = -(1/n) RT \ln k_2 . \tag{20}$$

For S-type isotherms, K must be used instead of k_2 in Eq. (20). The standard entropy of surface micellization (ΔS^0_{sm}) can be found from

$$\Delta S^0_{sm} = -d\Delta G^0_{sm}/dT , \tag{21}$$

and the standard enthalpy of surface micellization (ΔH^0_{sm}) can be obtained from

$$\Delta H^0_{sm} = \Delta G^0_{sm} - T\Delta S^0_{sm} . \tag{22}$$

Table 2 shows the values of n, k_2, ΔG^0_{sm}, ΔS^0_{sm}, and ΔH^0_{sm}, for LS-type adsorption isotherms of DEMS on carbon black. It is evident that the main contribution to the negative ΔG^0_{sm} is the large positive value of ΔS^0_{sm}, whereas ΔH^0_{sm} is positive. In other words, the surface micellization is an entropy-driven process. Table 2 also shows that the n of DEMS increases with increasing temperature, which is similar to the micellization of nonionic surfactants in the bulk phase. In addition,

Table 2. Surface micellization and micellization of decylmethylsulfoxide on carbon black [5]
a) On carbon black

T/K	n	K_2	ΔG^0_{sm} (kJ · mol^{-1})	ΔS^0_{sm} (JK^{-1} · mol^{-1})	ΔH^0_{sm} (kJ · mol^{-1})
298	11.6	4.0×10^{28}	− 13.6		4.8
303	12.7	8.8×10^{31}	− 14.6	64	4.8
313	13.3	3.5×10^{33}	− 15.2		4.8

b) In bulk solution*)

T/K	CMC (mol · dm^{-3})	ΔG^0_m (kJ · mol^{-1})	ΔS^0_m (JK^{-1} · mol^{-1})	ΔH^0_m (kJ · mol^{-1})
288	1.95×10^{-3}	− 15.0		1.1
303	1.86×10^{-3}	− 15.8	56	1.2
313	1.82×10^{-3}	− 16.4		1.1

*) ΔG^0_m = RT In CMC

Table 1 also shows that the value of ΔG^0_{sm} for S-type isotherms of nonionic surfactants on silica gels is close to that of ΔG^0_{sm} in bulk phase. From ΔG^0_{sm} values of OMS and DEMS the $\Delta G^0_{sm}/CH_2 = -3.5$ kJmol^{-1} is found, which coincides with that of micellization in bulk solution [3, 24]. All these results reflect the analogy between the nature of surface micellization and bulk micellization.

Surface micelle concentration (SMC) [5]

Similar to the definition of critical micelle concentration (CMC), the critical surface micelle concentration or simply surface micelle concentration (SMC) is defined as the concentration range at which the adsorption dramatically increases as surface micelles form on the solid surface. However, usually the SMC estimated from the adsorption isotherm is empirical and always somewhat arbitrary. Therefore, a mathematical approach is helpful.

We take two different definitions to assign the SMC as follows: 1) SMC is the concentration at which the straight line with the maximum $d\Gamma/dc$ on the adsorption isotherms intersects the line of $\Gamma = 0$ or $\Gamma = \Gamma_\infty/n$; 2) SMC is defined as being the point at which $d^3\Gamma/dc^3 = 0$. Unfortunately, neither of the definitions can provide a simple analytical solution for SMC from Eq. (16). However, as the surface micelle formation is usually involved in the S-shape portion of the isotherms, it is reasonable to apply the above definitions to the S-type and LS-type isotherms. For an S-type Eq. (6), the results are

$$\mathrm{SMC} = \left(\frac{n-1}{n+1}\right)^{(n+1)/n} K^{-1/n}, \tag{23}$$

and

$$\mathrm{SMC} = \left[\frac{3(n-1) - [3(n^2-1)]^{1/2}}{3(n-1) + [3(n^2-1)]^{1/2}} \cdot \frac{1}{K}\right]^{1/n}, \tag{24}$$

with definitions 1 and 2, respectively. For LS-type isotherms, with $k_1 c \gg 1$, Eq. (19),

$$\mathrm{SMC} = \left[\frac{(n-2)}{n}\right]^{n/(n-1)} k_2^{1/(1-n)}$$

and

$$\mathrm{SMC} = \left[\frac{3(n-2) - [3n(n-2)]^{1/2}}{3(n-2) + [3n(n-2)]^{1/2}} \cdot \frac{1}{k_2}\right]^{1/(n-1)}, \tag{26}$$

with definitions 1 and 2, respectively. A number of results has been cited in [5, 18] which show that no significant differences can be found between the values of SMC calculated from Eqs. (23) and (24) for S-type isotherms, or from Eqs. (25) and (26) for L-type isotherms. Both calculated SMC values agree with the empirical values. A useful rule is that usually the SMC value roughly equals half of the CMC value for the same surfactant [3, 5, 16, 18], although the SMC is certainly affected by the nature of the solid surface.

Reverse surface micelle formation [6]

The essential point of the two-step model is to attribute the motive power of the adsorption to two kinds of interactions: one is between the solid surface and surface active species, another between the surface active species adsorbed and those to be adsorbed. Hydrophobic effect should be one of the latter, but not the unique one. Actually, the two-step model has been applied successfully to the adsorption of amphiphilic compound at solid/nonpolar solvent interface. Obviously, there is no hydrophobic effect in this case.

In highly nonpolar solvent such as hydrocarbons the polar group of the amphiphilic molecules becomes solvophobic and, in such media, aggregates form in which the polar groups form a core [25, 26]. Such species are generally referred to as reverse micelles. Obviously, one may use the term reverse surface micelle to describe the aggregates which form at the solid/nonpolar solvent interfaces. It is reasonable to suggest that the two-step model can also be accepted in this case, although the nature of interactions and the structure of surface aggregate may be entirely different from that in the surfactant adsorption from aqueous solution. In the first step the amphiphiles can be adsorbed as individual molecules on the solid through van der Waals interaction, while in the second step the adsorption increases dramatically as the reverse surface micelles form through dipole-dipole interaction and/or hydrogen bonding between the head groups (e.g., hydroxyl for alcohols) of the amphiphiles, and each molecule adsorbed in the first step provides a possible center for reverse surface micelle formation. The difference between the adsorption of surfactants from aqueous solution onto solid and that of amphiphiles from nonpolar solvent onto nonpolar solid is that the physical basis for micelle formation in the former case is hydrophobic interaction, while that in the latter case is polar interactions. Therefore, the two-step model and the general Eq. (16) can also be used in this case.

It has been shown [6] that Eq. (16) can fit the adsorption data of *n*-decanol on carbon black from heptane [27]. By using the procedures described previously, the optimum values of n, k_1, and k_2 of Eq. (16) can be evaluated (Table 3). The results show that the constant k_1 is quite small and no significant change is observed as the temperature increases from 15° to 35° C, while the constant k_2 dramatically decreases with increasing temperature, indicating that the motive forces for the two steps are different. Table 3 shows that ΔH^0_{rsm} is large and negative, with ΔS^0_{rsm} quite small and also negative. Obviously, in contrast to the surface micellization from aqueous solution, the reverse surface micelle formation from nonpolar solvent is not an entropy-driven process. The results qualitatively support the proposed mechanism of reverse surface micellization.

Composition for surfactant adsorbed film-individual isotherms

Obviously, the interface of solid/surfactant solution, in general, consists of monomers, surface micelles, and unoccupied sites. According to Eqs. (12–15), the following individual isotherms equations can be derived:

$$\Theta_1 = \frac{\Gamma_1}{\Gamma_\infty/n} = \frac{k_1 C}{1 + k_1 C + k_1 k_2 C^n}\,, \tag{27}$$

$$\Theta_{sm} = \frac{\Gamma_{sm}}{\Gamma_\infty/n} = \frac{k_1 k_2 C^n}{1 + k_1 C + k_1 k_2 C^n}\,, \tag{28}$$

Table 3. Reverse surface micelle formation of 1-decanol from heptane on graphilized carbon black [6]*)

T/K	n	k_2	$-\Delta G^0_{rsm}$ kJ·mol^{-1}	$-\Delta S^0_{rsm}$ kJ·mol^{-1}	$-\Delta H^0_{rsm}$ kJ·mol^{-1}
288	6.0	3.5×10^8	7.85		65.5
293	5.9	2.2×10^7	6.98	0.17	65.6
298	5.8	2.0×10^6	6.20	0.16	65.8
303	5.2	3.5×10^4	5.07	0.23	65.7
308	3.0	1.0×10^2	3.93	0.23	65.5
			AV	0.20	65.5

*) $k_1 \sim 1$

and

$$\Theta_s = \frac{\Gamma_s}{\Gamma_\infty/n} = \frac{1}{1 + k_1 C + k_1 k_2 C^n} , \qquad (29)$$

where Θ_1, Θ_{sm}, and Θ_s represent the surface fraction occupied by the monomers, surface micelles, and unoccupied sites, respectively. Generally, as shown in Fig. 7, as concentration increases, Θ_s decreases monotonously; Θ_1 increases initially, then decreases; Θ_{sm} is nearly nil in the very low concentration region, but increases steeply when the SMC is reached, then approaches to a limiting value, depicting an S-shape curve. It should be noted that the analysis of individual isotherms could be of significance, since it makes possible to control the surface concentrations and fractions of monomers, surface micelles, and unoccupied sites through adjusting of bulk concentration, and from Θ_{sm}–C curves, the SMC could be determined more accurately.

Spreading pressure of surfactant adsorbed film [23]

It is well known that the spreading pressure (π) can be evaluated from adsorption data and, more conveniently, from the isotherm equation, if available. The Gibbs equation for adsorption of a single solute from a dilute solution onto an insoluble solid can be written as

$$- d\gamma = RT\Gamma d\ln C , \qquad (30)$$

where γ is the interfacial free energy of solid/solution interface. Noting that $\pi = \gamma_0 - \gamma$ (where γ_0 is

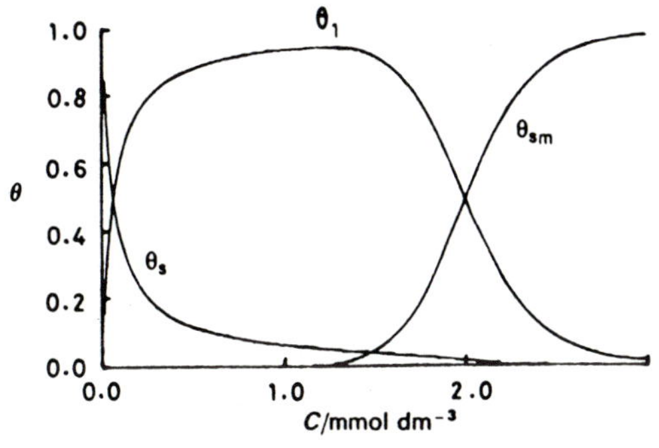

Fig. 7. Individual isotherms of monomers (Θ), surface micelles (Θ_{sm}), and unoccupied sites (Θ_s) for DEMS on carbon black [23]. $k_1 = 1.5 \times 10^4$, $k_2 = 4 \times 10^{28}$ and $n = 11.8$

the γ at $C = 0$) and $d\pi = - d\gamma$, integration of Eq. (30) yields

$$\pi = RT \int_{C=0}^{C} \Gamma d\ln C . \qquad (31)$$

Inserting general Eq. (16) into Eq. (31) and integrating the resulting equation gives

$$\pi = (\Gamma_\infty/n)\, RT \ln(1 + k_1 C + k_1 k_2 C^n) . \qquad (32)$$

As mentioned previously, in the limiting cases the general Eq. (16) will reduce to Langmuir Eq. (17) and S-type Eq. (6), respectively. The corresponding π–C relations can be derived as

$$\pi = \Gamma_\infty RT \ln(1 + k_1 C) , \qquad (33)$$

and

$$\pi = (\Gamma_\infty/n)\, RT \ln(1 + KC^n) ; \qquad (34)$$

the former is the Szyszkowski equation.

Since A can be evaluated from the surface area of adsorbent and Γ, the π–A curves can be obtained through combining π–C and A–C curves.

Individual small surface micelles – various evidences from experiments [17]

Low saturated adsorption value

The adsorption of cationic surfactants onto negatively charged polar adsorbents usually shows typical two-plateau character or LS-type isotherm [16, 21, 28, 29]. At concentrations higher than CMC, the nearly constant saturated adsorption value Γ_2 (i.e. Γ_∞) indicates the upper limit of the surface micellization. From the Γ_2 values, the mean surface area occupied by one surfactant ion in the adsorbed layer can be calculated. These values indicate that, even at high concentrations, the adsorbed layer is far from being arranged as close-packed mono- or bilayers, as depicted by Harwell et al. [15] (Fig. 8a and b). This typical feature of polar adsorbents with a low number of primarily charged adsorption sites can be reasonably explained by the existence of small, isolated surface micelles (Fig. 8c) [4, 5, 16–18, 20, 21]. The aggregation numbers of these surface micelles, calculated from the ratio Γ_2/Γ_1 [16, 17] or from fitting the general equation for the adsorption isotherm [5, 18], indicate small surface micelles, with aggregation numbers ($n = 3$–20) distinctly smaller than that in micelles ($n > 50$). It seems that the (spherical

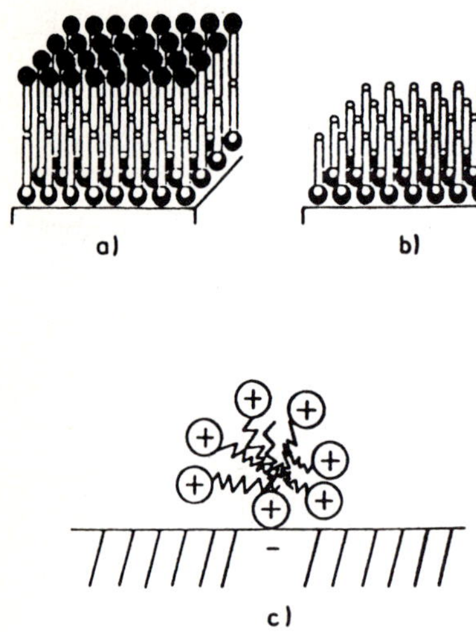

Fig. 8. Models of adsorbates for ionic surfactants on hydrophilic surfaces. a) Admicelle, according to [15]; b) Hemimicelle, according to [15]; c) Small surface micelle, according to [16, 21]

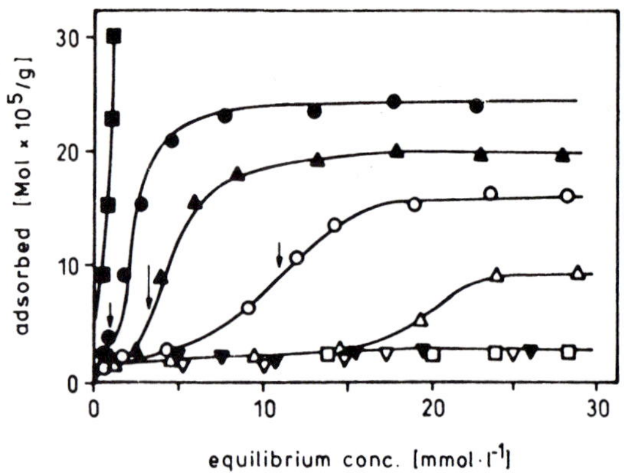

Fig. 9. Adsorption of cationic aggregating and non-aggregating compounds on silica (aerosil 200), at pH~4 and 20 °C [17, 28]. □, Acanil; ▽, Chlopromazine-HCl; ▼, Codeine; △, Decylpyridinium chloride; ○, Dodecylpyridinium chloride; ▲, Tetradecylpyridinium chloride; ●, Hexadecylpyridinium chloride; ■, Octadecylpyridinium chloride; Vertical arrows mark the CMC for surfactants

more or less) surface micelle must be isolated on the surface, and the reasonable diameter of surface micelle corresponds to the length of the surfactant ion (L) roughly [30], if the mean distance between the two closest adsorption sites on the solid surface is larger than L.

Influences of alkyl chain length of surfactant

The influence of surfactant alkyl chain length on the adsorption is demonstrated in Fig. 9 [28]. Corresponding to the shift of the CMC, the vertical step of the isotherm shifts towards lower concentrations as the alkyl chain length of surfactant increases [16, 28]. The saturated adsorption Γ_2 increases slightly with the hydrocarbon chain length. As a consequence, the aggregation number also increases.

For the establishment of the model the adsorption isotherms in Fig. 6 and the adsorption isotherms of the short chain surfactants (C_{10} and C_{14}) in Fig. 9, for example, show a decisive feature: the second step of adsorption is clearly separated from the first adsorption plateau, where saturation due to the electrostatic interaction with the solid surface (Γ_1) is clearly observed. Figure 9 also shows that surfactants with longer hydrocarbon tails and, thus, a stronger tendency for aggregation, begin the second step of adsorption (i.e., surface micellization) before saturation is reached for the first step. The two mechanisms overlap and may, therefore, no longer be clearly distinguished by the shape of the isotherm. However, it is obvious that the second step requires the first step. All these features have been quantitatively interpreted above by the general adsorption Eq. (16) based on the two-step model and the mass-action treatment.

Influences of counterions

Counterions are attracted to the Stern layer formed by the surfactant. They reduce the repulsion between the head groups by charge compensation. In this way, they favor the input of surfactant ions into the surface micelles. Their effectiveness is obviously determined by the size of the counterions, including their hydration shells. Weaker hydrated counterions can approach closer to the oppositely charged head groups. Therefore, they are more effective in charge compensation and can enhance the adsorption (i.e., surface aggregation) of the surfactant ions. This is clearly shown by the different values of Γ_2 and aggregation numbers of alkylpyridinium cations adsorbed on silica in presence of Cl^- and NO_3^- [17, 31]. The Γ_2 value of tetradecylpyridinium cations in the presence of weakly hydrated NO_3^- is more than twice the value in the presence of more strongly hydrated Cl^- ions [31]. Generally, the counterions obey the lyotropic series in promoting adsorption [22]. In addition, the divalent sulfate ion increases the Γ_2 value of tetradecylpyridinium cations on silica by a factor 4 in comparison with Cl^- [31]. Charge compensation of two surfactant head groups by one counter-ion obviously favors surface micellization.

Influences of neutral electrolytes

The presence of a neutral electrolyte has important consequences for the adsorption of ionic surfactants. According to the shift of the CMC, the steep increase of the adsorption occurs at lower concentrations (i.e., lower SMC) [16, 34]. The second effect is the increase in the saturation value of adsorption [16, 29, 31, 34]. With increasing neutral electrolyte concentration, the electrical potential on the adsorbed layers decreases. As a consequence, the mutual repulsive forces of the surfactant head groups are reduced and, thus, the incorporation of more surfactant ions into the adsorbed layer is facilitated. It is reasonable to assume that the number of primary adsorption sites at the solid surface is not significantly changed by the addition of neutral electrolytes, provided specific surface reactions can be excluded. Therefore, the increase in the adsorption must be accompanied by an equivalent increase in the aggregation number and an expansion of the volume of the surface micelles [16, 34].

Influence of pH

When the adsorbents carry surface groups that behave as electrolytes, the number of available adsorption sites depends on the surface density of the corresponding functional groups, their dissociation constants, and the pH of the medium. For example, for silicas the number of adsorption sites on a negatively charged surface is increased with increasing pH and, accordingly, the adsorption of surfactant in the first step increases [28, 34]. The total adsorption (Γ_2) also rises, because the number of surface micelles is enhanced. However, the surface aggregation numbers show a reversed tendency. These can be seen clearly from Table 4 [17]. The results can be explained by assuming that one surface micelle develops around more than one anchor chain if the density of sites is high enough [17]. Finally, at a very high density of primary adsorption isolated micelles may no longer exist

Table 4. Effect of pH on the saturation values of cationic surfactants in the first and second adsorption plateaus and surface aggregation number
(Adsorbent: Silica Aerosil 200)[17]

Surfactant	pH	Γ_1 (10^{-6}) mol·g^{-1}	Γ_2 (10^{-6}) mol·g^{-1}	n
Tetradecyldimethyl benzylammonium chloride	3.8	20	180	9.0
	4.5	30	230	7.7
	5.5	50	350	7.0
	6.5	100	440	4.4
	8.5	200	560	2.8
Tetradecylpyridinium chloride	3.5	25	180	7.2
	4.5	35	220	6.3
	5.6	80	370	4.6
	6.7	140	500	3.6
	7.0	150	550	3.6
	7.5	220	600	2.7
	8.3	210	670	2.3
	9.5	360	750	2.1
Dodecylpyridinum chloride	4.0	25	150	6.0
	4.4	40	220	5.5
	5.6	120	380	3.1
	6.7	200	490	2.5
	7.5	270	570	2.1
	8.3	300	620	2.1
	9.5	400	650	1.6

and relatively homogeneous double layer structures may be found [34].

Influences of porosity of adsorbent

The adsorption of surfactant on porous adsorbents is usually influenced by pore exclusion effects [3, 35, 36]. The saturation values of adsorption indicate that the surface area available to surfactant adsorption decreases with decreasing pore radius in silica with pores smaller than 10 nm. For silicas with pores smaller than 4 nm the adsorption-enhancing power of neutral electrolytes is very low, while the increase of the adsorption sites on the surface induced by a pH-shift provides an effective procedure for enhancing surfactant adsorption. Obviously, the establishment of surface micelles in narrow pores is more hindered than the adsorption into the first step adsorption layer.

Influences of surface polarity

It is well known that the hydrophilic matrix of silica can be rendered more or less hydrophobic by chemisorbed alkyl silanes. Very recently, the adsorption of laurylsulfate anions was studied on surfaces of hydrophilic silica (Aerosil Ox 50) surface modified by chemisorbed alkyl chains (introduced by alkyldimethyl-chlorosilanes) [37]. The surface densities of the chemisorbed alkyl chains (C_8, C_{12} and C_{18}) was adjusted between 1 and 20% of the surface silanol groups, leaving the silica still hydrophilic and negatively charged in aqueous suspension. On these materials, even laurylsulfate anions can be adsorbed in spite of the repulsive forces of the negatively charged surface, if the chemisorbed alkyl chains are not long enough (e.g., C_{12} and C_{18}). Obviously, only hydrophobic interactions contribute to the adsorption; they are restricted to the alkyl chains of the surfactant anions and the surface-grafted silanes, and the latter serve as anchors. On dodecyl-anchor groups laurylsulfate anions form aggregates with $n = 2$, on octadecyl-anchors aggregates with $n = 3$. The total amount of adsorbed laurylsulfate increases linearly with the surface density of alkyl chain anchors up to 10%. Adsorption on these anchors can be strongly increased by the addition of neutral electrolytes, which reduce the repulsive forces. In the presence of 0.1 M NaCl the size of the laurylsulfate aggregates increases to $n = 4$ at dodecyl chains and $n = 6$ at octadecyl chains. While on bare silica there is still no adsorption of laurylsulfate anions, they are weakly attached, even to octylgroups ($n = 1$–2). These facts, once more, strongly support the existence of well-defined small surface micelles and the viewpoint of surface micelle formation on isolated anchors.

More experimental results which support the surface micelle formation and the two-step adsorption model can be found in [17, 37, 38].

Acknowledgement

T. Gu wishes to thank the Norwegian Research Council for Sciences and the Humanities (NAFV) for financial support. He also wishes to thank professor J. Sjøblom, University of Bergen, for his kind hospitality.

References

1. Clunie JS, Ingram BT (1983) In: Parfitt GD, Rochester CH (eds.) Adsorption from Solutions at the Solid/Liquid Interface. Academic, London, Chap 3
2. Hough DB, Rendall HM (1983) In: Parfitt GD, Rochester CH (eds.) Adsorption from Solutions at the Solid/Liqud Interface. Academic, London, Chap 6
3. Gu T, Zhu B-Y (1990) Colloids Surfaces 44:81
4. Zhu B-Y, Gu T (1989) J Chem Soc, Faraday Trans 1 85:3813
5. Zhu B-Y, Gu T, Zhao X (1989) J Chem Soc Faraday Trans 1 85:3819
6. Zhu B-Y, Gu T (1990) Colloids Surfaces 46:339
7. Everett DH (1964) Trans Faraday Soc 60:1803; (1965) 61:2478
8. Koganovski AM, Klimenko NA, Chobanu MM (1977) Kolloidn Zh 39:358
9. Klimenko NA (1978) Kolloidn Zh 40:1105; (1979) 41:781
10. Koopal LK, Lyklema I (1979) J Electroanal Chem 100:895
11. Shumann D, Bennes R, Bellostas D (1979) J Electroanal Chem 101:73
12. Chander S, Fuerstenau DW, Stigter P (1983) In: Ottewill RH, Rochester CH, Smith AL (eds.) Adsorption from Solution. Academic, London, p 197
13. Rudzinski W, Dominko A, Partyka S, Brun B (1985) Adsorption Sci Techn 2:207
14. Gaudin AM, Fuerstenau DW (1955) Trans Am Inst Min Metall Pet Eng 202:66; (1955) 205:958
15. Yeskie MA, Harwell JH (1988) J Phys Chem 92:2346
16. Gao Y, Du J, Gu T (1987) J Chem Soc Faraday Trans 1 83:2671
17. Rupprecht H, Gu T (1991) Colloid Polym Sci 269:506

18. Zhu B-Y, Gu T (1991) Adv Colloid Interface Sci 37:1
19. Zhu B-Y, Zhao X, Gu T (1988) J Chem Soc Faraday Trans 1 84:3957
20. Gu T, Gao Y, He L (1988) J Chem Soc Faraday Trans 1 84:4471
21. Gu T, Huang Z (1989) Colloids Surfaces 40:71
22. Gu T, Huang Z, Sun Z, Forsling W (1991) Acta Chem Scand 45:526
23. Zhu B-Y, Gu T (1991) J Chem Soc Faraday Trans 87:2745
24. Zhao G-X, Zhu D (1988) Colloids Surfaces 33:289
25. Kertes AS, Gutman H (1975) In: Matijevic E (ed.) Surface and Colloid Sciences, Vol 8. Wiley, New York, p 193
26. Mittal KL, Mukerjee P (1977) In: Mittal KL (ed.) Micellization, Solubilization and Microemulsions, Vol 1. Plenum, New York, p 1
27. Findenegg GH, Koch C, Liphard M (1983) In: Ottewill RH, Rochester CH, Smith AL (eds.) Adsorption from Solution. Academic, London, p 87
28. Rupprecht H (1968) Dissertation. University of Munich
29. Rupprecht H (1976) Acta Pharm Techn 22:37
30. Gu T, Rupprecht H (1990) Colloid Polym Sci 268:1148
31. Rupprecht H, Ullmann E, Thoma K (1971) Fortschr. Kolloide Polymere 55:45
32. Rupprecht H, Biedermann M (1973) Pharm Ind 35:640
33. Rupprecht H (1976) Cosmetics Toiletries 31:38
34. Huang Z, Ma J, Gu T (1989) Acta Chim Sinica (Engl Ed) 2:105
35. Vitzthum J, Strnadova M, Bernard R, Rupprecht H (1990) Progr Colloid Polym Sci 83:110
36. Rupprecht H, Unger K, Biersack MJ (1977) Colloid Polym Sci 255:276
37. Sigg J (1991) Dissertation. University of Regensburg
38. Vitzthum J (1990) Dissertation. University of Regensburg

Authors' address:

Prof. Dr. H. Rupprecht
Lehrstuhl für Pharmazeutische Technologie
Universität Regensburg
Universitätsstr. 31
D-W-8400 Regensburg
FRG

Critical dissociation of ferritin during adsorption at a liquid-solid interface

H. Nygren

Department of Histology, University of Göteborg, Sweden

Abstract: Adsorption of ferritin to a hydrophobic solid surface was studied by transmission electron microscopy. At low bulk concentrations adsorption is initiated at a few sites and continues by growth of clusters at these sites. The nucleation-and-growth-like kinetics gives an acceleration of the apparent binding rate which is damped by mass transport limitation. The clusters grow to a critical size and dissociate rapidly from the surface. This process of accelerated binding and critical dissociation can be seen repeatedly during adsorption of the protein.

At high bulk concentrations, the surface is covered with protein during the first seconds of incubation, but the adsorbed layer then becomes unstable and dissociation is seen within the first minute of incubation, resulting in a clustered spatial distribution of adsorbed protein.
These findings suggest that protein adsorption reaches an equilibrium, but that the kinetics of the process is discontinuous.

Key words: Discontinuous kinetics; protein adsorption; ferritin; electron microscopy; oscillations

Introduction

The kinetics of protein adsorption at a liquid-solid interface is characterized by an initial rapid, often diffusion-rate-limited reaction, followed by a continuous decrease of the forward reaction rate [1]. Adsorbed protein films are stable when rinsed with salt solutions, which has lead to conclusions that protein adsorption is an irreversible process [2]. Still, the amount of adsorbed protein is concentration dependent, adsorbed protein layers can be rapidly exchanged in the presence of protein in the bulk [3] and this process can be facilitated by shearing flow [4] These experimental findings indicate equilibrium properties of the adsorbed film and are paradoxical to the apparent irreversibility of the adsorption process. Furthermore, data may often be fitted into a Langmuir isotherm. This behavior is similar to the kinetics of antigen-antibody reactions at liquid-solid interfaces and may reflect important common properties of heterogeneous macromolecular reactions [5].

Most experimental data on the kinetics of protein adsorption are from integrating measurements with low lateral resolution. However, a few reports present data on protein adsorption at the molecular level studied by transmission electron microscopy (TEM). Ferritin, a plasma protein with a molecular weight of 880 kD, consisting of 24 similar subunits and a central iron core with a diameter of 5 nm, has been used as a model protein due to its electron density [6].

In studies of protein adsorption by TEM using ferritin as model system, it has been shown that the kinetics of ferritin adsorption to a hydrophobic surface is strongly dependent on the supramolecular structure of the adsorbed protein layer, and that a stable plateau level of adsorption is formed below a monolayer [7, 8].

The present study was undertaken in order to elucidate the relationship between the formation of stable levels and the kinetics of adsorption.

Material and methods

Sample preparation

Hydrophobic, methyl silanized quartz grids were made as described in [9]. Horse spleen ferritin

(Fluka, Switzerland. Crystallized ×3) was dissolved in phosphate buffered saline (PBS; 0.02 M phosphate buffer pH 7.2; 0.15 M sodium chloride). The protein solution was applied to the grids with a pipette for short incubation times (< 100 s). For longer incubations, quartz grids were placed inverted on drops of PBS in a humified chamber and ferritin solution was injected into the drops to a final concentration of 10 μg/ml or 10 mg/ml. The adsorption was terminated by rinsing with PBS for 5 s. The grids were finally dried in a dry air current.

Electron microscopy

The grids were examined with an electron microscope at 60 kV accelerating voltage. Electron micrographs were taken at a magnification of ×20 000 and copies were made at higher magnification. The pictures were scanned and digitized. The number of adsorbed molecules/cm^2 was counted for 2–4 image fields (9–100 cm^2) on each picture.

Results and discussion

The adsorption of ferritin from a solution containing 10 mg/ml onto a methylized quartz surface is shown in Fig. 1. After 5 s of adsorption the surface is covered with a rather uniform protein layer (Fig. 1a). After 15 s, scattered areas of empty space can be seen (Fig. 1b), indicating instability of the adsorbed layer and dissociation of bound protein. After 45 s incubation time, the adsorbed protein molecules are distributed as clusters with large confluent uncovered surface area in between (Fig. 1c).

The experimental finding presented in Fig. 1 is consistent with data presented by Feder and Giaever [6], who reported that adsorbed ferritin was found heterogeneously distributed over a polymer surface, showing clustering and large uncovered areas. This distribution was interpreted as a result of loss of material during sample preparation since it was possible to obtain a more even distribution of the protein at high bulk concentration (10 mg/ml) and short incubation time (10 s).

At a lower bulk concentration, the process of adsorption and dissociation can be followed in more detail. The kinetics of adsorption are graphically presented in Fig. 2. An initial acceleration is seen from 5–10 s of adsorption. The amount of protein bound after 10 s of adsorption is close to the amount of protein that can reach the surface by mass transport (see [7] for details). After 40 s, 100 s, and 250 s of adsorption, some experiments showed small amounts of adsorbed ferritin on the surface, indicating rapid dissociation of protein during sample preparation (Fig. 3). In other experiments, dense clusters were seen at these adsorption times [7, 10]. The sample preparation routines were extremely standardized and little or no variation is seen at other adsorption times (Fig. 2). This

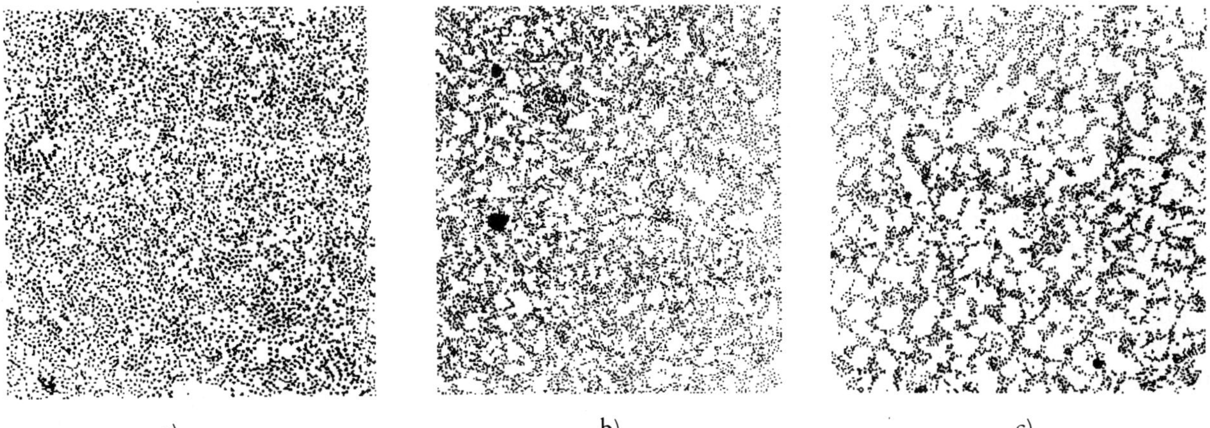

a) b) c)

Fig. 1. Electron micrograph of ferritin molecules adsorbed on a methylized quartz grid from a bulk concentration of 10 mg/ml of ferritin dissolved in phosphate-buffered saline (PBS). The picture has been digitized and printed on a laser printer. Magnification: Electron dense particles = 5 nm. Adsorption time: a) 5 s; b) 15 s; c) 45 s

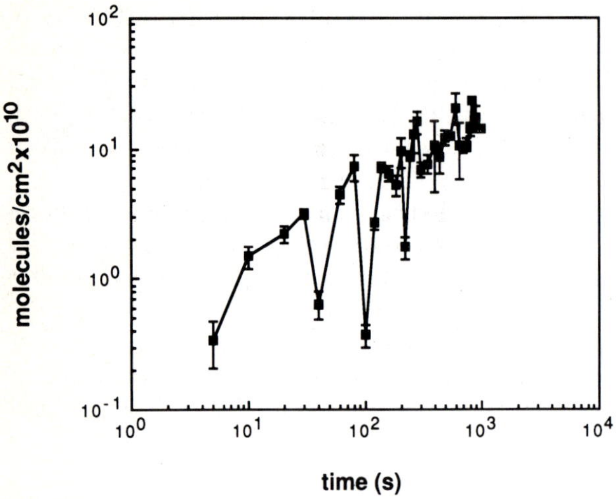

Fig. 2. A plot of the surface concentration of ferritin (molecules/cm^2) vs time (s). Bulk concentration of ferritin was 10 μg/ml dissolved in PBS. Error bars indicate ± SD

Fig. 3. Electron micrograph of ferritin molecules adsorbed on a methylized quartz grid from a bulk concentration of 10 μg/ml of ferritin dissolved in PBS. The picture has been digitized and printed on a laser printer. Magnification: Electron dense particles = 5 nm. Adsorption time: a) 80 s; b) 100 s

experimental finding is interpreted as an instability of large clusters seen in some experiments, and that the bulk/surface relation of protein may be more than one distribution in this system (Figs. 3a, b).

It has been shown previously that the initial adsorption of ferritin from lower bulk concentration occurs by diffusion-rate-limited, two-dimensional aggregation without any loss of material from the surface during preparation of samples [7, 10]. The initial aggregation can be computer-simulated by assuming reversible diffusion-limited aggregation [11]. The formation of dense clusters with low fractal dimensions (Fig. 3a) can be computer-simulated with a set of rules assuming higher probability of adsorption as a near neighbor, lateral diffusion of adsorbed molecules, and dissociation of molecules with many neighbors [12]. In the present study, experimental evidence is found indicating that the process of adsorption is not continuous with this set of rules, but that the clusters formed may become critical and dissociate rapidly. This may be an important mechanism for the establishment of an equilibrium at the interface, and also demonstrates that protein adsorption may have discontinuous kinetics.

Acknowledgement

This study was supported by grants from the Swedish Medical Research Council (grant 06235) and the Board of Technical Development.

References

1. Cuypers PA, Willems GM, Kop JM, Corsel JW, Jansen MP, Hermens WT (1987) In: Brash JL, Horbett TA (eds.) Proteins at Interfaces. Physicochemical and Biochemical Studies. ACS Symposium Series 343:208–221
2. Norde W (1980) In: Lee LH (ed.) Adhesion and Adsorption of Polymers, Part B. Plenum, New York, pp 801–829
3. Vroman L, Adams AL, Fischer GC, Munoz PC (1980) Blood 55:156–159
4. Brash JL, Samak QM (1978) J Colloid Interface Sci 65:423–428

5. Nygren H, Stenberg M (1989) Immunology 66:321–328
6. Feder J, Giaver I (1980) J Colloid Interface Sci 78:144–148
7. Nygren H, Stenberg M (1990) J Biophys Chem 38:67–75
8. Nygren H, Stenberg M (1990) J Biophys Chem 38:77–85
9. Stenberg M, Stemme G, Nygren H (1987) Stain Technol 62:231–236
10. Nygren H, Stenberg M (1990) Progr Colloid Polymer Sci 82:15–18
11. Stenberg M, Nygren H (1990) Progr Colloid Polymer Sci 82:10–14
12. Stenberg M, Nygren H (1991) Biophys Chem 41:131–141.

Author's address:

Dr Håkan Nygren
Dept. of Histology
P.O.B. 33031
S-400 33 Göteborg
Sweden

Extraction and fractionation of lignosulfonate by a supported liquid membrane

A.-K. Kontturi, K. Kontturi, P. Niinikoski, and G. Sundholm

Laboratory of Physical Chemistry and Electrochemistry, Helsinki University of Technology, Espoo, Finland

Abstract: The paper presents a summary of our results of the supported liquid membrane extraction of a polydisperse polyelectrolyte, lignosulfonate. Trilaurylamine dissolved in decanol was found to be a suitable liquid membrane. The smallest molecules are transported first. A simple model for the transport of lignosulfonate is outlined and the results of the experiments are presented.

Key words: Supported liquid membrane; macromolecule; lignosulfonate; carrier-mediated transport

Introduction

It has been known for a long time that lignosulfonates can be extracted from acidic solutions by long-chain aliphatic amines and then stripped back into an aqueous phase by a base:

i) extraction:

$$R_3N(o) + H^+ + LS^- \rightleftarrows R_3NH^+LS^-(o) \qquad (1)$$

ii) stripping:

$$R_3NH^+LS^-(o) + OH^- \rightleftarrows R_3N(o) + H_2O + LS^-. \qquad (2)$$

The letter o refers to the organic phase, which is typically an aliphatic alcohol.

The liquid-liquid extraction of lignosulfonate (LS) mentioned above was studied in our laboratory some years ago [1]. Later, lignosulfonate was used as a model substance in studies of transport properties of polyelectrolytes [2–8]. From these studies we know the effective charge numbers and ionic diffusion coefficients of lignosulfonate molecules in various aqueous solutions. The molar mass of lignosulfonate varies between about 1000 and 100 000. The molar mass distribution of the extracted lignosulfonate depends on both the amine and the organic solvent [1]. The liquid membrane extraction combines the transport and the extraction processes.

This project on liquid membranes was started to find out details about the reactions at the liquid-liquid interface and to find out what determines the molar mass distribution of the extracted lignosulfonate. The procedure is familiar from other liquid membrane studies: namely, a well-known liquid-liquid extraction is more or less directly transformed to a liquid membrane process. In our case, however, the extracted species was a polydisperse polyelectrolyte.

In liquid membrane extraction of lignosulfonate the amine dissolved in an organic solvent is impregnated into a hydrophobic porous membrane. The extraction and the back-extraction proceed continuously at the different sides of the membrane (Fig. 1). The amine acts as a carrier. The process in Fig. 1 is called co-transport because hydrogen ions are transported with the lignosulfonate molecules in the same direction across the membrane.

It is, of course, essential that both the amine and the organic solvent stay in the membrane. As a carrier, we have used trilaurylamine (TLA), which has a very low solubility in water. As an organic solvent, we have used decanol, which is in many ways a poor solvent for the liquid membrane application, mainly because of the low interfacial tension between water and decanol. However, other solvents that were used earlier in liquid-liquid-extraction studies were found to be unsuitable for liquid membranes [9]. In laboratory-scale

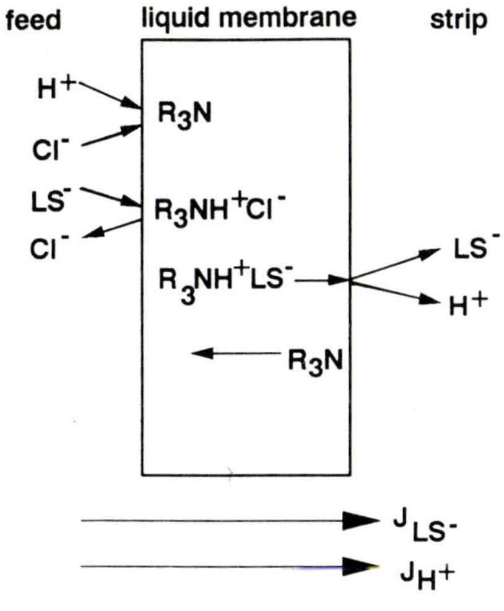

Fig. 1. Co-transport of lignosulfonate (LS) across a liquid membrane with an amine (R_3N) carrier

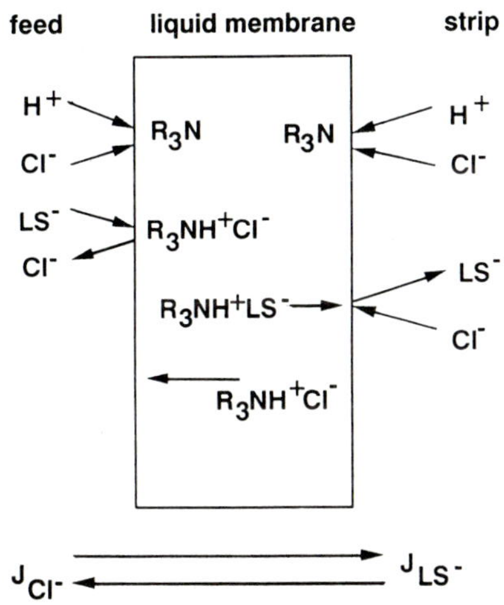

Fig. 2. Counter-transport of lignosulfonate (LS) across a liquid membrane with an amine (R_3N) carrier

experiments decanol can be used if the pressure differences across the liquid membrane are carefully balanced.

An alternative stripping reagent to sodium hydroxide is chloride ion. The stripping side is then an acidic solution with an excess of sodium chloride so that the ion-exchange reaction of the feed side goes backwards at the stripping side (Fig. 2). The process in Fig. 2 is called counter-transport because the coupled fluxes of lignosulfonate and chloride have opposite directions.

In this study, we show that the extraction of a polydisperse polyelectrolyte, lignosulfonate, can be performed as a steady-state carrier-mediated process, either in the co- or counter-transport mode.

Experimental

We used a steady-state set-up in the experiments with a flow-channel-type cell. The steady-state was achieved in 3 h and the flux of lignosulfonate could be determined from the steady-state concentration of the strip side. This concentration was determined by UV absorption. The lignosulfonate was from Rauma-Repola. The molar mass distribution is presented in Table 1. The determination method has been described elsewhere [1]. The effective charges measured in 0.1 M NaCl vary between -20.7 (50 000 $\mathrm{g\,mol^{-1}}$) and -2.6 (4000 $\mathrm{g\,mol^{-1}}$) [3]. Millipore FGLP membranes were used as porous support. The details of the experiments were described earlier in [9, 10].

Results

The influence of the concentrations of the species present in the system (Cl^-, amine, etc.) on the flux of lignosulfonate was studied. The primary assumptions made in this study were that the diffusion of the lignosulfonate-amine compound in the liquid membrane is the rate-determining step in the overall transport process, and the chemical reactions occurring are at local equilibrium.

Co-transport

In the case of co-transport the local equilibrium assumption was convincingly supported by the experimental results. A large amount of chloride in the feed suppressed the flux of lignosulfonate. An increase of either the amine or the lignosulfonate concentration in the feed increased the flux, as was expected for reactions at equilibrium.

The strip side reaction in the co-transport mode is simply an acid-base reaction. Our results showed that once the sodium hydroxide concentration was high enough for the reaction to occur, it had little effect on the flux of lignosulfonate. The concentration of lignosulfonate in the strip had no effect at all. The above primary assumption that the chemical reactions occurring were at local equilibrium was therefore replaced by the assumption that the

Table 1. The molar mass distribution of the extracted lignosulfonate in co-transport as percentage of total sample

Concentration	Molar mass fractions				
	> 40 000	> 20 000	> 10 000	> 5000	< 5000
Feed LS			% of total sample		
0.1 g/l	10	22	38	59	41
1.0 g/l	4	8	19	40	60
5.0 g/l	0	1	5	22	78
10.0 g/l	2	3	6	17	83
20.0 g/l	1	1	2	10	90
40.0 g/l	0	1	2	8	92
Trilaurylamine			% of total sample		
0.05 M	0	1	3	15	85
0.1 M	4	8	19	40	60
0.15 M	5	13	30	54	46
0.2 M	8	18	37	60	40
Original sample			% of total sample		
	20	34	55	75	25

stripping reaction in the co-transport case was fast and irreversible.

Counter-transport

In counter-transport the reaction at the feed side and the behavior of the flux of lignosulfonate is the same as in co-transport. The stripping reaction is the reverse of the feed side reaction.

However, the results of the experiments in which the strip side concentrations were changed turned out to be unpredictable. Firstly, the concentration of lignosulfonate had no effect. Secondly, there was an interesting function between the flux of lignosulfonate and the strip side chloride concentration. Based on these results we dropped the equilibrium assumption for the strip side reaction and treated it as a rate-determining step.

Discussion

On the basis of the experimental results described above, we developed a model for the transport of lignosulfonate.

Co-transport

The reactions at the feed side are now written in more detail:

$$\mathrm{TLA} + \mathrm{HCl} \rightleftarrows \mathrm{TLAH^+Cl^-} \tag{3a}$$

$$x\mathrm{TLAH^+Cl^-} + \mathrm{LSM}_n \rightleftarrows (\mathrm{TLAH})_x\mathrm{LSM}_{n-x} + x\mathrm{MCl}\ , \tag{3b}$$

where M is a cation. The stoichiometric numbers x and n are needed, because the effective charge number of lignosulfonate varies according to molar mass [3]. All the charges have to be neutralized by a cation from the organic or the aqueous phase. The assumptions made lead to the concentration gradients in the liquid membrane shown in Fig. 3. The

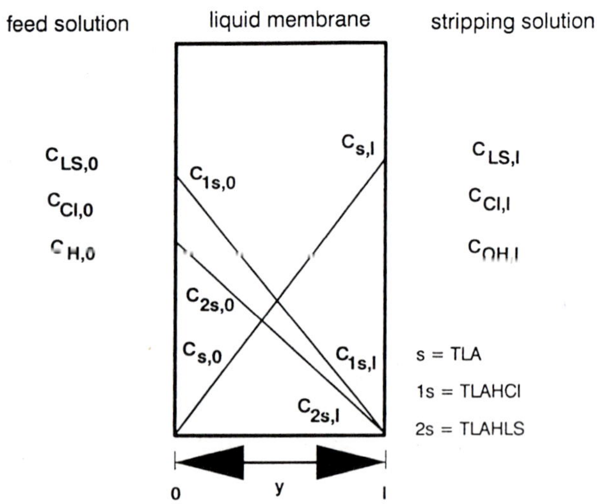

Fig. 3. The concentration gradients of the species in the membrane in co-transport as a result of the approximations in the model

concentration gradients are linear because there are only neutral components in the membrane. The concentrations of the amine compounds are zero at the stripping side. Also, the concentration of free amine is assumed to be zero at the feed side of the liquid membrane.

The fluxes of the components in the liquid membrane can now be written in the form

$$J_{2s} = J_{\mathrm{LS}} = D_{2s} \frac{C_{2s,0}}{l} , \tag{4}$$

where the symbols are explained in Fig. 3. In addition to the flux equations, three conditions are posed; they are:

$$- J_s = J_{1s} + xJ_{\mathrm{LS}} . \tag{5}$$

Further, the total amine concentration in the membrane remains constant:

$$C_{\mathrm{tot}} = \frac{1}{l} \int_0^l (C_s + C_{1s} + xC_{2s})\, dy . \tag{6}$$

Finally, the equilibrium condition at the feed side is

$$K = \frac{C_{2s,0} C^x_{\mathrm{Cl},0}}{C^x_{1s,0} C_{\mathrm{LS},0}} . \tag{7}$$

We now assume that only one amine per lignosulfonate molecule is needed to draw lignosulfonate into the liquid membrane ($x = 1$). The flux of lignosulfonate is then obtained in the form

$$J_{\mathrm{LS}} = \frac{C_{\mathrm{tot}}}{l\left(\frac{1}{D_1} + \frac{1}{D_2}\frac{C_{\mathrm{Cl},0}}{KC_{\mathrm{LS},0}}\right)} , \tag{8}$$

where

$$D_1 = \frac{2D_s D_{2s}}{(D_s + D_{2s})} , \qquad D_2 = \frac{2D_s D_{2s}}{(D_s + D_{1s})} .$$

(Subscripts are defined in Fig. 3). In fact, reasonable results could only be obtained by putting x equal to 1. Equation (8) gives a simple relationship between the flux of lignosulfonate, and the concentration of the species on the feed side and in the liquid membrane. According to Eq. (8), the flux of lignosulfonate should be directly proportional to the total amine concentration. As is shown in Fig. 4, the results of the experiments agree with this prediction when the amine concentration is less than 0.2 M.

The limiting flux may be caused by aggregation of amine compounds, or it may indicate that diffusion in the water phase becomes rate-limiting. However, our calculations based on the hydrodynamics of channel flow [10] show that the diffusion layers have only a small effect. The inverse value of the flux of lignosulfonate should be directly proportional to the chloride concentration in the feed solution. This is also in accordance with the experiments (Fig. 5).

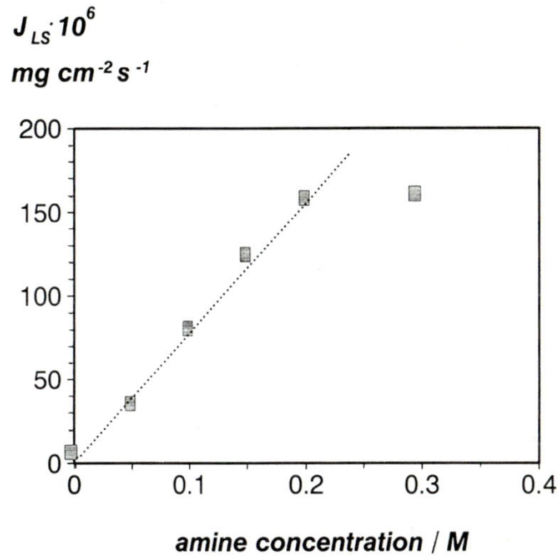

Fig. 4. The flux of lignosulfonate in co-transport as a function of amine concentration in the membrane. The points are experimental values

Counter-transport

In the case of counter-transport, the reactions at the feed side are the same as in co-transport, except that the amine hydrochloride formed remains as hydrochloride. Here, the concentration of lignosulfonate-amine-complex is not zero at the strip side of the membrane (Fig. 6). Therefore, the gradients in the membrane are smaller than in co-transport, which is also seen in the smaller fluxes in this transport mode. The reaction at the strip side is not in equilibrium. Instead, we have postulated a reaction mechanism by which the rate-determining step is the reaction of adsorbed chloride ion with the adsorbed amine-lignosulfonate-complex [10].

The flux of lignosulfonate through the interface can now be written in the form

$$J_{\mathrm{LS}} = \frac{(k_f b_{2s} C_{2s,l} b^x_{\mathrm{Cl}} C^x_{\mathrm{Cl},l})}{(1 + b_{\mathrm{Cl}} C_{\mathrm{Cl},l})^x} , \tag{9}$$

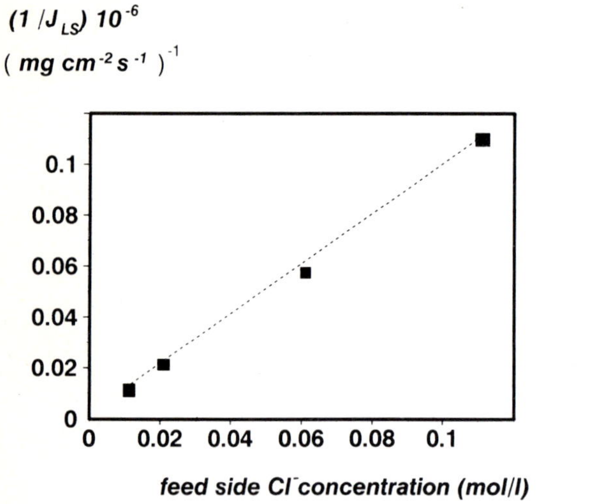

Fig. 5. The inverse value of the flux of lignosulfonate in co-transport as a function of the feed side chloride concentration. The points are experimental values

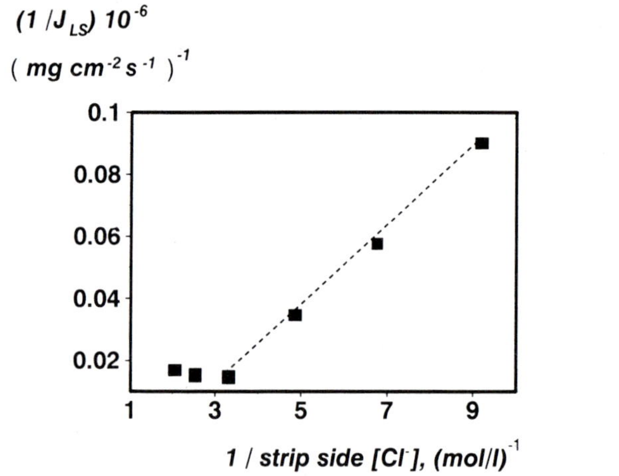

Fig. 7. The inverse value of the flux of lignosulfonate in counter-transport as a function of the inverse value of the strip side chloride concentration. The points are experimental values

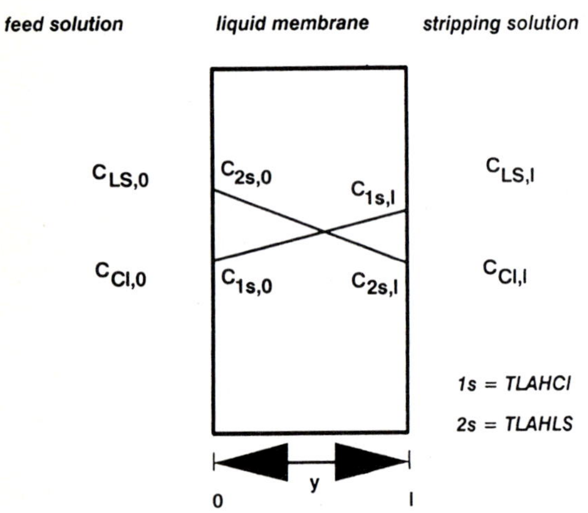

Fig. 6. The concentration gradients of the species in the membrane in counter-transport

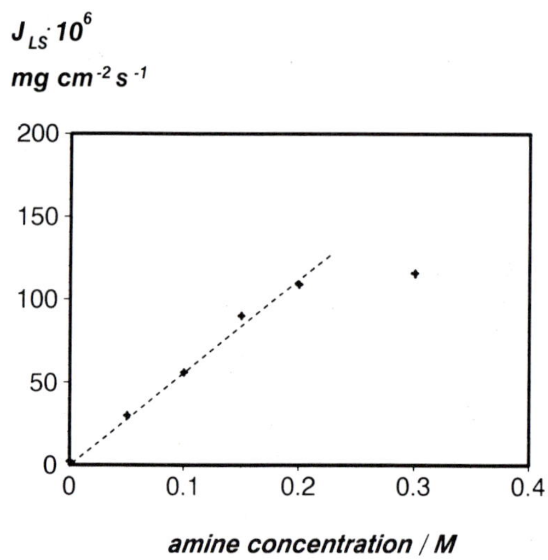

Fig. 8. The flux of lignosulfonate in counter-transport as a function of the amine concentration. The points are experimental values

where b's are coefficients in a Langmuir isotherm, and k_f is the rate-constant of the rate-determining step. When this equation is combined to the transport equations, and other conditions, which are similar to the co-transport case, the flux of lignosulfonate is obtained in the form

$$\frac{1}{J_{LS}} = \left(\frac{1}{k_f A' b_{2s}} + \frac{B}{A'}\right) + \frac{1}{k_f A' b_{2s} b_{Cl}} \frac{1}{C_{Cl,l}}, \tag{10}$$

where

$$A' = \frac{C_{tot} K C_{1s,0}}{C_{Cl,0}}, \quad B = \frac{1}{D_{2s}}.$$

We have assumed that x is equal to one and that the total amine concentration is equal to the concentration of the amine hydrochloride at the feed side of the liquid membrane. Equation (10) predicts that the inverse of the LS flux should be directly proportional to the inverse value of the strip side chloride concentration. The experiments confirmed this dependence up to a chloride concentration of about 0.3 mol/l (Fig. 7). According to our model, surface saturation causes the constant flux after this concentration. The small change after saturation could be explained by a salting-out effect. The flux of lignosulfonate should be directly proportional to the total amine concentration. This is verified in Fig. 8. A limiting flux is observed at 0.2 M.

Conclusions

From the analysis of both transport modes, we found some evidence that approximately one amine per lignosulfonate is needed to transport the macromolecule through the liquid membrane. This is, at first sight, a surprising result, because the mean molar mass of our lignosulfonate is about 22 000. However, analysis of the molar mass distribution of the extracted lignosulfonate shows that mainly the smallest molecules are transported. This is demonstrated clearly in Table 1, where the molar mass distribution of the extracted lignosulfonate (co-transport) is presented as a function of feed side lignosulfonate and amine concentration.

To get a more detailed model, we need diffusion coefficients of lignosulfonate molecules in the organic phase. The first approximations to the diffusion coefficients were obtained by Walden's rule, so that the known values in water were corrected by the viscosity of decanol. However, these values were an order of magnitude greater than the approximation obtained by the present model [10]. Also, our measurements in mixed solvents indicate that the conformation of lignosulfonates changes in low dielectric media [7]. Therefore, Walden's rule cannot be applied. It should be pointed out that it is, in principle, possible to measure the ionic diffusion coefficients in the organic phase, in the same way that we have measured them in the aqueous phases.

In addition to the diffusion coefficients, we need information about the reactions at the two interfaces. For this purpose, we have set up a rotating disc equipment where the mass transport to the liquid-liquid interface is controlled by rotating the supported liquid membrane. A study of the reactions at the interfaces using this system is in progress.

Acknowledgement

Financial support from the Academy of Finland and Acta Chemica Scandinavica is gratefully acknowledged.

References

1. Kontturi A-K, Sundholm G (1986) Acta Chem Scand Ser A 40:121
2. Kontturi A-K, Kontturi K (1987) J Coll Interface Sci 120:256
3. Kontturi A-K, Kontturi K (1988) J Coll Interface Sci 124:328
4. Kontturi A-K, Kontturi K (1987) Acta Polytechn Scand 178:143
5. Kontturi A-K (1988) J Chem Soc Faraday Trans 1 84:4033
6. Kontturi A-K (1988) J Chem Soc Faraday Trans 1 84:4043
7. Kontturi A-K, Kontturi K, Niinikoski P (1990) J Chem Soc Faraday Trans 86:3097
8. Kontturi A-K, Kontturi K, Niinikoski P (1991) J Chem Soc Faraday Trans 87:1779
9. Kontturi A-K, Kontturi K, Niinikoski P, Sundholm G (1990) Acta Chem Scand 44:879–882
10. Kontturi A-K, Kontturi K, Niinikoski P, Sundholm G (1990) Acta Chem Scand 44:883

Authors' address:

Anna-Kaisa Kontturi
Laboratory of Physical Chemistry and Electrochemistry
Helsinki University of Technology
Kemistintie 1A
SF-02150 Espoo
Finland

Intermolecular interaction and ordering of fibrinogen at a liquid-solid interface

H. Nygren and C. Karlsson

Department of Histology, University of Göteborg, Göteborg, Sweden

Abstract: The adsorption of fibrinogen to a hydrophilic quartz surface was studied by transmission electron microscopy using a negative staining technique to visualize the protein. The distribution of adsorbed fibrinogen was random at low surface concentrations and low bulk concentrations (0.29 nM) when analyzed as the pair correlation function $g(r)$. At higher concentration (0.73 nM) an end-to-end polymer polycrystalline structure was seen at the surface, indicating attractive interaction between the ends of the rod-shaped molecule. The polycrystalline pattern was broken up at higher concentrations and snowflake-like crystals were formed.

The results indicate the presence of weak attractive forces between the ends of fibrinogen molecules.

Key words: Protein adsorption; supramolecular structure; pair correlation; fibrinogen; interface

Introduction

Platelet adhesion and aggregation are key phenomena in blood coagulation. Fibrinogen, a rod-shaped plasma protein with a molecular weight of 340 000 D, has been shown to promote platelet adhesion when adsorbed at solid surfaces, and it participates in surface-induced thrombosis [1, 2]. Recent studies have demonstrated that fibrinogen acts as an adhesin for cells by a tripeptide, Arg-Gly-Asp, interacting with amembrane glycoprotein, GP IIb-IIIa, present at the platelet surface [3]. Since the tripeptide active site is only a small part of the protein macromolecule, the molecular and supramolecular orientation of fibrinogen may be important for the platelet reaction to immobilized protein.

Supramolecular patterns of adsorbed fibrinogen have been described [4, 5] that show formation of ridges and walls on the surface. In higher resolution, formation of protein polymers has been reported [6, 7] suggesting intermolecular interaction at the interface. This interaction may induce a long-range ordering of the adsorbed protein [8]. When increasing the surface concentration towards a laying monolayer, molecules adsorbed side-on showed a polycrystalline structure [8].

The present study was undertaken in order to elucidate the ordering of fibrinogen molecules at a liquid-solid interface and the supramolecular structure of the adsorbed fibrinogen layer as related to a phase transition of the fibrinogen molecules.

Experimental

Chemicals

Human fibrinogen (Kabi, Stockholm, Sweden) was dissolved in 0.15 M NaCl to a concentration of 10 mg/ml and was kept frozen in aliquots. For the experiments, fibrinogen stock solution was diluted in 0.02 M PBS pH 7.2. Uranyl acetate (Fluka Buchs, Switzerland) and buffer salts (Merck, Darmstadt, FRG) were analytical grade.

Electron microscopy

Negative staining: Negative staining of adsorbed fibrinogen molecules was performed on quartz grids. Silicon wafers (Wacker Chemitronic, Frankfurt, FRG) were heat oxidized and the silicon was etched to form sample supporting grids as described previously [9]. The grids were made hydrophilic by oxidation with dichrome sulfuric acid. The contact angle with water was 24° at the hydrophilic surface.

Quartz grids were placed inverted on droplets of PBS in a moist chamber. Fibrinogen solution was injected into the droplets with a micropipette to give final concentration as indicated in the figures. The incubation time was 20 h at room temperature. The adsorption was terminated by rinsing with PBS followed by a short rinse in 1–2% uranylacetate. The excess of uranylacetate solution was blotted off with a filter paper and the grids were dried in air.

Examination: The specimen were examined in a JEOL 100 CX electron microscope at an accelerating voltage of 60 kV. Photographs were taken at primary manifications 33 000 × and 50 000 × and were further enlarged on paper copies to 150 000 ×. The size of molecules and the intermolecular distance was measured with a ruler on photographs magnified 300 000 ×.

Pair distribution

The pair correlation function $g(r)$ was evaluated as the probability of finding a neighbor at a distance r from a given molecule. The distribution of molecules as a function of the radial distance $M(r)$ was measured using a circular grid. The distribution found was related to a random distribution by normalizing with the density probability [10]:

$$A^{-1}[N(N-1)]^{0.5}\,, \tag{1}$$

where A is the area and N the number of molecules.

Results and discussion

Figures 1a–d show electron micrographs of the adsorbed layer of fibrinogen on hydrophilic quartz surface after 20 h adsorption time with bulk concentrations of fibrinogen in the range 0.29–29.0 nM. The dried and negatively stained fibrinogen molecules are seen as 45 ± 4 nm-long and 5 nm-wide nodular rods. This size and shape indicates a drying artefact which reduces the thickness of the rod by about 50% compared with the fully hydrated molecule which is described as a 9 nm × 45 nm rod [11].

At a bulk concentration of 0.29 nM (Fig. 1a), adsorbed fibrinogen is found to be distributed as single molecules with occasional formations of end-to-end dimers and trimers. The pair correlation function $g(r)$ shown in Fig. 2 reveals a distribution of molecules close to random, indicating that the fibrinogen molecules are adsorbed without any significant intermolecular interaction.

Adsorption of fibrinogen from a bulk concentration of 0.73 nM is shown in Fig. 1b. An orderly layer of end-to-end polymerized fibrinogen molecules can be seen with a polycrystalline structure where different domains are visible with parallel orientation of the molecules. The average intermolecular distance between the parallel molecules is 16 nm. This distance is significantly longer than the diameter of a fully hydrated molecule, indicating the presence of solvent in the protein layer.

Adsorptionof fibrinogen from a bulk concentration of 2.9 nM (Fig. 1c) show a heterogeneous distribution of molecules with scattered sites of end-to-end polymerized clusters. An increase of the bulk concentration of fibrinogen to 29.0 nM (Fig. 1d) resulted in snow-flake-like crystals of fibrinogen molecules. The individual molecules are faintly visible as end-to-end polymers within the crystals.

The data presented above indicate that attractive intermolecular interaction is present between fibrinogen molecules adsorbed at a surface. The forces involved are probably weak, since no interaction can be seen at low surface concentrations (Figs. 1a, 2). Attractive interaction has been shown previously between ferritin molecules during adsorption to a surface [12, 13], but the interaction between ferritin molecules is stronger than the interaction between fibrinogen molecules, indicated by higher probabilities of finding neighbors at short distances from a given molecule, even at low surface concentrations [13].

The polycrystalline layer seen in Fig. 1b breaks up at higher bulk concentrations of protein, but is rigid enough to cause a stable level in the adsorption isotherm at low surface concentrations [6]. The mechanism has also been described during adsorption of ferritin, where high bulk concentrations of protein reduce the amount adsorbed at the surface [13]. The isotherm of fibrinogen adsorption [6] and the breaking up of supramolecular structures seen in the present study indicate equilibrium

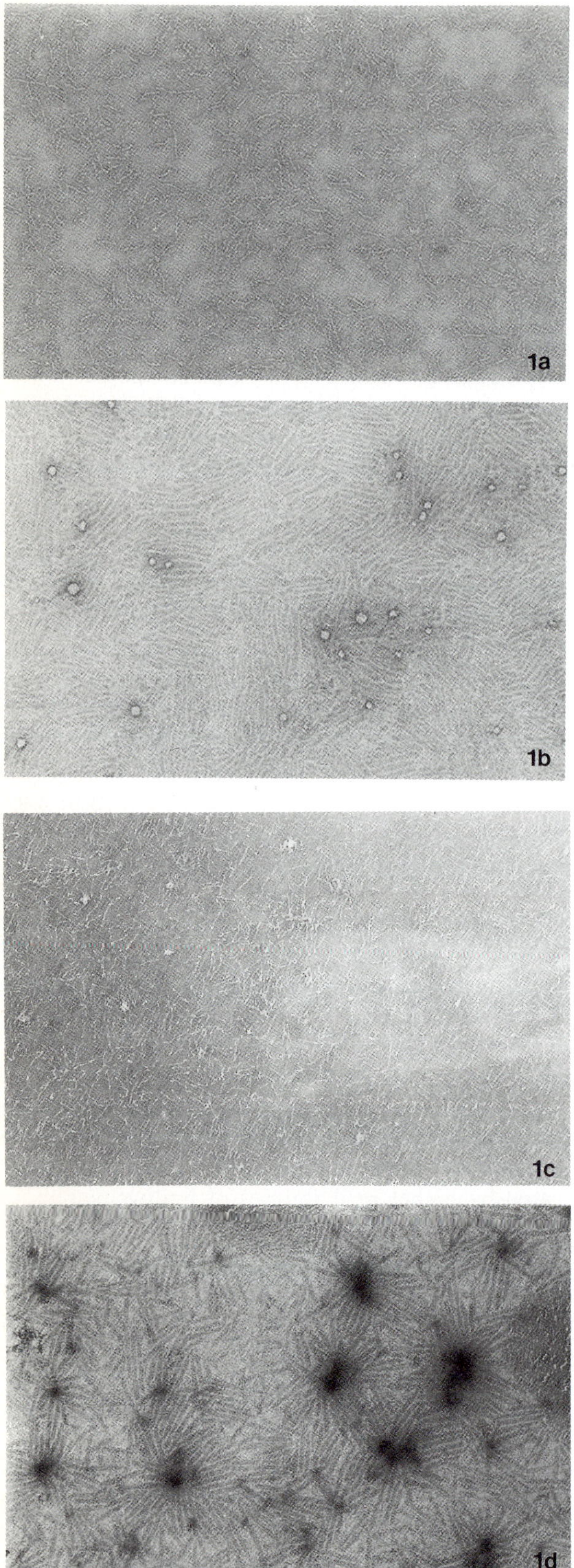

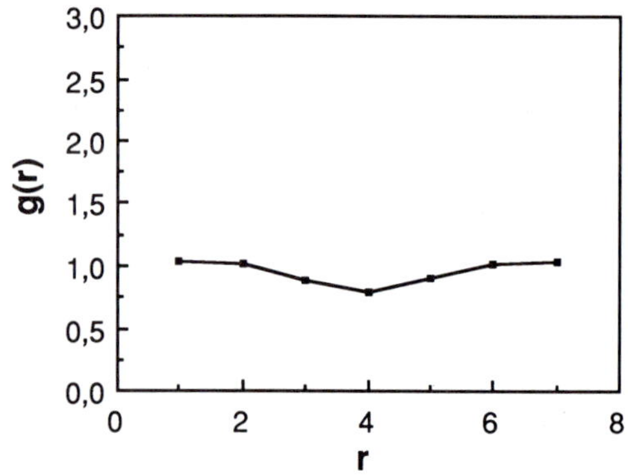

Fig. 2. Pair correlation function $g(r)$ vs r (molecular radius; $1r = 23$ nm) measured for fibrinogen molecules adsorbed to a hydrophilic quartz surface. The data are based on correlation analysis of Fig. 1a. Random distribution is indicated by $g(r) = 1$

properties of fibrinogen adsorption, although the adsorbed layer is stable during rinsing with buffer.

Polycrystalline samples and long-range ordering have also been described in an emulsion of oil droplets at glass-water interfaces [14], indicating that attractive interaction between colloidal particles at interfaces may be a general phenomenon. Equilibrium properties are also suggested for the oil in water system [14].

The nature of the attractive forces acting between fibrinogen molecules is not known. One possible attractive force is the entropic force first measured between hydrophobic mica surfaces [15]. The highly directional properties of the interaction between fibrinogen molecules indicate different polarities in different parts of the molecule with, most probably, hydrophobic ends. The end-to-end interactions of fibrinogen are possibly also responsible for the stacked end-on adsorption of fibrinogen at higher surface concentrations that results in the formation of ridges and walls at the surface [5]. The different orientations of fibrinogen molecules result in different reactivities of antibodies against fibrinogen at surfaces (unpublished observations), and may possibly also effect cellular reactions at the interface.

Figs. 1a–d. Electron micrographs of the adsorbed layer of fibrinogen to hydrophilic quartz surfaces after 20 h adsorption time, with different bulk concentrations of fibrinogen. a) 0.29 nM; b) 0.73 nM; c) 2.90 nM; d) 29.0 nM. Magnification = × 150000

Acknowledgement

This study was supported by grants from the Swedish Medical Research Council (grant 12X-06235) and by grants from The Board of Technical Development.

References

1. Packham MA, Evans G, Glynn MF, Mustard JF (1969) Lab Clin 73:686–693
2. Zucker MB, Vroman L (1969) Proc Soc Exp Biol Med 131:318–324
3. Ginsberg M, Loftus JC, Plow EF (1988) Thrombosis and Haemostasis 59:1–6
4. Eberhart RC, Prokop LD, Wissenger J, Wilkov MA (1977) Trans Am Soc Artif Intern Organs 23:134–136
5. Brynda E, Houska M, Lednicky F (1986) J Colloid Interface Sci 113:164–169
6. Nygren H, Stenberg M (1988) J Biomed Mater Res 22:1–11
7. Wigren R, Elwing H, Erlandsson R, Welin S, Lundström I (1991) FEBS Lett 280:225–228
8. Nygren H, Stenberg M, Karlsson C (1992) J Biomed Mater Res 26: 77–91
9. Stenberg M, Stemme G, Nygren H (1987) Stain Technol 62:231–236
10. Berry RS, Rice SA, Ross J: Physical Chemistry. Thermodynamic properties of fluids. Chapter 23:pp 830
11. Bachmann L, Schmitt-Fumian WW, Hammel R, Lederer K (1975) Macromol Chemie 176: 2603–2618
12. Nygren H (1988) J Immunol Meth 114:107–114
13. Nygren H, Stenberg M (1990) Biophys Chem 38:77–85
14. Bibette J, Roux D, Nallet F (1990) Phys Rev Lett 65:2470–2473
15. Israelachvili JN, Pashley RM (1982) Nature 300:341–342

Authors' address:

Dr Håkan Nygren
Department of Histology
P.O.B. 33031
S-400 33 Göteborg
Sweden

Progress in Colloid & Polymer Science Progress Colloid Polym Sci 88:100–109 (1992)

Organic xanthates adsorbed on gold surfaces: an infrared and photoelectron study

N. O. Persson, K. Uvdal, B. Liedberg, and M. Hellsten[1])

Laboratory of Applied Physics, IFM, University of Linköping, Sweden
[1]) Berol AB, Stenungsund, Sweden

Abstract: Infrared reflection absorption spectroscopy (IRAS) and x-ray photoelectron spectroscopy (XPS) have been used to study the adsorption of potassium *p*-methyl benzyl xanthate and potassium *p*-trifluoromethyl benzyl xanthate on gold surfaces. Monolayer structures were formed by adsorption from 1 to 10 μM aqueous solutions of the xanthates for different adsorption times. The reflection-absorption (R-A) spectra are compared with calculated R-A spectra, which are based on the optical constants $n(\nu)$ and $k(\nu)$ derived from KBr transmission spectra of the corresponding Au(I) salts. The experimentally obtained peak positions are in good agreement with the calculated ones, indicating that the chemical structure of the surface complexes are closely related to those of the metal salts. However, the relative intensities vary considerably between experimental and calculated R-A spectra. We discuss these intensity changes in terms of molecular orientation and packing. A detailed comparison of the relative ratios for the peaks belonging to the OCS_2 and CF_3 groups reveals that ordered structures are formed on the gold surface, and that a reorientation from a flat to an almost vertical orientation occurs with increasing immersion time and concentration of the xanthate solution. These observations are confirmed with angle-dependent XPS(Θ).

Key words: Xanthates; adsorption; gold surfaces; infrared spectroscopy; x-ray photoelectron spectroscopy

Introduction

Organic xanthates are widely used as collectors in mineral flotation [1]. In order to improve this process, it is of great interest to gain insight into the adsorption behavior of xanthates on solid surfaces. Such studies have been performed on pure metals and sulphide mineral surfaces by, for example, electrochemical methods [2–4], infrared techniques [5–8], and x-ray photoelectron spectroscopy [9–12]. The electrochemical results have been explained with the formation of surface compounds [4] followed by formation of dixanthogen, especially on metal surfaces [2–4]. Recent infrared investigations on sulphide minerals indicate formation of metal xanthates on the mineral surface, and that the presence of soluble oxides is of great importance for the hydrophobization process [5].

We present in this paper a combined infrared reflection-absorption spectroscopy (IRAS) [13, 14] and x-ray photoelectron spectroscopy (XPS) [15] study on the adsoption of potassium *p*-methyl benzyl xanthate (KMBX, Fig. 1a) and potassium *p*-trifluoromethyl benzyl xanthate (KFBX, Fig. 1b) on evaporated gold surfaces. Most of the experimental work is performed on the trifluoromethyl substituted analog KFBX (Fig. 1b), because of the strong and unique spectral signature of the CF_3 group in both the infrared and XPS spectra. Our main attention is focused on the coordination behavior and molecular organization (packing) within the very first adsorbed layer as a function of xanthate concentration in solution and adsorption time. Of particular interest is to establish if reorganization phenomena occur during the initial phase of adsorption. Recent infrared studies on copper [7]

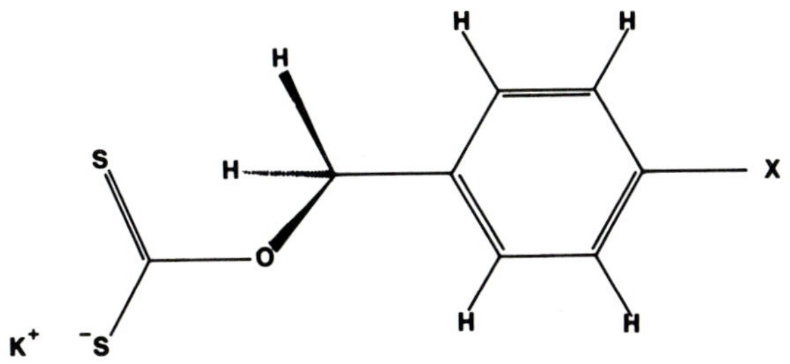

Fig. 1. Molecular structure of potassium *p*-methyl benzyl xanthate, KMBX (a) and p-(trifluoromethyl)benzyl xanthate KFBX (b)

reveal that xanthates react with copper through a dissolution precipitation mechanism and easily form multilayer structures of cuprous xanthates on the surface, even at fairly low exposures. In order to avoid that this type of multilayer formation occurs, we decided to use gold as the substrate material, because gold is not really expected to follow a dissolution precipitation mechanism.

The IRAS technique gives, through peak shifts and intensity changes, pertinent information about the molecular structure and active parts of the adsorbed molecule. It is also possible to extract information about the orientation of the molecular skeleton and/or functional groups in relation to the metal surface by utilizing the surface selection rule [13, 14]. Such analysis requires reliable intensity data from a proper reference compound, and that the transition dipole moments are known for the vibrational modes in the molecule. We have in this study chosen to compare the experimental R-A spectra with calculated R-A spectra [16], which are based on the optical constants $n(\nu)$ and $k(\nu)$ obtained from the transmission (KBr) spectrum of the corresponding Au(I) xanthate salts. The main reason for using this method, instead of a direct comparison with the transmission spectrum, is that it accounts for so-called optical effects [16, 17] in the R-A spectrum. These optical effects may give rise to very large peak shifts (distortions) in regions with strong absorption, and must be compensated for before the spectral interpretation. We have successfully used this method for the identification of Cu(II)-amino acid complexes on copper surfaces [18].

Complementary information about the adsorbed layer can be obtained from angle dependent x-ray photoelectron spectroscopy, XPS(Θ) [19, 20]. By analyzing the photoemitted electrons at different take-off angles Θ in relation to the surface normal, it is possible to obtain a depth profile of atoms within the adsorbed layer. Thus, XPS(Θ) can tell us if certain parts of the molecule are preferentially oriented toward or away from the metal surface, i.e., it can distinguish between up and down in an adsorbed molecule.

Experimental

The potassium *p*-methyl xanthate was synthesized according to the following procedure: 30 g (0.246 moles) of *p*-methyl benzyl alcohol and 150 g acetone were added to a flask equipped with stirrer, thermometer, and nitrogen sparging, and then 15.6 g (0.278 moles) KOH which had been ground in a mortar was also added. When the main part of the KOH had dissolved, 18.3 g (0.240 moles) of carbon disulphide was added dropwise and the reaction temperature was kept below 30 °C by cooling with an ice bath. The reaction mixture was then kept at room temperature for 2 h. Then 80–90% of the acetone was removed by distillation and 150 mls of petroleum ether (b.p. 60–70 °C) was added to the residue. The xanthate precipitate was removed by filtering and the filter cake was washed twice with petroleum ether. After drying the solid, the assay was determined to be 94% by titration with HCl in ethanol. The fluorinated compound KFBX was prepared in an analogous way. Here, the purity of the salts varied between 83 and 98% in the batches used.

The purity of the potassium xanthates was always checked before use by recording their IR spectra in KBr according to the procedure described below, and compared with spectra from newly synthetized salt. If needed, the salt was worked up by dissolution in hot acetone, warm filtration of this solution to remove inorganic impurities, cooling, and recrystallization by adding diethyl ether or diisopropyl ether [21].

The gold(I) xanthates were precipitated from an excess of the appropriate potassium xanthate in water solution with $HAuCl_4$ according to the method of Denko and Anderson [22]. The $HAuCl_4$ was purchased from J. M. Speciality Products, England (assay 48.5% Au) and the water used came from a Milli-Q unit. The yellow solid was separated by centrifugation, washed several times with analytical grade diethyl ether, and dried in vacuo for several days before spectra in KBr were recorded.

KBr pellets were prepared from ca. 2 mg of gold xanthate and 600 mg of KBr, which was mixed and crushed in a mortar; 250 mg of this mixture was pressed in a circular pellet form ($\phi = 13$ mm) at 10-ton pressure. Transmission spectra were recorded with pure KBr as reference.

The gold films ca. 1500 Å-thick were electron-beam evaporated at a base pressure of 10^{-9} torr and an evaporation rate of 2 Å/s onto silicon (100) wafers. The silicon wafers had been primed with a 10 Å-thick chromium layer prior to the gold evaporation in order to improve the adhesion between gold and silicon.

X-ray diffraction (XRD) measurements on the gold surfaces were performed on a Philips x-ray Diffractometer PW 1050/25 [18, 23a]. The texture coefficient $\gamma_{hkl} = I_{hkl}/(\sum I_{hkl})$, where I_{hkl} is the intensity of the (hkl) peak in the diffractogram, is used to characterize the degree of orientation of the gold films. For polycrystalline gold, γ_{111} is calculated to be 0.43 [23b] on the basis of the following intensities: I_{111}; I_{200}; I_{220}; I_{311}; I_{222}. The value 0.95 obtained in the present work, suggests therefore that the gold films have a strong (111) texture.

The gold surfaces were always washed in TL1 solution, i.e., one part 25% NH_3 solution + one part 30% H_2O_2 solution + five parts distilled water, at 80 °C for at least 5 min before use unless taken directly from the evaporator unit. All glassware used in adsorption experiments was washed in TL1 according to the same procedure.

The adsorption of xanthate on gold surfaces was achieved by immersing the surfaces in a water solution of the xanthate at a concentration of 1–10 μM and an adsorption period of 1–120 min. The solutions were prepared in Milli-Q water, through which nitrogen had been bubbled for at least 30 min to remove oxygen. The adsorption vessel was stirred during the immersion, and covered with parafilm. Before the surface was lifted off, the solution surface was drained. The gold surface was then rinsed with water or xanthate solution and then blown dry with nitrogen and put into the spectrometer vacuum (cf. below) as soon as possible.

The IR unit was a Bruker IFS 113 v Fourier Transform spectrometer equipped with a DTGS detector. The R-A spectra were obtained with a Bruker GIR (Grating angle Incidence Reflection) accessory [18] aligned at 85 °. All the R-A spectra were recorded with radiation polarized parallel to the plane of incidence. The KBr and R-A spectra were obtained by averaging 200 and 500 interferograms, respectively, at 4 cm^{-1} resolution, and all R-A spectra presented in this paper are plotted according to

$$\left(\frac{\Delta R}{R_0}\right)_{\parallel} = \left(\frac{R_0 - R}{R_0}\right)_{\parallel}, \tag{1}$$

where R and R_0 are the reflectivity from the metal with and without the layer present.

The XPS measurements were carried out with unmonochromatized Mg(K_α) photons (1253.6 eV). Spectra were taken with an analyzer resolution of 0.2 eV, and the resolution was determined from the full width at half maximum (FWHM) of Au 4f(7/2) line to be 0.9 eV. This Au line was also used to calibrate the binding energy scale at 83.8 eV. The spectrometer used has been described in detail previously [20, 24]. When measuring in the "surface sensitive" mode, or in other words, when collecting electrons with a take-off angle of 80 ° with respect to the surface normal, the signal from the outermost surface is enhanced. This enhancement is correlated to the limited escape depth of the photoelectrons [19]. In the bulk sensitive mode the photoelectrons exit with an angle of 30 ° with respect to the normal, and hence more information from the "bulk" of the sample can be obtained.

Calculation of R-A spectra

The procedure used to calculate R-A spectra from a transmission spectrum of an isotropic model compound (KBr pellet) is schematically outlined in Fig. 2. The first part of the procedure involves an iterative Kramers–Kronig algorithm [25] where the input extinction coefficient $k(\nu)$ is determined from the Beer–Lambert law. In this step the homogeneously dispersed model compound in the KBr pellet is transformed into an isotropic layer with effective thickness d_{eff}, which can be determined from the concentration and density of the model compound, and the physical dimensions of the KBr pellet. The input $k(\nu)$ is then used to calculate the refractive index $n(\nu)$ from the

Kramers–Kronig transform [25]:

$$n(\nu_i) = n_\infty + \frac{2}{\pi} \int_0^\infty \frac{\nu k(\nu)\, d\nu}{\nu^2 - \nu_i^2}, \tag{2}$$

where n_∞ refers to the refractive index at high frequencies. The obtained $k(\nu)$ and $n(\nu)$ sets are used to generate the absorbance spectrum $A^c(\nu)$ from the exact Fresnel formalism. The calculated spectrum $A^c(\nu)$ is then compared with the experimental spectrum $A(\nu)$. This procedure is iterated by refining $k(\nu)$ (cf. Fig 2) and thereby also $n(\nu)$, until an acceptable match is obtained between $A^c(\nu)$ and $A(\nu)$. The iterative Kramers–Kronig procedure is performed on a VAX computer with a program IRXA written by us. The finally accepted $k(\nu)$ and $n(\nu)$ are later transferred to a separate simulation program, SIMSPEC, for generation of R-A spectra. Figure 3 shows the results from the iterative Kramers–Kronig procedure for gold(I) *p*-trifluoromethyl benzyl xanthate, Au(I)FBX, and a simulated R-A spectrum. The R-A spectrum is based on a three-layer model and an isotropic layer thickness of 10 Å. It is obvious from Fig. 3 that the band shapes and relative intensities differ significantly between the original transmission (bulk) spectrum $A(\nu)$ and the simulated R-A spectrum. Note especially the relative intensity changes for the main absorptions near 1330, 1180, and 1070 cm^{-1}, respectively. These changes further illustrate the importance of adopting the present or similar [16] approaches in cases where infrared intensities are used in more rigorous orientation analyses of molecular adsorbates and thin films.

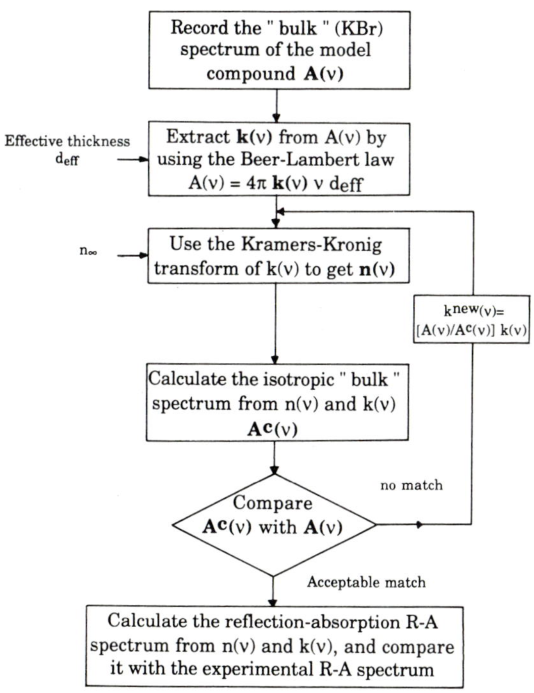

Fig. 2. Block diagram illustrating the iterative Kramers–Kronig procedure used for extraction of optical constants $n(\nu)$ and $k(\nu)$ from "bulk" infrared spectra

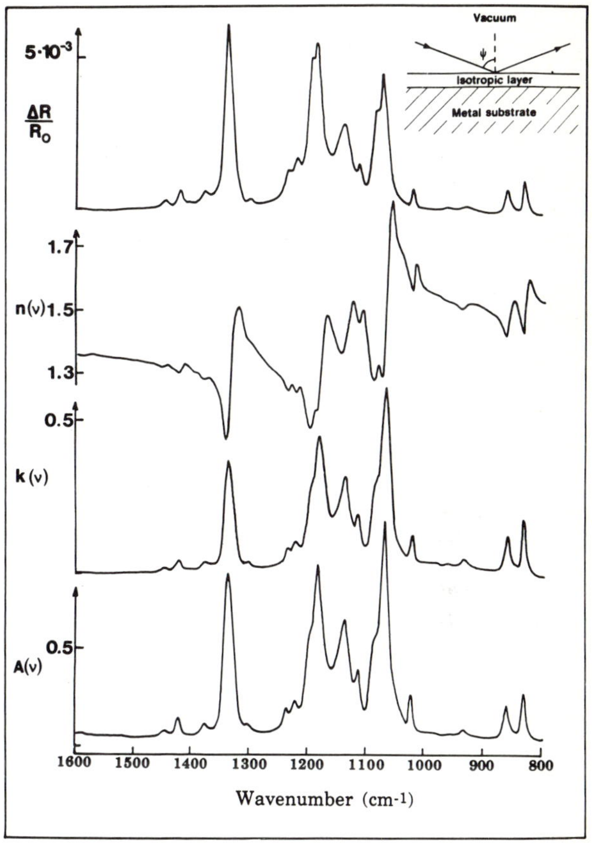

Fig. 3. Infrared $A(\nu)$, $n(\nu)$, and $k(\nu)$ spectra of AuFBX, together with a simulated R-A spectrum of a 10 Å-thick AuFBX layer on gold. The insert shows the optical model used in the simulation of the R-A spectrum with $\Psi = 85°$. $n_\infty = 1.5$ in the Kramers–Kronig transform

Results and discussion

Infrared spectra

In Fig. 4 are shown a simulated (cf. above) R-A spectrum of a Au(I)FBX layer on gold, together with R-A spectra of adsorbed KFBX after various

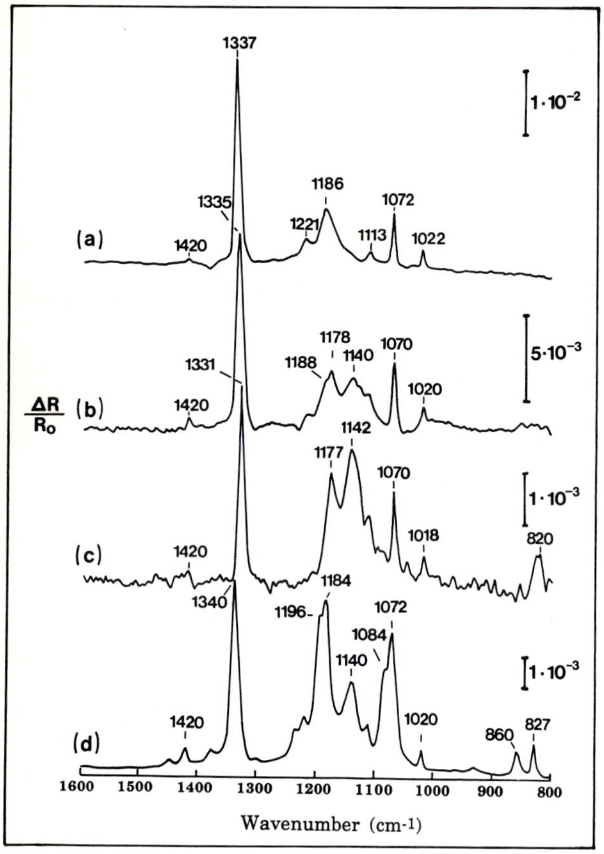

Fig. 4. Experimental and simulated R-A spectra of KFBX on gold. The adsorption time for the experimental spectra were a) 90 min, b) 15 min, and c) 5 min, and the xanthate concentration was $1.0 \cdot 10^{-6}$ M. The simulated spectrum in (d) was obtained as in Fig. 3

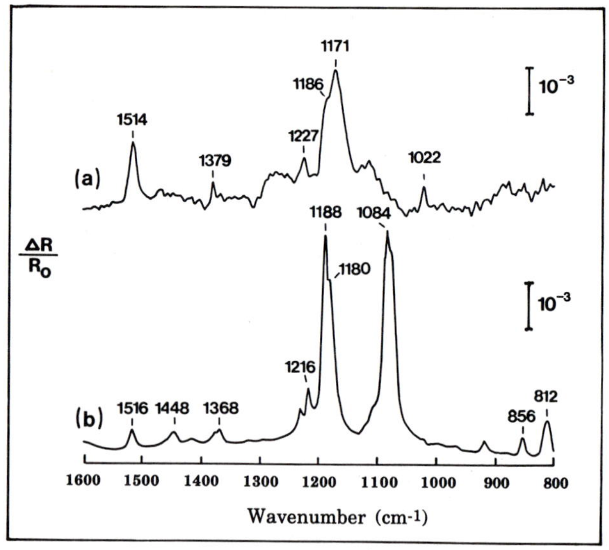

Fig. 5. Experimental and simulated R-A spectra of KMBX on gold. The adsorption time was 120 min and the xanthate concentration was $10 \cdot 10^{-6}$ M. $n_\infty = 1.5$ and the thickness of the layer of AuMBX was 5 Å for the simulated spectrum

immersion times in $1.0 \cdot 10^{-6}$ M solution, and in Fig. 5 a simulated spectrum for gold(I) *p*-methyl benzyl xanthate, Au(I)MBX, with a R-A spectrum from a surface immersed 2 h in $10 \cdot 10^{-6}$ M solution of KMBX. The R-A spectrum of KFBX obtained after 90 min most likely respresents a full monolayer. This conclusion is based on the observation that the overall spectral signature and intensity remain unaffected upon increasing the adsorption time or xanthate concentration in solution. R-A spectra obtained for shorter adsorption times than 5 min give essentially the same R-A spectral signatures as the one obtained after 5 min, but with a lower intensity. The situation is different for KMBX because we are not able to prepare any stable layer of that molecule at the concentration used for KFBX ($1.0 \cdot 10^{-6}$ M), suggesting that the affinity to gold is different for the two molecules.

There is generally a very good agreement between experimental and calculated frequencies (cf. Figs. 4 and 5), suggesting that the chemical structure of the surface complexes are closely related to those of the Au(I) salts. However, it is equally obvious from Figs. 4 and 5 that the relative intensities of the obtained bands are drastically altered in the R-A spectra of the adsorbed layers, and that the effects are dependent on the adsorption time (Fig. 4). The most pronounced differences occur in the 1100–1050 cm^{-1} region where the $O{-}CS_2$ group is expected to absorb strongly. In order to investigate these intensity changes in detail and the possibility of orientation effects according to the surface selection rule, an assignment of the strongest bands was made. The results are shown in Table 1, together with the transition dipole moments (M_i) and the full widths at half maximum (FWHM). The vibrations of various metal xanthates have been reviewed by Winter [26] and the vibrations of the aromatic ring are assigned according to Katritzky and Simmons [27]. The assignments of the C-F vibrations were taken from Bellamy [28].

The assignment of the 1066 cm^{-1} band (Table 1) in the KFBX spectra on gold to an in-plane vibration of the benzene ring must be discussed in more

Table 1. Strong bands in the KBr-spectra of Au-xanthates and their full width at half maximum (FWHM) and transition dipole moment directions M_i in molecular coordinates

AuFBX (cm^{-1})	AuMBX (cm^{-1})	FWHM (cm^{-1})	Assignment	M_i
1337	–	20	$\nu_s CF_3$	$\parallel$ C-CF_3
1178, 1136	–	30–40[a])	$\nu_{as} CF_3$	$\perp$C-CF_3
1185	1186	20	νC-O	$\parallel S_2$C-O
1080	1070	20	ν_{as}S-C-S	$\perp S_2$C-O
1066	–	~ 10[a])	X-C_6H_4-CF_3	$\parallel C_6H_4$[b])
1020	1024	8	βC-H (ipd)[c])	$\parallel$ C-CF_3
827	808	10	γC-H(opd)[d])	$\perp C_6H_4$[b])

[a]) Taken from reference spectra of *p*-(trifluoromethyl) benzyl alcohol [29]
[b]) Refer to the molecular plane of the benzene ring
[c]) ipd = in-plane deformation
[d]) opd = out-of-plane deformation

detail because of its importance in the following discussion. This band was not listed by Katritzsky and Simmons [27] for any of their para-substituted benzene derivatives. The observed band, however, was too narrow to be ascribed to a $\nu_{as}(C\text{-}S_2)$ stretching vibration in the xanthate group. Furthermore, this band occurs in the spectra of *p*-trifluoromethyl benzyl alcohol and other *p*-substituted trifluoro benzene derivatives [29], but not in the spectrum of *p*-methyl benzyl alcohol [29] or in the spectrum of the non-fluorinated compound KMBX. The presence of the 1066 cm^{-1} band in the spectrum of the fluorinated compound KFBX, but not in that of KMBX, may reflect the differences in electron donating (accepting) properties of the CF_3 and CH_3 groups, respectively. These properties are clearly visible in the Hammet substitution constants for parasubstituted benzene derivatives, which change from − 0.17 (donating) for the CH_3 group to 0.54 (accepting) for the CF_3 group [30]. Such a large change will certainly influence the electronic structure of the benzene moiety, and thereby also its vibrational spectrum. We therefore assign the sharp 1066 cm^{-1} band in the R-A spectrum of KFBX (Fig. 4) to an in-plane vibration of the benzene ring, rather than to a $\nu_{as}(C\text{-}S_2)$ vibration of the O–CS_2 group. This assignment is supported by the absence of a similar band in the R-A spectrum of KMBX, Fig. 5. The above assignment is also confirmed by recent studies of KFBX on silver and copper [31], where the 1066 cm^{-1} band appears well-separated from the $\nu_{as}(C\text{-}S_2)$ vibration near 1040 cm^{-1}.

No IR-spectra of gold(I) xanthates in the literature are known to the authors, but the xanthate group vibrations listed in Table 1 are in good accordance with those of copper(I) xanthates given by Winter [26]. However, we still need to find an explanation for the disappearance of the $\nu_{as}(C\text{-}S_2)$ in the spectra of KFBX and KMBX on gold. In the electrochemical study of the adsorption of ethyl xanthate to gold [4], Lezna et al. found that the adsorption started with the binding of the two sulphur atoms to the metal surface. This is in accordance with our results, because the S-C-S asymmetric stretching vibrations at 1080 and 1070 cm^{-1} for AuFBX and AuMBX, respectively, disappear even for the shortest adsorption time (cf. Figs. 4 and 5). If the xanthate group binds in this way to the gold surface, the dipole moment of the $\nu_{as}(C\text{-}S_2)$ vibration will be oscillating parallel to the surface and the corresponding band will thus disappear in an R-A spectrum. A bridging or chelating coordination to the gold surface, regardless of coverage, is therefore the most plausible explanation to the experimental results. (The nature of the gold-xanthate interaction will be discussed further on.)

Comparing the spectra after adsorption for 5, 15, and 90 min (Fig. 4) reveals other features. The most striking detail is the drastic increase of the intensity of the symmetric C-F stretching at ca. 1330 cm^{-1} relative to the other bands for increasing adsorption time. This vibration has an oscillating dipole moment parallel to the C-CF_3 axis of the molecule (cf. Table 1). On the contrary, the asymmetric CF_3

stretching bands near 1180 and 1135 cm^{-1} are weakened considerably with adsorption time. The corresponding dipole moments are aligned perpendicular to the C-CF_3 axis of the molecule. The observations concerning the C-F strectching vibrations unambiguously indicate that the CF_3 group orients itself with its C_3(C-CF_3) axis nearly perpendicularly to the gold surface as the packing density increases.

Now, let us turn to the benzene ring. A look at Figs. 4 and 5 reveals that the C-C stretching vibrations [27] at 1300 to 1600 cm^{-1} are present, but weak or hidden behind other bands, e.g., the symmetric C-F stretch at 1330 cm^{-1}. This is also the case with the in-plane C-H bending or ring vibrations [27] near 1000 to 1200 cm^{-1}.

The out-of-plane C-H bending vibration [27] at 800–850 cm^{-1} is left for analysis. This vibration is associated with an oscillating dipole moment perpendicular to the ring plane. It is evident from Figs. 4 and 5 that this band is weakened and ultimately disappears after a sufficiently long adsorption time, an observation that supports the conclusion that the benzene ring successively orients itself with the ring plane perpendicular to the gold surface, when the packing density on the surface is increased.

X-ray photoelectron spectroscopy

The infrared spectroscopy measurements indicate that the adsorbed xanthate molecules bind to the gold surface with both sulphur atoms at a relatively low degree of adsorption and, then, the organic part of the molecule erects itself perpendicular to the surface as the degree of adsorption increases. In Fig. 6 are shown the F(1s), C(1s) and S(2p) core level spectra of KFBX on gold after immersion in $1.0 \cdot 10^{-6}$ M solution for an adsorption time of 90 min at two take-off angles, $\theta = 30°$ and $80°$.

Table 2 summarizes the binding energies for the spectra in Fig. 6, as well as for a spectrum of a submonolayer prepared by adsorption from a $1.0 \cdot 10^{-6}$ M solution for 5 min (cf. Fig. 4c). The S(2p) binding energy at 161.8 eV is in good agreement with those obtained from chemisorbed thiols [32] and thiol containing amino acids [33] on gold, revealing that the binding to gold occurs through the sulphur atoms. The fact that the binding energies are identical for the two preparations (Table 2) further supports our previous proposal that the bonding to gold is essentially the same in the two cases, i.e., independent of surface coverage. Angle-

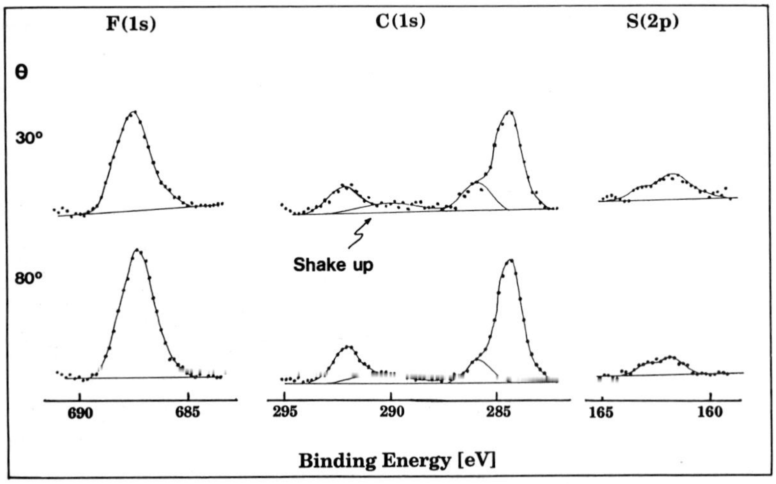

Fig. 6. XPS(Θ) core level spectra of KFBX on gold. The adsorption was performed from a $1.0 \cdot 10^{-6}$ M solution for 90 min

Table 2. XPS(Θ) binding energies (eV)

	C_{Ph}(1s)	C_{CO}(1s)	C_{CF_3}(1s)	F(1s)	S(2p)
Submonolayer	284.5	285.8	292.0	687.4	161.8
Full monolayer					
$\Theta = 30°$	284.5	285.8	292.0	687.4	161.8
$\Theta = 80°$	284.5	285.8	292.0	687.4	161.8

dependent data for the full monolayer in Fig. 6 is summarized in Table 3. It should immediately be pointed out that it was impossible to see any angle-dependent effects for the submonolayer (Fig. 4c), an observation not in conflict with the infrared results, which predicted a flat orientation at low coverages. The relative intensity ratios for the full monolayer exhibit, however, more pronounced differences when the take-off angle increases from 30° to 80°.

A comparison between the C_{tot}/S and F/S intensity ratios shows that the experimental values are substantially higher than those expected stoichiometrically. This behavior may have several explanations, including, for example, cross-section changes of the S(2p) orbital caused by interaction with the substrate. Another more fundamental explanation to the observed effect is related to the symmetry properties of the electronic orbitals. When the p-states are analyzed, subshell cross-sectional effects must be accounted for [34]. If the p-orbitals are assumed to be ordered relative to the surface, the cross-sections can be expected to vary with the electron take-off angle. This change in cross-section complicates the analysis, and the angle dependency resulting from p-states should therefore be handled with care. Thus, it is generally more reliable to compare intensity ratios from orbitals with spherical symmetry, and to avoid intensity data from strongly interacting atoms (orbitals) like S(2p). The following discussion will therefore focus on s-orbitals. The C_{CF_3}/C_{CO} and F/C_{CO} intensity ratios at $\Theta = 30°$ are in good agreement with stoichiometric values. It is also clear that the above ratios increase strongly with increasing take-off angle, indicating that a highly ordered structure is formed with the trifluoromethyl group oriented away from the metal surface. A similar trend is observed for the C_{Ph}/C_{CO} intensity ratio. Thus, the angle-dependent XPS(Θ) results give an excellent confirmation of the infrared observations that a transition from a flat to an almost vertical orientation of the xanthate molecules takes place with increasing packing density.

Table 3. XPS(Θ) relative intensities of a monolayer of KFBX on gold, prepared by adsorption from a $1.0 \cdot 10^{-6}$ M solution for 90 min

Case	C_{tot}/S	C_{Ph}/C_{CO}	C_{CF_3}/C_{CO}	F/C_{CO}	F/S
KFBX	4.5	3	0.5	1.5	1.5
Full monolayer					
$\Theta = 30°$	9.6	4.0	0.6	1.6	2.7
$\Theta = 80°$	14.3	5.2	1.2	2.6	4.5

Nature of the gold-xanthate interaction

The two most probable ways of binding to the surface of the xanthate ions are a bidentate bridging between two gold atoms, or a bidentate chelating to one atom [26]. Both coordinations would give the same orientation of the molecule relative to the surface and thus lead to the disappearance of the asymmetric $\nu_{as}(CS_2)$ stretching band at 1070–1080 cm^{-1}. To the authors' knowledge, no structural studies of any solid gold xanthate have been performed. However, the structures of some Au(I) complexes with similar dithio ligands have been determined.

Chiari et al. [35] have studied gold dithioacetate with x-ray and infrared techniques, and they concluded that the structure is a tetrameric one with four gold atoms in a rhomboid with a side length of 3.013 Å and a unique Au(2)-Au(4) distance of 3.296 Å. The S-Au-S angles are nearly 180°, and the mean Au-S bond length is 2.296 Å. The xanthate groups form bridges between two Au atoms in the rhomboid, and the eight S-atoms form two planes parallel to the rhomboid. From the IR results (CSS asymm str = 1160 cm^{-1} and CSS sym str = 899 cm^{-1}) the same authors concluded that gold dithioacetate has a bridging interaction in line with the x-ray results. Chiari et al. [35] also made IR measurements on gold dithiobenzoate. Here, the signals attributed to C-S stretching were different (CSS asymm str = 1039, CSS symm str = 911 cm^{-1} and CSS bend = 656 cm^{-1}), and it was then concluded that the xanthate group chelated the gold atom.

Åkerström [36] studied gold(I) salts with some *N,N*-dialkyl dithiocarbamates. He proposed more or less polymeric structures, where the units are dimers of two gold atoms that form bridges between two xanthate groups.

Hesse and Jennische [37] made more extensive x-ray studies on Au(I) *N-N*-dipropyl dithiocarbamate. They found that the structure consisted of isolated dimers of the Au salt and that the unit cell is formed by two dimers. Each dimer is bridged together by the two Au atoms which coordinates linearly to one S-atom from each ligand. The Au-

Au distance in a dimer is 2.76 A. The dimers are ordered with the Au atoms in linear coordination, and the distance between two Au atoms from two adjacent dimers is 3.40 Å.

Melnik and Parish [38] have reviewed structural data for gold compounds and it is seen that gold(I) compounds mostly coordinate linearly. Molecular orbital calculations by Schwerdtfeger et al. [39] gave the same results.

Our IR bands ascribed to the xanthate group (cf. Table 1) give no definitive indication of which type of coordination is at hand between the gold atoms on the surface and the sulphur atoms of the xanthate group. However, no band corresponding to a chelate coordination [35] appears near 650 cm^{-1}, which may give an indication of a bridging interaction. The bridge structure also seems probable from structural data of xanthates [26]. X-ray data on the xanthate ion estimate the S-C-S angle to be 124 °, the S_1-C and the S_2-C bond lengths to be 1.67 and 1.70 Å, respectively. The S_1-S_2 distance can then be calculated to 2.97 Å, which is near to the Au-Au distance in a (111) surface of Au (2.88 Å) [40]. An Au-Au distance of 2.88 Å gives an S-C-S angle of 117.4 °, which is tolerable deformation [26]. From the above considerations and the fact that a chelate interaction is hard to combine with a linear coordination to the metal, we conclude that the interaction is bridge-like in our case. The proposed structures of adsorbed KFBX and KMBX, at low and high packing densities are schematically depicted in Fig. 7.

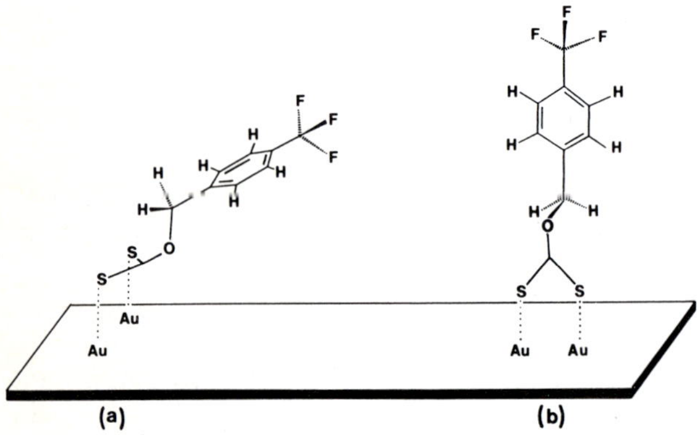

Fig. 7. Proposed arrangement and coordination of xanthate molecules on gold. In (a) is shown the molecule at low packing density after the short adsorption time, and in (b) the more perpendicular conformation at a higher packing density after prolonged adsorption

Conclusions

We have investigated the adsorption of potassium *p*-(trifluoromethyl) benzyl xanthate and potassium *p*-methyl benzyl xanthate on gold surfaces by means of Fourier transform infrared reflection absorption spectroscopy and x-ray photoelectron spectroscopy. The close resemblance between experimental and simulated R-A spectra of Au(I) xanthates, and a comparison with structurally well characterized model compounds like Au(I) dithioacetates or Au(I) dithiocarbamates have led us to propose a bridge-like coordination for the chemisorbed xanthates. From the variations of the relative intensities of the IR bands with adsorption time and the angular dependence of the XPS spectra, we further conclude that the organic part of the molecule lies down on the gold surface at low packing densities, but erects itself, approaching a conformation normal to the surface when the packing density increases. An orientation transition of the type presented here (Fig. 7) will influence the hydrophilic/hydrophobic properties of the interface, and certainly also the ability for the collector molecules to act more effectively in a flotation process. Similar studies of KMBX and KFBX on silver and copper will be published separately.

Acknowledgements

This work was supported by the Swedish Board for Technical Development (STU). We also wish to thank Helene Roth at Berol AB for the synthesis of the two potassium xanthates.

References

1. Aplan FF (1980) In: M. Grayson (ed.) Flotation. Kirk-Ottmer, Encyclopedia of Chemical Technology. John Wiley & Sons, New York X:523
2. Woods R (1971) J Phys Chem 75:354
3. Hyunh Thi O, Lamache M, Bauer D (1981) Electrochim Acta 26:33
4. Lezna RO, de Tacconi NR, Arvia AJ (1988) J Electroanal Chem 255:251
5. Persson P (1990) Thesis, Uppsala
6. Mielczarski JA, Yoon RH (1989) J Phys Chem 93:2034
7. Mielczarski JA, Leppinen J (1987) Surface Sci 187:526.
8. Cases JM, De Donato P, Kongolo M, Michot L (1989) Colloids and Surf 36:323
9. Mielczarski J, Suonninen E, Johansson L-S, Laajalehto K (1989) Intern J Mineral Process 26:181

10. Ranta L, Minni E, Suonninen E, Heimala S, Hintikka V, Saari M, Rastas J (1981) Appl Surf Sci 7:393
11. Mielczarski J (1987) J Coll Interf Sci 120:201
12. Mielczarski J, Suonninen E (1984) Surf Interf Anal 6:34
13. Francis SA, Ellison AH (1959) J Opt Soc Amer 49:131
14. Greenler RG (1969) J Chem Phys 50:1963
15. Siegbahn K, Nordling C, Fahlman A, Nordenberg R, Hamrin K, Hedman J, Johansson G, Bergman T, Karlsson S-E, Lindgren I, Lindberg B (1967) In: ESCA-Atomic Molecular and Solid State Structure Studied by means of Electron Spectroscopy, Nova Acta, R Sci Ups Ser IV, Vol. 20.
16. Allara DL, Nuzzo RG (1985) Langmuir 1:52
17. Allara DL, Baca A, Pryde CA (1978) Macromolecules 11:1215
18. Ihs A, Liedberg B, Uvdal K, Törnkvist C, Bodö P, Lundström I (1990) J Coll Interf Sci 140:192
19. Fadely CS, Baird RJ, Siekhaus W, Novakov T, Bergström SAA (1974) J Electron Spec and Rel Phenomena 4:93
20. Uvdal K, Bodö P, Ihs A, Liedberg B, Salaneck WR (1990) J Coll Interf Sci 140:207
21. Du Rietz C (1957) Sv Kem Tidskr 69:310
22. Denko CW, Anderson AK (1945) J Am Chem Soc 67:2241
23. a) Bertilsson L, Uvdal K, unpublished work
    b) Powder Diffraction File, (Berry LG, ed) Sets 1–5, publ. by the Joint Committee of Powder Diffraction Standards, Philadelphia 1974, p 571
24. Salaneck WR, Stafström S, Bredas JL, Andersson S, Bodö P, Ritsko J (1988) J Vac Sci Technol A 6:3134
25. Bertie JE, Eysel HH (1985) Appl Spectroscopy 39:392
26. Winter G (1980) Rev Inorg Chem 2:253
27. Katritzky AR, Simmons P (1959) J Chem Soc 1959:2051
28. Bellamy LJ (1975) The Infrared Spectra of Complex Molecules, Vol I, 3rd ed. Chapman and Hall, London, p 372
29. The Aldrich Library of FT-IR Spectra, ed I by CJ Pouchert (1985), Vol 1
30. Pine SH, Hendrickson JB, Cram DJ, Hammond GS (1981) Organic Chemistry (4th ed.), McGraw-Hill International Student Ed. Singapore, p 623
31. Persson N-O, Liedberg B (to be published)
32. Bain CD, Biebuyck HA, Whiteside GM (1989) Langmuir 5:723
33. Uvdal K, Bodö P, Liedberg B (1992) J Coll Interf Sci 149:162
34. Kemeny PC, Leckey RCG, Jenkin JG, Liesgang J (1974) J Electron Spectr and Rel Phenomena 5:881
35. Chiari B, Piovesana O, Tarantelli T, Zanazzi PF (1985) Inorg Chem 24:366
36. Åkerström S (1959) Arkiv Kemi 14:387
37. Hesse R, Jennische P (1972) Acta Chem Scand 26:3855
38. Melnik M, Parish RV (1986) Coord Chem Rev 70:157
39. Schwerdtfeger P, Boyd PDW, Burell AK, Robinson WT, Taylor MJ (1990) Inorg Chem 29:3593
40. Kittel C (1976) Solid State Physics, 5th ed, John Wiley & Sons, New York p 32

Authors' address:

N. O. Persson
Laboratory of Applied Physics, IFM
University of Linköping
S-581 83 Linköping
Sweden

Lipid arrangements in human stratum corneum

A. Bouwstra*), G. S. Gooris*), J. A. van der Spek#), and W. Bras+)

*) Pharmaceutical Technology, Center for Bio-Pharmaceutical Sciences, Leiden University, Leiden, The Netherlands
#) SERC Daresbury Laboratory/NWO, Warrington, United Kingdom
+) Technical Department, Leiden University, Leiden, The Netherlands

Abstract: The structure of human stratum corneum was investigated with small-angle x-ray scattering (SAXS). At room temperature the scattering curve was characterized by a strong intensity at low scattering vector ($Q < 0.8$ nm^{-1}) and two complicated diffraction peaks originating from a lamellar structure of the lipids. The lamellar lipid structure in the stratum corneum transformed to a disordered structure between 65° and 75°C, the same temperature region at which a thermal lipid transition occurred. After cooling down to room temperature a recrystallization of at least a part of the lipids took place, after which only one unit cell with a repeat distance of 13.4 nm could be detected. Comparison of the scattering curve of the stratum corneum after crystallization with the scattering curve of the stratum corneum before recrystallization leads to the conclusion that, in the original curve, the lipids are arranged in two unit cells with repeat distances of 6.4 and 13.4 nm. From model calculations it appears that the latter unit cell consists of more than one bilayer. The scattering curves of stratum corneum hydrated to various levels were measured. A change in the water content of stratum corneum between 6% w/w and 60% w/w (fully hydrated) did not result in swelling of the bilayers, but the scattering curve obtained with stratum corneum hydrated to 60% w/w differed from those at lower hydration levels: the scattering curve at 60% w/w showed only the diffraction peaks corresponding to a unit cell with a repeat distance of 6.4 nm.

Key words: Human stratum corneum; SAXS; lamellar lipid structure; scattering curve

List of abbreviations

d – repeat distance
λ = wave length
I = Intensity
Θ = scattering angle
Q = scattering vector
n = order of the diffraction peak
N = number of unit cells

Introduction

The stratum corneum is the outermost layer of the skin, which acts as the main barrier for diffusion of substances through the skin. The stratum corneum consists of flattened cells embedded in a matrix of lipids. The lipids are arranged in a lamellar phase and at least a part of its lipid bilayers are in the gel or the crystalline state.

Several studies have been carried out in order to gain information about the structure of lipids and proteins in stratum corneum. The reason for this being that lipid regions in the stratum corneum are supposed to act as permeation channels for drugs. Studies carried out on the structure of the stratum corneum includes, for example, thermal analysis [1–5], small angle x-ray scattering (SAXS) [6–10], wide angle x-ray scattering (WAXS) [10, 11], and Fourier transform infrared spectroscopy [12, 13].

With thermal analysis several transitions could be detected [1–5]. The transitions at lower temperature originated from lipids, while the transitions at higher temperatures were influenced by proteins or originated from denaturation of proteins. Most of the transitions are influenced by the hydration level in the stratum corneum: an increase in water content resulted in an increase in the average ratio peak height to full width at half maximum for the thermal transitions.

Friberg et al. [6] used (SAXS) to study the lipid arrangement in human stratum corneum. They found a very broad diffraction peak, from which they calculated a corresponding repeat distance in the range of 5.0 to 8.0 nm. White et al. [7] studied the lipid structure of murine stratum corneum with WAXS and SAXS. They observed a repeat distance of 13.1 nm. In a very recent study, Hui et al. [8] observed that upon hydration no swelling of the bilayers occurred in murine stratum corneum. With WAXS the lipid structure of murine stratum corneum was characterized by two diffraction lines located at 0.412 nm and 0.375 nm. Increase in temperature resulted in a disordering of the lipid structure. Bouwstra et al. [9] studied human stratum corneum with SAXS and observed a repeat distance of 6.5 nm, but could not exclude the existence of a repeat distance of approximately 13 nm. They also studied the effect of hydration on the lipid structure in the stratum corneum. No swelling of the bilayers was observed upon varying the hydration level between 6% to 40% w/w. Until now the differences in repeat distances between murine stratum corneum and human stratum corneum are not understood. Wilkes [11] studied the structure of human and rat stratum corneum with WAXS as a function of temperature. Depending on the orientation of the stratum corneum with respect to the primary beam, they found various diffraction peaks; positioning the neonatal rat stratum corneum normal to the primary beam resulted in two lipid bands at 0.42 nm and 0.37 nm, while halos were present at 0.46 nm and at 0.98 nm when passing the x-rays parallel to the surface of the stratum corneum. These halos are common for protein structures. In the case of human stratum corneum a third lipid reflection was observed corresponding to a repeat distance of 0.46 nm. At higher temperatures rat and human stratum corneum also showed a disordering of the lipid structure. In a very recent publication an extensive study has been published by J-C. Garson et al. [10]. They observed an oriented structure in human stratum corneum and could detect many diffraction peaks in the wide-angle region. One of the most important conclusions is the existence of a protein which is very similar to β keratin. They used a high flux synchrotron radiation source. In the small-angle region they observed repeat distances of 6.5 nm and 4.5 nm, and attributed these bands to lamellar structures consisting of different types of lipids.

Since detailed information of the lamellar structure in human stratum corneum is still lacking, a study has been carried out with SAXS. Measurements were carried out after heating the stratum corneum to various temperatures. To investigate the reversibility of the transitions, diffraction patterns were obtained after heating/cooling cycles. Because water is a very potent penetration enhancer, and the organization of the lipids at various water content is of fundamental interest, the lipid structure was also studied as a function of the hydration level in the stratum corneum up to 60% w/w.

Experimental

Preparation of stratum corneum samples

Abdomen skin obtained after surgical operation was dermatomed to a thickness of 200 μm. The stratum corneum was separated from the epidermis by digestion of a 0.1% tripsin in phosphate buffer saline (PBS) solution (pH = 7.4) at 37 °C for 12 h. The stratum corneum was subsequently treated with a 0.2% solution of trypsin inhibitor (Type 11-S from Soybeam; Sigma chemicals, The Netherlands) in PBS. The stratum corneum was dried and stored in a desiccator over silica gel. Before use, the stratum corneum was hydrated over salt solutions or water. The stratum corneum was hydrated to water contents varying between 6% w/w and 60% w/w. The water contents were measured by weighing the stratum corneum before {sc(hydr)} and after drying {sc(dry)}. The hydration levels were calculated as follows: {sc(hydr)} − sc(dry)]/sc(hydr) × 100%.

Small angle x-ray scattering

All measurements were carried out at the Synchroton Radiation Source at Daresbury Laboratories using station 8.2. This station has been built as part of a NWO/SERC agreement. The camera produces a highly collimated beam with a cross-section of $0.4 \times 4\ mm^2$ at the sample position. With the SRS operating at 200 mA and 2 GeV the x-ray intensity is approximately 4 1011 photons/s with λ (wave length) = 0.15 nm at the sample position. Smearing of the diffraction pattern due to the finite size of the x-ray beam is negligible. The sample to detector distance can be set between 0.2 to 4.5 m, thus enabling studies of systems with repeat distances $0.4 < d < 100$ nm. For data collection a multiwire quadrant detector was used. This detector can handle count rates up to 250 000 c/s. The detector system spatial resolution is 0.5 mm. The resulting resolutions in the repeat distance depend on the scattering angle Θ. Examples of the obtained resolution expressed in d are: 0.13 nm at $Q = 1\ nm^{-1}$ ($d = 6.3$ nm), 0.06 nm at $Q = 0.15\ nm^{-1}$ ($d = 4.2$ nm), and 0.015 nm at $Q = 0.3\ nm^{-1}$ ($d = 2.1$ nm), in which Q is the scattering vector defined as $4\pi \sin\Theta/\lambda$.

The detector definitely improves the signal-to-noise ratio at higher diffraction angle, compared with the previously used linear detector [9]. For all the experiments the sample-to-detector distance was set to 2.0 m. The diffraction patterns were normalized with respect to synchrotron beam decay and absorption of the sample. Corrections for positional inhomogeneity in the detector sensitivity were performed as well. No smoothing algorithms were applied to the data. In order to correct the curve for background scattering the scattering curve of the empty cell was subtracted from the scattering curves obtained with the sample cell filled with stratum corneum.

Calibrations were performed with the help of a wet rat tail collagen sample with a repeat distance of 67 nm.

One or two sheets of stratum corneum, the total weight being approximately 5 mg, were approximately randomly packed in a special designed temperature controlled sample cell. The volume of the inner compartment of the sample cell was $1 \times 1.5 \times 8\ mm^3$. The sample cell was sealed by a lid using two small screws. Between the lid and the sample cell a rubber ring was placed to prevent water loss during the measurement. The sample cell was placed in a sample holder. The temperature was measured and controlled by a thermocouple, which was placed in the sample holder near the sample cell. The sample was heated by an electric wire, which was located in the sample holder. The sample cell was equipped with two mica windows.

The scattering intensities are plotted as a function of the scattering vector Q. The exposure time for all measurements was 15 minutes. The stratum corneum sample was checked for damage which might occur from the primary beam. For that purpose the scattering data of stratum corneum were collected in 15 cycles, in each cycle data collection took place during 1 min. Comparing the various scattering curves no differences were detected. Therefore, no damage of the stratum corneum caused by x-rays was observed.

In order to obtain information about structural changes as function of temperature, stratum corneum samples were measured in a temperature range between 25° and 120 °C. In a second series of experiments the stratum corneum was heated to various temperatures, equilibrated for 1 h, and subsequently cooled to room temperature in order to crystallize the lipids. The stratum corneum was equilibrated for at least 48 h after which the scattering curves were measured. Possible degradation of the lipids was checked with thin-layer chromatography. In a third series of experiments the stratum corneum was reheated after recrystallization of the lipids from 120 °C. These experiments have been carried out with stratum corneum sheets oriented parallel to the incident beam.

Finally, the scattering curves of stratum corneum were measured as function of the water content. The hydration level in the stratum corneum varied between 6% w/w and 60% w/w. All the experiments were carried out with stratum corneum from one donor, except for the curves measured at various temperatures. These measurements have been carried out with stratum corneum from another donor.

Results

In Fig. 1a the scattering profile of human stratum corneum hydrated to 40% w/w is shown. The scattering curve is characterized by a large intensity at $Q < 0.8\ nm^{-1}$ and a broad diffraction peak at $Q = 0.98\ nm^{-1}$. The broad diffraction peak clearly exhibits a shoulder on the righthand side, indicating that it actually consists of two partially unresolved peaks. Sometimes a second very small shoulder has been detected at higher Q-value. At $Q = 1.85\ nm^{-1}$ a weak diffraction peak could be detected. The weak diffraction peaks exhibit a shoulder on the righthand side as well. In order to determine the source of the various peaks and the source of the strong scattering at low angle, the scattering pattern was compared with the pattern obtained with stratum corneum after extraction of the lipids with a chloroform/methanol mixture. The results are shown in Fig. 1b. After extraction both diffraction peaks disappeared while the strong scattering at low angle was still present. It seems that the diffraction peaks are caused by the lipids in the stratum corneum, while the strong scattering at low angle is due to the proteins in the corneocytes.

According to Braggs law, $n\lambda = 2d \sin \Theta$, a reciprocal relationship exists between the dimensions of the structure and the scattering angles.

In the case of a lamellar phase the Bragg reflections are located at equidistant positions in Q space, i.e., the reflections are located at Q_1, $Q_2 = 2Q_1$, $Q_3 = 3Q_1, \ldots, Q_n = nQ_1$, in which n is the order of the diffraction peak. The position of the nth order diffraction peak is directly related to the repeat distance d (length of the unit cell) according to

$$Qn = 2\pi\, n/d \,. \tag{1}$$

The scattering curve of human stratum corneum exhibits one strong diffraction peak. From the position of this peak the repeat distance can be calculated according to Eq. (1). Assuming that the main position of this peak is the first-order diffraction peak a repeat distance (d) of 6.4 nm was found. Substituting this value in Eq. (1), a Q-value of 1.96 nm^{-1} for the second-order peak was found, which actually is the Q-value at which the shoulder of the weak diffraction peak is positioned. It seems that the main position of the strong diffraction peak and the shoulder of the weak diffraction peak are caused by a lamellar structure with a repeat distance of 6.4 nm. A unit cell with a repeat distance of 6.4 nm cannot explain the source of the shoulder of the strong diffraction peak and the main peak of the weak diffraction doublet. These peaks should be based on another repeating unit; this point will be discussed below.

Another explanation for the diffraction peaks might be the presence of a unit cell with a repeat distance of 12.8 nm. In that case, the main position of the strong diffraction peak should be of second order. The main position of the weak diffraction peak should then be explained by the presence of cholesterol crystals, which exhibit a reflection at $Q = 1.88\ nm^{-1}$ ($d = 3.35$ nm). However, the third-order diffraction peak is calculated at $Q \approx 1.47\ nm^{-1}$, which is not at the position of the shoulder of the strong diffraction peak located at $Q = 1.38\ nm^{-1}$. It seems that this discrepancy is too large to justify a unit cell with a repeat distance of 12.8 nm.

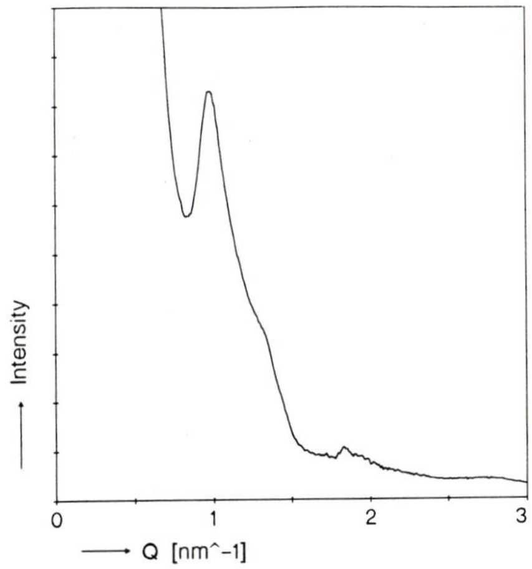

Fig. 1a. The scattering curve of human stratum corneum hydrated to 40% w/w. A strong diffraction peak and a weak diffraction peak could be detected, both consisting of a main position and a shoulder on the righthand side

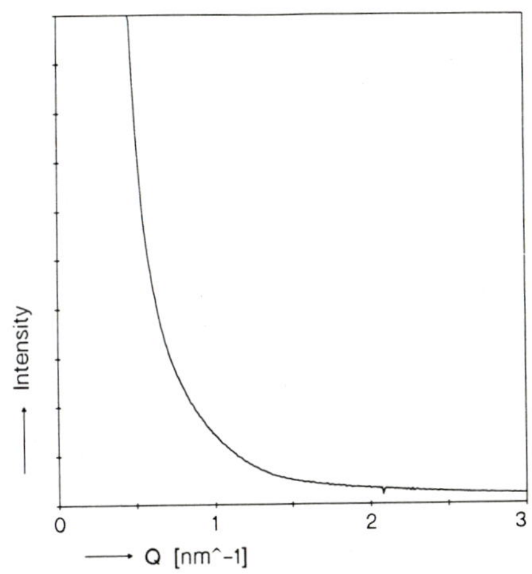

Fig. 1b. The scattering curve of human stratum corneum after extraction of the lipids

The structure of the stratum corneum as function of temperature

Various studies are known in which the thermal behavior of stratum corneum have been investigated. Although it is known that the occurring transitions are due either to lipids or to proteins in the stratum corneum, it is not exactly known whether the lipid transitions are due to melting of the alkyl chains or that a phase transition occurs in which the lamellar phase transforms to another phase. The thermal behavior of human stratum corneum, which is shown in Fig. 2, has been published before [5]. The first and second thermal transition appeared at 37 °C and 70 °C. Both peaks were ascribed to reversible phase transitions of the lipids in the stratum corneum. The third transition (85 °C) is probably due to lipids associated with the proteins in the stratum corneum. The transition was only reversible in the cases that heating did not exceed the denaturation temperatures. Finally, the fourth irreversible transition is due to the denaturation of the proteins in the stratum corneum.

In order to obtain more information about the origin of the thermal transitions of the lipids the scattering curves were measured as a function of temperature. The results are shown in Figs. 3a and b. The scattering curve measured at 60 ° does not differ significantly from that measured at 25 °C, indicating that no detectable disordering of the lamellae or a change in *d* occurred during the thermal transition.

The scattering curve obtained at 67 °C, which is in the middle of the temperature range of the second thermal transition, differed from those obtained at lower temperatures. Both the main diffraction peak and its shoulder were significantly reduced in intensity. At 75 °C the main diffraction peak and the weak diffraction peak completely

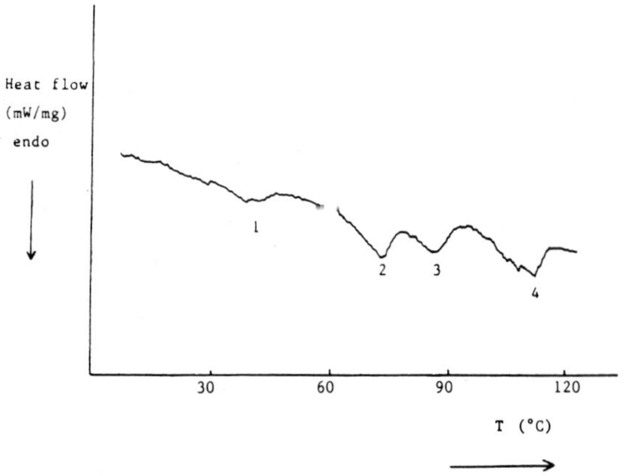

Fig. 2. The thermal transitions of human stratum corneum hydrated to 20% w/w

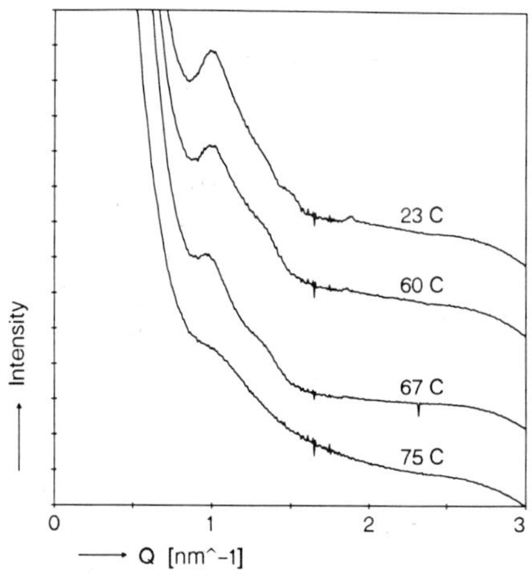

Fig. 3a. The scattering profile of human stratum corneum hydrated to 20% w/w at various temperatures. The temperatures have been indicated in the figure

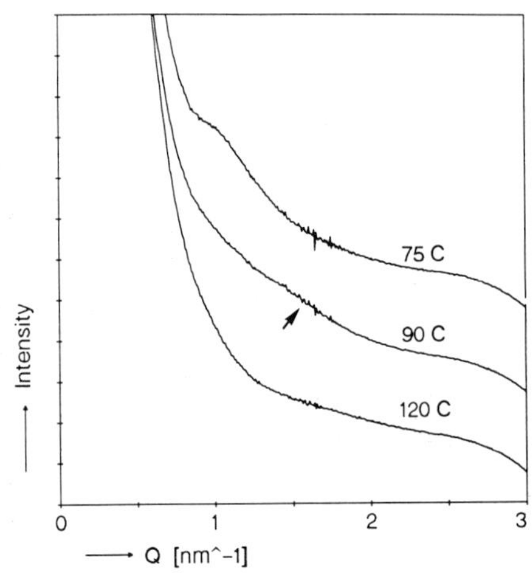

Fig. 3b. The scattering profile of human stratum corneum hydrated to 20% w/w at various temperatures. The temperatures have been indicated in the figure. At 90 °C the scattering curve exhibits a very poor "peak", as indicated by an arrow

disappeared. Only a shoulder on the descending scattering curve remained at $Q = 1\ nm^{-1}$. It seems that the system underwent a phase change in this temperature region. Since no new diffraction peaks appeared a disordering of the lamellar phase is the most likely explanation; whether this is disorder of the first or second kind, or whether it is based on undulations of the bilayers cannot be determined from the data.

The origin of the shoulder on the descending curve still present at 75 °C is not fully understood, but from the thermal analysis results it could be concluded [5] that it can be caused by lipids linked to proteins in the stratum corneum. As in the case of the third thermal transition [5] the shoulder on the scattering curve is sensitive to protein denaturation. After recrystallization of the lipids from 120 °C the shoulder is not present on the scattering curve at 75 °C, as is shown in Fig. 5.

The presence of only a shoulder on the scattering curve and the absence of higher order diffraction peaks implies that the long-range order completely disappeared. In fact the shoulder in the diffraction curve could possibly be caused by the presence of one well-ordered lipid layer; this could be the corneocyte lipid envelope [14]. These lipids, which are linked to the corneocyte envelope, should still be present after extraction of stratum corneum. However, the shoulder on the descending scattering curve is not observed after methanol/chloroform extraction. This might be caused by a disordering of the lipid envelope during extraction.

Between 75 ° and 90 °C, the temperature region of the third thermal transition, the shoulder of the descending scattering curve disappeared, but a small very broad peak appeared at higher Q-value (see arrow in Fig. 3b). The disappearance of the shoulder confirms the suggestion that this shoulder in the scattering curve can be correlated with the third thermal transition. At 120 °C, which is above the denaturation temperature of the proteins, only a descending scattering curve was observed.

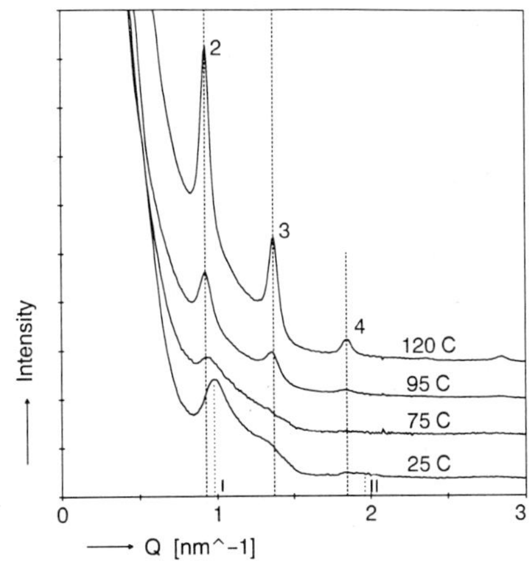

Fig. 4a. The scattering curve of human stratum corneum after heating and cooling to room temperature. The temperatures to which the stratum corneum have been heated are indicated in the figure. $n = 1, 2$, etc., refers to the orders of diffraction peaks of the cell with a repeat distance of 6.4 nm. $n =$ I, II refers to the diffraction orders of the unit cell with a repeat distance of 6.4 nm

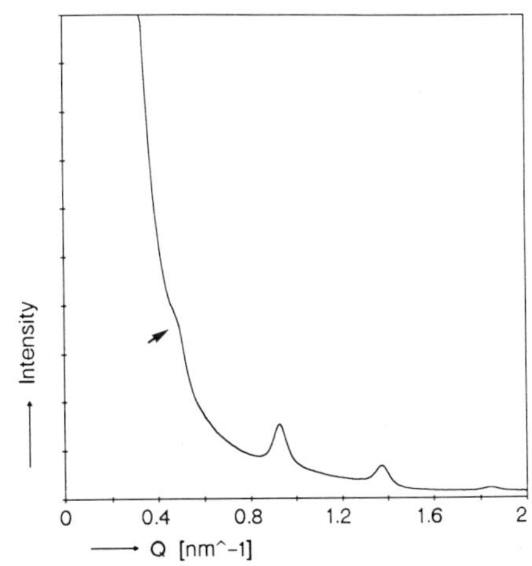

Fig. 4b. The scattering curve of human corneum after heating to 120 °C and cooling to room temperature. The arrow indicates the first-order diffraction peak

Lipid structure in stratum corneum after recrystallization

In order to find a more detailed interpretation for the scattering curve, the lipids were allowed to recrystallize after heating to various temperatures and subsequent cooling to room temperature. The changes in structure of the stratum corneum lipids was checked after heating to 120 °C and then cooling to room temperature; no changes in the lipid structures could be observed using TLC. After recrystallization the stratum corneum was equilibrated for at least 48 h. The resulting scattering curves are shown in Fig. 4a. The curve labeled 25 °C is the original scattering curve of human stratum corneum without heating. The main position of the peak at $Q = 0.98\ \text{nm}^{-1}$ and the shoulder of the weak diffraction peak at $Q = 1.96\ \text{nm}^{-1}$ can be identified as, respectively, the first- and second-order reflection from a unit cell with a length of 6.4 nm. The source of the shoulder on the righthand side of the strong diffraction doublet and the source of the main position of the weak diffraction peak are not known.

First, the results obtained after heating to 120 °C and cooling to room temperature will be interpreted. In this curve at least five diffraction peaks were observed, of which three are quite strong.

These peaks are located at equidistant positions. The exception is the distance between the position of the primary beam (located at $Q = 0\ \text{nm}^{-1}$) and the first diffraction peak. This distance is twice the other interpeak distances. Moreover, the position of the first peak, located at $Q = 0.94\ \text{nm}^{-1}$ has significantly shifted with respect to the position of the main diffraction peak in the original scattering curve. It was concluded that after heating to 120 °C and cooling the lipids recrystallized in a lamellar structure which differed from that obtained from the original scattering curve, and that the diffraction peak located at $Q = 0.94\ \text{nm}^{-1}$ is a second-order diffraction peak. For the length of the unit cell calculated from this second-order diffraction peak, a value of 13.4 nm was found. In Fig. 4b the scattering curve after heating to 120 °C has been plotted at another scale. In this curve a slight change in the slope of the scattering curve can be observed at approximately $Q = 0.5\ \text{nm}^{-1}$. This is approximately the position at which the first order diffraction peak was expected. The change in the slope at $Q = 0.5\ \text{nm}^{-1}$ is also observed in the original scattering curves, as shown in Fig. 6b. It seems that the first-order diffraction peak is present, but the intensity of this peak is very low, which is quite remarkable. An explanation for this phenomenon is given below, using a model with several bilayers in one unit cell. The steeply descending background makes it difficult to distinguish the peak from the background. After comparing the scattering curve obtained after heating to 120 °C with the original scattering curve, the following observations can be made. The third-order diffraction peak of the 120 °C curve is located at the same position as the shoulder on the right-hand side of the main diffraction peak. In addition, the fourth-order diffraction peak in the 120 °C curve is located at the main position of the weak diffraction peak of the original scattering curve. From these observations it was concluded that the two peaks (doublets) in the original scattering curve are related to two unit cells, with repeat distances of 6.4 and 13.4 nm, respectively. The scattering at the main position of the strong diffraction peak originates from a unit cell with a repeat distance of 6.4 nm, while the intensity at the shoulder of the strong diffraction peak and the main position of the weak diffraction peak results from a unit cell with a repeat distance of 13.4 nm.

From the full width at half maximum the mean number of bilayers has been calculated with the Scherrer Equation [15]. The average number of unit cells appeared to be 6. In the second-, third-, and fourth-order diffraction peaks the full width at half maximum does not differ within the experimental error. Since broadening of the peaks caused by disorder of the second kind increases as a function of Q^4 [14], it seems that this disordering is absent after recrystallization.

In Fig. 3a it is shown that the shoulder as well as the main position of the main diffraction peak decreased in intensity between 60 ° and 75 °C. This indicates that in both unit cells the lipid arrangement changed in the same temperature range. This directly implies that it is impossible to distinguish between the lipids in the two unit cells by thermal analysis.

The scattering curve obtained after heating to 95 °C and subsequent cooling to room temperature showed a similar scattering profile as that obtained after heating to 120 °C and cooling, although the intensities of the individual peaks of the former curve are lower. Heating to 75 °C and cooling to room temperature resulted in another profile of the scattering curve: only one diffraction peak was observed, of which the main position was already shifted towards the position of the second-order diffraction peak in the 120 °C curve. This indicates that the lipids crystallized in a lattice with the same unit cell as the 120 °C, but that the long-range order is lower compared to stratum corneum heated to higher temperatures. Comparing the curve in Fig. 3a obtained after heating to 75 °C with the corresponding scattering curve in Fig. 5a after the cooling process leads to the following observation. It is likely that the scattering profile caused by the recrystallization of the lipids is superimposed on the scattering curve measured at 75 °C. The latter curve exhibits the shoulder on the descending scattering curve. If this is true, then the shoulder on the scattering curve observed at 75 °C is already present at lower temperatures, but detection at lower temperatures is impossible due to the fact that it will be obscured by the strong diffraction peak in the same Q region.

After recrystallization of the lipids from 120 °C, the scattering curves have been measured as a function of temperature. The lamellar structure became disordered between 55 ° and 75 °C, which is in the same temperature range as before recrystallization. However, the shoulder still present at 75 °C during the first heating procedure disappeared. This is in excellent agreement with the results obtained using thermal analysis after recrystallization of the lipids from 120 °C. During reheating the third thermal transition is shifted to the same temperature region at which the second thermal transition was located [5].

The relative intensities of the various peaks after recrystallization of the lipids from 120 °C depicted in Figs. 4a and 5 are different. This is probably due to a difference in orientation of the stratum corneum. The scattering curves depicted in Fig. 4 are measured with stratum corneum in a parallel orientation with respect to the primary beam. This implies that the exponent of the Lorenz factor is smaller than 2, whereas in a completely disoriented structure the Lorenz factor is Q^2. A decrease in the exponent of the Lorenz factor results in relatively higher intensities of the diffraction peaks at higher Q values. This is in agreement with the observed differences between Figs. 4a and 5.

Structure of stratum corneum as function of the hydration level

Since water is an effective penetration enhancer [16] the changes in the lipid structure induced by changes in the water content of the stratum corneum are of fundamental interest. For this reason, scattering curves were obtained of stratum corneum samples hydrated between 6% w/w (almost dry) and 60% w/w (fully hydrated); the curves are shown in Fig. 6a. The position of the main diffraction peak does not change between 6% w/w and 40% w/w hydration level. A small shift in the top of the peak is observed in the case of 60% w/w hydration, but this could also be due to the lower intensity of the peak, since the peak is located on a very steep descending scattering curve.

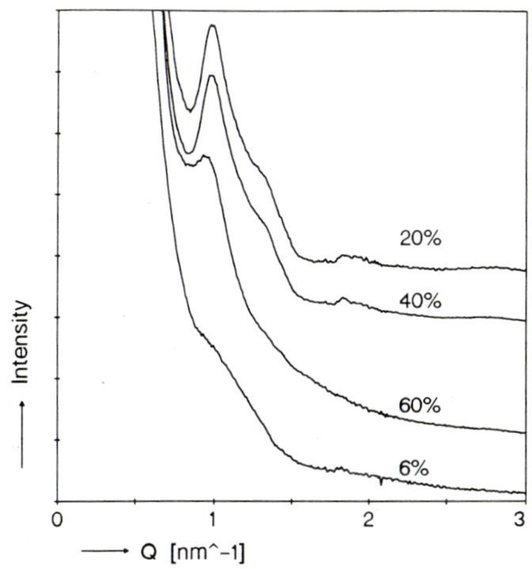

Fig. 6a. The scattering curve of human stratum corneum hydrated to levels varying between 6% w/w and 60% w/w. At 60% w/w hydration the shoulder on the strong diffraction peak disappeared

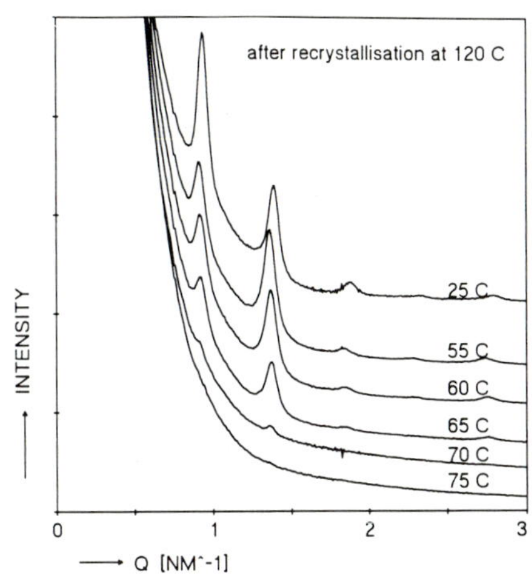

Fig. 5. The scattering curves after recrystallization of the lipids from 120 °C. After heating to 120 °C the stratum corneum was cooled to room temperature and heated to the temperatures indicated in the figure. During the experiments the stratum corneum was oriented with respect to the primary beam

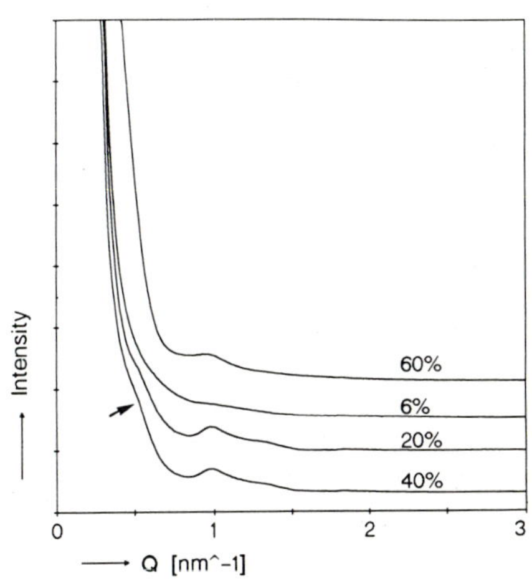

Fig. 6b. The scattering curve of human stratum corneum hydrated to various levels varying between 6% w/w and 60% w/w. The curves have been plotted at another scale to show the first-order diffraction peak (shoulder at the descending scattering curve, see arrow) of the unit cell with a repeat distance of 13.4 nm. This shoulder disappeared at 60% w/w hydration

This indicates that the repeat distance does not change upon hydration and that no swelling of the bilayers occurs. This is quite remarkable, since the hydration level varied between 6 and 60% w/w. It seems that water is intercallated into the bilayers for only very small amounts, and that most of the water is absorbed by the corneocytes in the stratum corneum. This is in accordance with recent findings of Mak et al. [17], who did not observe any influence of water content in the statum corneum in the infrared C-H absorbance determined by FTIR. Another possibility is the forming of water-rich regions in intercellular regions in the stratum corneum (phase separation), but this is not confirmed by electron microscopy observations. The intensity of the peak increases if the hydration level increases from 6% w/w to 40% w/w, indicating a stronger ordering of the lamellae at higher water contents. A part of the answer can be given after examining the scattering curves drawn in Figs. 6a and 6b. It was found that the curve of 60% w/w hydrated stratum corneum does not have the shoulder on the righthand side of the main diffraction peak (Fig. 6a), and that the change in the slope of the scattering curve at $Q = 0.5\ \mathrm{nm}^{-1}$ is not present (Fig. 6b). Additionally, the intensity of the main position of the weak diffraction peak decreased upon hydration from 40 to 60% w/w. From these observations, it was concluded that the unit cell with a repeat distance of 13.4 nm disappeared upon hydration of the stratum corneum from 40 to 60% w/w. Due to the low intensity of the main diffraction peak in the 60% w/w curve it is more likely that the lipids originally arranged in the unit cell of 13.4 nm transformed to a disordered lipid structure instead of undergoing a phase transition into a lipid arrangement of the 6.4 nm unit cell.

Model calculations

To obtain a better understanding of the arrangement of the lipids in the unit cell, model calculations were carried out for the lipids located in a unit cell with a length of 13.4 nm, since in this case the relative intensities of six diffraction orders are known. For the calculations, the scattering curve measured after recrystallization of the lipids from 120 °C depicted in Fig. 5a was used, since the curve of untreated stratum corneum does show only partly resolved peaks. In the model calculations disorder of the first and second kind [17] was neglected. Moreover, it was assumed that the lipids are randomly orientated in the sample cell. According to Levine [18], in general, the scattering intensity I as a function of the scattering vector Q from a stack of lamellae is given by

$$I(Q) = \left\{ \int_0^d \rho(x) \exp(-iQx)\, dx \right\}^2 \times \frac{\sin^2\left(\frac{1}{2}NQd\right)}{\sin^2\left(\frac{1}{2}Qd\right)} . \tag{2}$$

In this equation N is the number of bilayers, x the spatial coordinate, $\rho(x)$ is the electron density of an individual bilayer, and d the repeat distance. The first part of this equation is the square of the Fourier transform of the electron density of the bilayer, the so-called form factor. The second part of the equation is the interference factor, which only depends on the repeat distance and the number of bilayers.

The most striking characteristic of the scattering curve under consideration is the low intensity of the first-order diffraction peak. Model calculations in previous studies [7, 9] already showed that a low intensity of the first-order diffraction peak can only be achieved by assuming a unit cell with more than one bilayer, thus allowing only small differences in the electron densities of the bilayers. Therefore, in this particular case the form factor in Eq. (2) is the Fourier transform of the electron density of a unit cell which might consist of more than one bilayer, which is, in fact, the electron density of the repeating unit perpendicular to the lamellae. In the second term in Eq. (2) the number of bilayers N is replaced by the number of unit cells, which is the repeating unit. The calculated intensities were divided by Q^2, the Lorenz factor, to make the calculated intensities directly comparable with the measured intensities. Using Q^2, it is assumed that the lipids are randomly oriented in the sample cell. From the 120 °C curve in Fig. 5a the height of the peaks was calculated by subtraction of the baseline from these peaks. The resulting experimental intensities are given in Table 1. For the first-order diffraction peak, we assumed a low but non-zero intensity. In the calculations the mean number of unit cells $\langle N \rangle$ was set to 6, according to the number of unit cells obtained with the Scherrer equation. The width of the calculated peaks was in

Table 1. The calculated intensities are compared with the experimental data of stratum corneum after heating to 120 °C and cooling to room temperature. The intensities of the second-order diffraction peak of the calculated and experimental data are set to 1 (scaling), n is the order of the diffraction peak

n	Experimental intensities	Calculated intensities
1	–	0.42
2	1	1
3	0.49	0.45
4	0.08	0.08
5	0.01	0.04
6	0.02	0.02

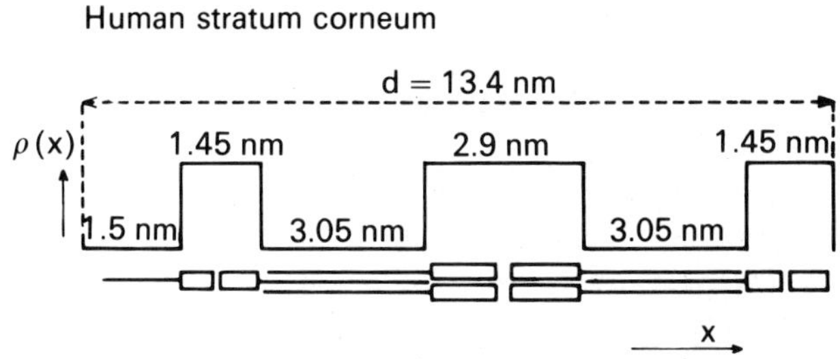

Fig. 7. The electron density profile of the unit cell with a repeat distance of 13.4 nm. Three alkyl chain regions are found, of which two are identical

good agreement with the experimentally obtained peak width. In the ideal electron density profile of the bilayers, used in the calculations, the number of unit cells does not influence the relative intensities of the diffraction peaks. By applying variations to the electron density profile the intensities of the diffraction peaks were fitted. The best fit was obtained with the electron density profile shown in Fig. 7. The calculated intensities can be found in Table 1. The obtained electron density model possesses three alkyl chain regions, of which two are identical. It was impossible to fit the diffraction peaks with a model with only two identical alkyl chain regions. A similar model for the arrangements of the alkyl chains with two different hydrocarbon regions was also proposed by Swarzendruber et al. [20] as a sequence of one narrow and two broad alkyl chain regions. The widths of the two different hydrocarbon regions in our model are 1.5 nm and 3.05 nm. These differences in length agree very well with the lipid arrangement proposed by Swarzendruber: the short alkyl chains of the ceramides (approximately 13 C atoms in the hydrocarbon region, completely extended 1.7 nm in length) are intercallated in the bilayers with a width of 1.5 nm, while the longer alkyl chains of the same molecules, with a mean length of approximately 26 C atoms [21] (completely extended 3.5 nm in length) are embedded in the hydrocarbon regions of adjacent bilayers with a width of 3.05 nm. Assuming that at least two alkyl chain layers are present in one bilayer, and that the width of the hydrocarbon regions is only 3.05 nm, this unequivocally shows that the alkyl chains in the lipid layers interpenetrate the bilayers.

However, the head group regions, which are characterized by a higher electron density, are very large, especially those which are located between the two adjacent hydrocarbon regions with equal width, i.e., 2.9 nm (see Fig. 7). It was impossible to simulate the intensities of the diffraction peaks with smaller head group regions, because that resulted in a too high intensity of the fifth-order diffraction peak. The width of the headgroup regions cannot be caused by the lipids present in the stratum corneum, since the main components, ceramides, free fatty acids, triglycerides, and cholesterol all consist of small head groups. The large head group regions (1.5 nm and 2.9 nm) may be explained by proteins intercallated in the lipid bilayers causing an extension of the high electron density regions.

Discussion

It is reasonable that the ultrastructure of the lipids in human stratum corneum consists of domains built up of two unit cells with repeat distances of 6.4 nm and 13.4 nm, respectively. Due to the various lipids present in the stratum corneum, of course, it is possible that more than two lamellar phases are present, but it seems that the scattering intensity is too low to detect these phases. Very recently, several higher order lamellar lipid reflections were found using WAXS, which can also be explained by the two unit cells [22]. A part of these results is shown in Fig. 8, which depicts the WAXS reflections of human stratum corneum. These results have been obtained with sheets of stratum corneum in parallel position. From the positions of the reflections at low angle (close to the beam stop), the corresponding distances have been calculated

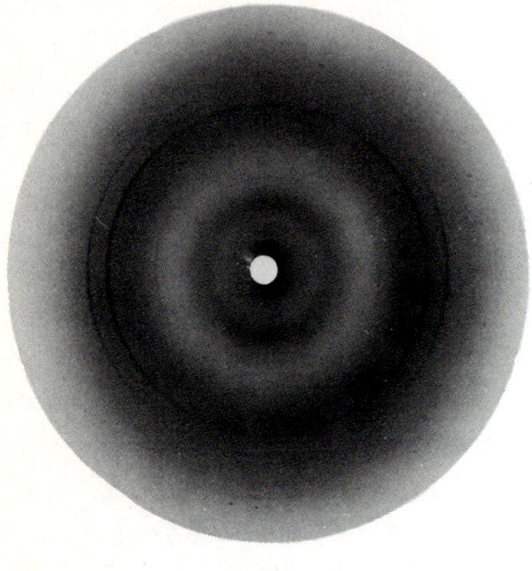

Fig. 8. The wide-angle reflections of human stratum corneum. The stratum corneum was oriented parallel to the primary beam. The two sharp reflections (indicated by an arrow) correspond to distances of 0.417 nm and 0.378 nm; these can be explained by an orthorhombic lateral packing of the lipids at room temperature [22].

for stratum corneum from two different donors. The results are shown in Table 2, together with the expected higher order reflection of the lamellar structures. In these calculations repeat distances of 6.8 nm (donor I) and 13.4 nm and 6.4 nm (donor II), respectively, were used. A repeat distance of 6.8 nm for donor I was used since the scattering curve of stratum corneum from donor I was an acceptional: no shoulder on the righthand side of the strong diffraction peak was observed, and the peak position corresponded with a repeat distance of 6.8 nm. Comparing the values of the observed reflections with those of the higher order reflections calculated from the lamellar structures with repeat distances of 6.8 nm, 6.4 nm and 13.4 nm (see Table 2), let us conclude that the reflections at low angle in Fig. 8 can be identified as higher order diffraction peaks that are at least partly due to the lamellar structures. These recent results confirm the results already obtained by SAXS that two lamellar structures are present in human stratum corneum, although there might be large interindividual differences between stratum corneum from different donors.

Whether both domains are randomly distributed in human stratum corneum or that each domain is located in certain regions of the stratum corneum is not yet clear. Garson et al. [10] revealed lamellar spacings of 6.5 and 4.5 nm for human stratum corneum. They did not find a repeat distance of 13.4 nm since the small first-order peak of this lamellar structure was not observed in their study, but their results showed that both unit cells are stacked perpendicular to the surface of the stratum corneum. They also stated that the 6.5 nm and 4.5 nm repeat distances are in accordance with the model proposed by Swarzendruber et al. [20]. In our opinion, this is open to discussion, since Swarzendruber did not propose stacks of lamellae with repeat distances of 6.5 nm and 4.5 nm, but instead, the so-called Landmann units which consist of a regular pattern of broad and narrow alkyl chain regions in the sequence broad, narrow, broad, broad, narrow, broad, broad, narrow, etc., which may result in longer repeat distances than shown in our model calculations in this paper.

The large head group regions in the electron density model are striking; they might be caused by the presence of proteins. The calculations presented in this paper should be considered as a first attempt to elucidate information about the arrangement of the lipids in one unit cell. In the future, more detailed calculations will be carried out.

Table 2. The reflections obtained at low angle from two different donors together with the explanations. s = strong, w = weak, m = medium intensity

d (nm)	Donor	Intensity	Explanation
2.28	I + II	arc, s	2.12: 6.4(3)*), 2.27: 6.8(3)*), 2.23: 13.4(6)*)
1.98	II	arc, vw	1.91: 13.4(7)
1.70	I + II	arc, w	1.60: 6.4(4), 1.70: 6.8(4), 1.68: 13.4(8)
1.51	II	arc, m	1.48: 13.4(9)
1.39	I + II	arc, w	1.28: 6.4(5), 1.36: 6.8(5), 1.34: 13.4(10)

*) For example, 2.13: 6.4(3) = the calculated diffraction line is 2.13 nm, which is the third-order diffraction peak of the lamellar phase with repeat distance of 6.4 nm

Garson et al. [10] also observed various peaks (shoulders) at the range from 4–4.5 nm. In their paper, it is not clear whether all these peaks are found in one piece of stratum corneum, or if the peak positions vary in stratum corneum from different donors. In our experiments the intensity of this shoulder varies to a large extent between stratum corneum from different donors; sometimes, a second shoulder is detected. This second shoulder is shown in Fig. 5a on the scattering curve obtained at 23 °C. These variations between different donors and the additional shoulder may explain the various peaks Garson et al. obtained.

In murine stratum corneum only one unit cell with a repeat distance of 13.1 nm was observed [7]. This undoubtly results in the conclusion that stratum corneum from different sources can possess quite different lipid structures. The unit cell in murine stratum corneum probably contains also more than one lipid bilayer [7]. In the case of murine stratum corneum, an increase in temperature already leads to a diffuse pattern at 40 °C. This was not observed with human stratum corneum at 45 °C. The differences in lipid structure between murine and human stratum corneum cannot be detected with thermal analysis, since both unit cells undergo a phase transition in the same temperature region. Another difference between the two types of stratum corneum is the recrystallization behavior of the lipids. In murine stratum corneum the lipids recrystallized in exactly the same lattice as before the heating/cooling cycle, while recrystallization of the lipids in human stratum corneum resulted in a different structure, as could be concluded from the diffraction pattern. The difference may originate from the fact that human stratum corneum lipids possess the two types of unit cells, and that the lipids only recrystallize in one type. Since remarkable differences exists between murine and human stratum corneum, it should be interesting to investigate other types of stratum corneum.

Concerning the thermal transition at 40 °C, it seems that this transition is only due to a change in the packing of the lipids in the lamellae. This is confirmed by investigations of Wilkes et al. [11]. In their studies, they observed a disappearance of the lipid band corresponding to a d of 0.37 nm. According to White et al. [7], this could be a change from an orthorhombic structure to a hexagonal packing. The orthorhombic structure was also observed in human stratum corneum (see Fig. 8).

Acknowledgement

We thank Dr. M. Ponec for performing the TLC experiments.

References

1. van Duzee BF (1975) Thermal analysis of human stratum corneum. J Invest Dermatol 65:404–408
2. Knutsen K, Potts RO, Guzek GM, McKie JE, Lambert WJ, Higuchi WI (1985) J Contr Rel 2:67–87
3. Golden GM, Guzek DB, Kennedy AH, McKie JE, Potts RO (1987) Stratum corneum lipid phase transitions and water barrier properties. Biochemistry 26:2382–2388
4. Golden GM, Guzek DB, Harris RR, McKie JE, Potts RO (1986) Lipid thermotropic transitions in human stratum corneum. J Invest Derm 86:255–259
5. Bouwstra JA, Peschier LJC, Brussee J, Boddé HE (1989) Effect of *N*-alkylazocycloheptane-2-ones including azone on the thermal behaviour of human stratum corneum. Int J Pharm 52:47–54
6. Friberg SE, Osborne DW (1985) Small-angle x-ray diffraction patterns of stratum corneum and model structure for its lipids. J Disp Sci Techn 6:486–495
7. White SH, Mirejovsky D, King GI (1988) Structure of lamellar lipid domains and corneocytes envelopes of murine stratum corneum. An x-ray diffraction study. Biochemistry 27:3725–3732
8. Hou SYE, Mitra AK, Menon GK, White SH, Ghadially R, Elias PM (1991) Membrane structure in normal and essential fatty acid deficient stratum corneum: Characterization by ruthenium tetroxide staining and x-ray diffraction. J Invest Dermatol 97:215–223
9. Bouwstra JA, de Vries MA, Gooris GS, Bras W, Brussee J, Ponec M, Thermodynamic and structural aspects of the skin barrier. J Contr Rel
10. Garson, J-C, Doucet J, Leveque J-L, Tsoucaris G (1991) Oriented structure in human stratum corneum revealed by x-ray diffraction. J Invest Dermatol 96:43–49
11. Wilkes GL, Nguyen A-L, Wildhauer R (1973) Structure properties relations of human and neonatal rat stratum corneum Biochim Biophys Acta 304:267–275
12. Klimish HM, Chandra G (1986) Use of Fourier transformed infrared spectroscopy with attenuated total reflectance for in vivo quantitation of polydimethylsiloxaness. J Soc Comet Chem 37:73–87
13. Potts RO, Guzek DB, Harris RR, McKie JE (1985) A noninvasive, in vivo technique to quantitatively measure water concentration of the stratum corneum using attenuated total-reflectance infrared spectroscopy. Arch Dermatol Res 277:489–495
14. Wertz PW, Swarzendruber DC, Kitko DJ, Madison KC, Downing DT (1989) The role of corneocyte lipid envelopes in cohesion of stratum corneum 93:169–172
15. Alexander LE (1969) X-ray diffraction methods in polymer science. Wiley-Interscience, New York

16. Barry BW (1983) Dermatologic formulations. Dekker, New York/Basel
17. Mak VHW, Potts RO, Guy RH (1991) Does hydration affect intercellular lipid organization in the stratum corneum? Pharm Res 8:1064–1065
18. Blaurock AE (1982) Evidence of bilayer structure and of membrane interactions from X-ray diffraction analysis. Biochim Biophys Acta. 167–207
19. Levine YK (1973) X-ray diffraction studies of membranes.Prog Surf Membr Sci 3:279–352
20. Swartzendruber DC, Wertz PW, Kitho DJ, Madison MD, Downing DT (1989) Molecular models of the intercellular lipid lamellae in mammalian stratum corneum. J Invest Dermatol 92:251–257
21. Wertz PW, Downing DT (1983) Ceramides of pig epidermis: structure determination. J Lipids Res 24:759–765
22. Bouwstra JA, Gooris GS, Salomons-de Vries MA, Bras W, A wide angle x-ray study (excepted in Int J Pharm)

Authors' address:

J. A. Bouwstra
Pharmaceutical Technology
Center for Bio-Pharmaceutical Sciences
Leiden University
P.O. Box 9502
NL-2300 RA Leiden
The Netherlands

Progress in Colloid & Polymer Science Progr Colloid Polym Sci 88:123–130 (1992)

Stabilization of fluorocarbon emulsions by sugar-derived perfluoroalkylated surfactants and co-surfactants

J. G. Riess, J. Greiner, S. Abouhilale, and A. Milius

Laboratoire de Chimie Moléculaire, Associé au CNRS, Université de Nice-Sophia Antipolis, Nice, France

Abstract: Fluorocarbon emulsions have considerable potential as injectable oxygen carriers (blood substitutes, in particular), contrast agents for diagnosis, and drug-delivery systems.

One approach to reducing the interfacial tension which opposes the dispersion of fluorocarbons in water, and to contributing to the stabilization of their emulsions, is to use fluorophilic surfactants. Several families of perfluoroalkylated, sugar-derived surfactants with modular molecular structures, including maltosides, trehalose fatty acid monoesters, and glucose phosphate esters were investigated to this end. Relatively minor structural differences were found to cause widely different behaviors. Very stable emulsions of perfluorodecalin (FDC) and perfluorooctylbromide (PFOB, perflubron) were obtained with some of them; the neutral ones led to considerable synergistic stabilization of such emulsions when a polaxomer (Pluronic F-68) was used as the primary surfactant, while the anionic ones developed such a synergistic effect with egg-yolk phospholipids (EYP). Some also had a significant impact on particle sizes.

Key words: Fluorocarbon emulsion; perfluoroalkylated surfactant; synergy; Pluronic F-68; phospholipids; maltoside; trehalose fatty monoester; glucose phosphate ester

Introduction

Our aim is to gain improved mastery over the characteristics and stability of injectable fluorocarbon emulsions. Such emulsions are destined to be used as in vivo oxygen-delivery systems (including "blood substitutes"), contrast agents, and drug-delivery systems [1, 2]. One of the key approaches to such mastery lies in the design and evaluation of highly fluorinated surfactants [3]. Constraints imposed by the intravascular destination of these compounds include high definition, purity, biocompatibility, and excretion. The surfactants used so far for this purpose are Pluronic F-68 and egg-yolk phospholipids (EYP).

A range of new highly fluorinated sugar-derived surfactants was developed, primarily for emulsifying fluorocarbons, but also, more generally, to answer a need for highly surface-active, yet biocompatible surfactants [4–7]. Sugar-derived surfactants have raised increased interest [8], yet there appears to have been no perfluoroalkylated sugar derivatives reported prior to those investigated here. The molecular design chosen for the latter is modular so as to allow stepwise variations in properties [3]. Their polar head is comprised of mono- or disaccharides and their hydrophobic tail consists of perfluoroalkylated chains and hydrocarbon spacers of various lengths. These two parts are connected either through a glycoside bond or through a carboxylic or a phosphoric ester linkage. This arrangement allows considerable versatility and stepwise adjustment of the surfactants' characteristics (Fig. 1).

The choice of sugars as polar heads was motivated by their biological tolerance, but also because we were looking for co-surfactants capable of acting synergistically with Pluronic F-68-type block polymers. While such polymers stabilize emulsions primarily through steric effects, the fluorinated sur-

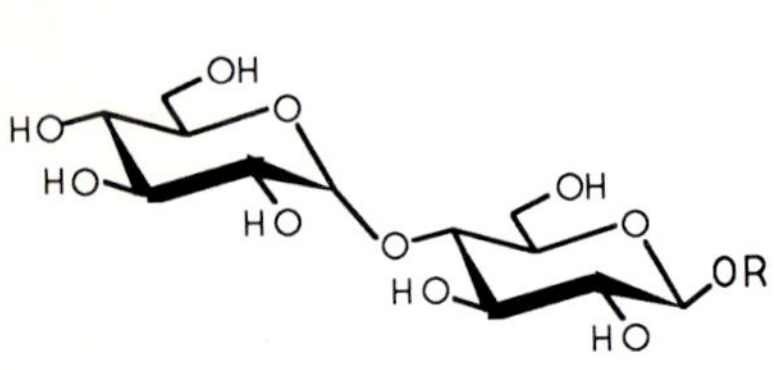

1 $C_8F_{17}(CH_2)_2$
2 $C_6F_{13}(CH_2)_2$
3 $C_6F_{13}CH{=}CH(CH_2)_3$
4 $C_6F_{13}CH{=}CH(CH_2)_9$
5 $CH_3(CH_2)_9$
6 $CH_3(CH_2)_{11}$

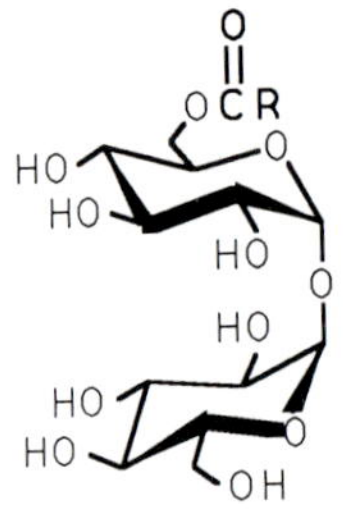

7 $C_8F_{17}(CH_2)_2$
8 $C_6F_{13}(CH_2)_4$
9 $C_4F_9(CH_2)_{10}$
10 $CH_3(CH_2)_9$

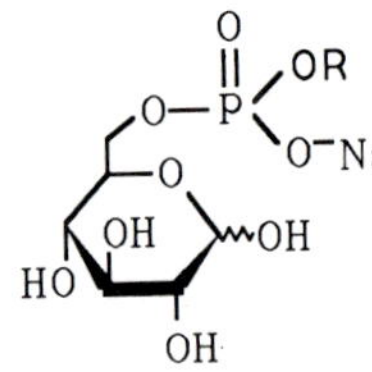

11 $C_6F_{13}(CH_2)_2$
12 $C_8F_{17}(CH_2)_2$
13 $CH_3(CH_2)_9$

Fig. 1. Molecular structure of some typical perfluoroalkylated sugar-derived surfactants

factant was expected to lower the interfacial tension and to improve the affinity of the interfacial film for the fluorocarbon phase. Interaction between the two surfactants was expected to occur through hydrogen bonding between the hydroxyl groups of the sugar molecule and the ether oxygens of the copolymer. Trehalose was selected for its symmetry, which made a one-step synthetic approach more feasible [6]. Phosphate diester linkages [7] were used to explore the effect of introducing a negative charge and in view of inducing synergistic stabilization effects with phospholipids.

This paper will emphasize, with some typical examples, the considerable differences in emulsifying and co-emulsifying abilities that were observed for some closely related members of these families of sugar-derived perfluoroalkylated surfactants. The surfactants were tested in three basic situations: as sole emulsifiers of fluorocarbons, as co-emulsifiers with Pluronic F-68, and as co-emulsifiers with EYP. Perfluorodecalin (FDC) was usually chosen as the fluorocarbon because of the known difficulty to obtain stable emulsions with this material, and perfluorooctylbromide (PFOB, perflubron) was used because it is presently the most promising oxygen-carrier for intravascular use [1]. Such a study was possible only because homologous series of well-defined, pure, monodisperse surfactants had been made available.

The syntheses of the materials used have been reported in the literature [4–7]. Their purity was checked by thin-layer chromatography and/or HPLC prior to use.

Results

Surface activities

All the water-soluble perfluoroalkylated amphiphiles tested display high surface activities (Table 1) with surface tensions at the critical micelle concentration (CMC) in the 20–25 mN/m range [$\gamma_s(H_2O) = 73$] and *F*-decalin/water interfacial tensions in the 1.7–6 mN/m range. These activities are, as expected, higher for the perfluoroalkylated amphiphiles than for their hydrocarbon analogues, especially for the neutral compounds.

Figure 2 illustrates the effect of replacing a 6- or an 8-carbon terminal segment of an 11-carbon fatty acid chain by perfluoroalkyl segments of the same lengths, on the efficiency (concentration of surfactant required to produce a given surface effect) and effectiveness (the maximum surface effect that the surfactant can produce whatever the amount used) [9] of a trehalose fatty acid monoester. Both of these parameters increase when an increasing portion of the hydrocarbon tail of compound **10** is replaced by a fluorinated one as in **7** and **8**.

The hydrocarbon spacer is also important. Thus, in the maltoside series, for example, for the same fluorinated chain length, C_6F_{13}, solubility decreases strongly with increasing spacer and total hydrophobic chain length (from over 100 g/L for **2**, to 25 g/L for **3**, and less than 0.01 g/L for **4**, i.e. for 2, 5 and 11 carbon spacers, respectively). Efficiency also increases significantly in the same series: for 0.01 g/L dispersions of **2**, **3** and **4** in 1 g/L solution of Pluronic F-68, γ_s decreases from 41.3 to 34.0 and

Table 1. Surface activities of typical perfluoroalkylated sugar derivatives

Compound	Hydrophobic tail *R*	CMC[a]) mM (g/L)	γ_s[b])	γ_i[b]) (FDC)
1	$C_8F_{17}(CH_2)_2$[c,d])	0.17 (0.13)	22.3	2.6
2	$C_6F_{13}(CH_2)_2$[c,d])	1.05 (0.72)	23.3	4.7
3	$C_6F_{13}CH{=}CH(CH_2)_3$	0.14 (0.10)	21.5	1.7
4	$C_6F_{13}CH{=}CH(CH_2)_9$	—	23.8[g])	2.3[g])
5	$CH_3(CH_2)_9$[e])	2.2 (1.06)	36.1	17.4
6	$CH_3(CH_2)_{11}$[c])	0.23 (0.12)	35.6	—
7	$C_8F_{17}(CH_2)_2$	0.007 (0.006)	21.7	2.8
8	$C_6F_{13}(CH_2)_4$	0.10 (0.075)	21.5	6.3[f])
9	$C_4F_9(CH_2)_{10}$	0.0075 (0.0055)	25.3	3.9
10	$CH_3(CH_2)_9$	0.39 (0.20)	38.3	15.8
11	$C_6F_{13}(CH_2)_2$	1.55 (1.00)	19.7	5.6
12	$C_8F_{17}(CH_2)_2$	1.07 (0.80)	24.2	5.1
13	$CH_3(CH_2)_9$	8.11 (3.57)	23.1	8.8

[a]) CMC determined by plotting γ_s vs Log[C] except [c]) spectrophotometrically [4]
[b]) γ_s (at CMC) and γ_i in mN/m ($\pm$ 0.3) at 20 °C, $\gamma_i(FDC/H_2O) = 56$; γ_i measured for 1 g/L aqueous solution except [f]) [C] = 0.1 g/L, [g] [C] = 1 g/L in 1 g/L Pluronic F-68 solution
[d]) anomeric mixture
[e]) [14]

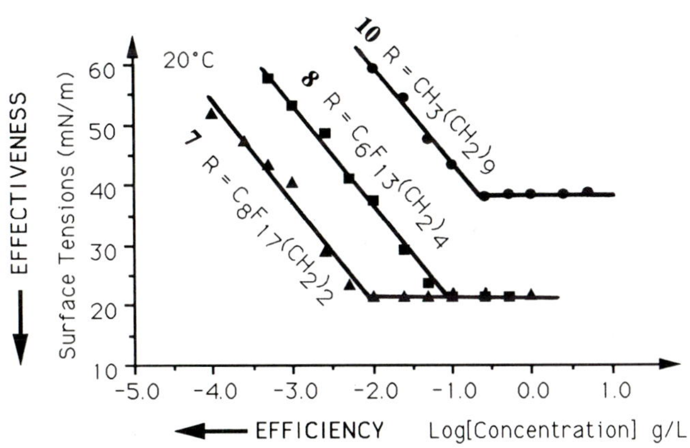

Fig. 2. Surface tensions (mN/m) vs Log[concentration] in g/L, for aqueous solutions of a series of trehalose fatty acid monoesters

25.6 mN/m, and γ_i FDC/H_2O from 23.9 to 5.3 and 5.5 mN/m, respectively.

For the same hydrophobic chain, for example $C_8F_{17}(CH_2)_2$, significant differences in CMC are observed, depending on the polar head; thus, the CMC is 1.07, 0.17, and 0.007 mM, respectively, for the phosphate **12**, maltoside **1**, and trehalose **7** derivatives (a 150/25/1 ratio).

For **3**, **7**, and **12**, some information on the degree of packing and orientation of the adsorbed surfactant molecules was obtained using the Gibbs Equation [10], $d\gamma = -2.203\chi RT\,\Gamma_m d(\log C)$ and the relationship $a_m = 10^{16}/N\Gamma_m$, where Γ_m and a_m are the surface concentration and area per molecule before the CMC with χ being the number of species (two for anionic compounds, one for neutral ones) in the surfactant molecule (Table 2). In the homologous trehalose ester series, for the same hydrophobic chain length (compounds **7**, **8**, and **10**), the area per molecule decreases with increasing fluorination of the chain; thus, the trehalose derivative **7**, with the longest perfluorooctyl segment has the highest degree of packing. Being anionic, the phosphate derivatives **11** and **12** display larger surface area; it is striking, however, that the $C_8F_{17}C_2H_4$ tail does not modify a_m significantly with respect to

Table 2. Surface concentration Γ_m and area a_m per molecule for typical perfluoroalkylated sugar derivatives

	$d\gamma/d(\log C)$	Regression coefficient	$\Gamma_m \cdot 10^{-10}$ (mole/cm^2)	a_m (Å^2)
3	– 16.5	1.000	2.9	56
7	– 24.7	0.989	4.4	38
8	– 17.1	0.993	3.1	55
10	– 15.6	0.990	2.8	60
11	– 26.6	0.999	2.4	70
12	– 15.8	0.997	1.4	118
13	– 14.3	0.998	1.3	130

$C_{10}H_{21}$, and that the highest packing is found for the $C_6F_{13}C_2H_4$ tail.

Hemolytic activity

It is noteworthy that these highly amphiphilic and surface-active molecules show essentially no detectable hemolytic activity provided the ratio of the fluorinated to hydrocarbonated segments' length in the hydrophobic tail is large enough (Table 3). It should be noted that, while the hemolytic activity of surfactants with hydrocarbon tails usually increases within a given family when the chain lengths and consequent surface activity increase, it *decreases*, on the contrary, when the hydrophobic tails are perfluoroalkylated. In this case, prolonging the perfluoroalkyl chain increases surface activity, but decreases hemolytic activity [11].

Neutral perfluoroalkylated maltoside-derived surfactants

When the perfluoroalkylated maltoside-derived surfactants were first investigated as emulsifiers for fluorocarbons, using sonication or microfluidization, the first noteworthy observation was that it turned out to be impossible to obtain any stable emulsion.

On the other hand, considerable stabilization could be achieved when these surfactants were used in conjunction with Pluronic F-68. This stabilization effect is illustrated in Fig. 3 for a 50% w/v concentrated *F*-decalin emulsion prepared with 5% w/v of a mixture of maltoside **1** and Pluronic F-

Table 3. Hemolytic activity of perfluoroalkylated vs hydrocarbon sugar-derived surfactants

Compound		Hydrophobic tail *R*	C g/L[a])	(Result)[b])
Maltoside	1	$C_8F_{17}(CH_2)_2$[c])	37/7	(0)
	2	$C_6F_{13}(CH_2)_2$[c])	100	(0)
	3	$C_6F_{13}CH{=}CH(CH_2)_3$	50/10	(0)
	4	$C_6F_{13}CH{=}CH(CH_2)_9$	0.05/0.5	(+ +)
			0.01/0.1	(0)
	5	$CH_3(CH_2)_9$	1	(+ +)
			0.1	(0)
	6	$CH_3(CH_2)_{11}$	0.1	(+ + +)
			0.05	(0)
Trehalose ester	7	$C_8F_{17}(CH_2)_2$	30/8	(0)
	8	$C_6F_{13}(CH_2)_4$	50	(0)
	9	$C_4F_9(CH_2)_{10}$	0.1	(+ + + +)
			0.005	(0)
	10	$CH_3(CH_2)_9$	1	(+ + + +)
			0.1	(0)
Glucose Phosphate	11	$C_6F_{13}(CH_2)_2$	100	(0)
	12	$C_8F_{17}(CH_2)_2$	30/20	(0)
	13	$CH_3(CH_2)_9$	5	(+ + + +)
			1	(0)

[a]) [C] g/L (concentration of surfactant in 0.9% NaCl solution) or [C]/[C′] (surfactant in 0.9% NaCl solution dispersed in Pluronic F-68 solution);
[b]) NaCl 0.9% as negative control (0) and H_2O positive control (+ + +);
[c]) anomeric mixture

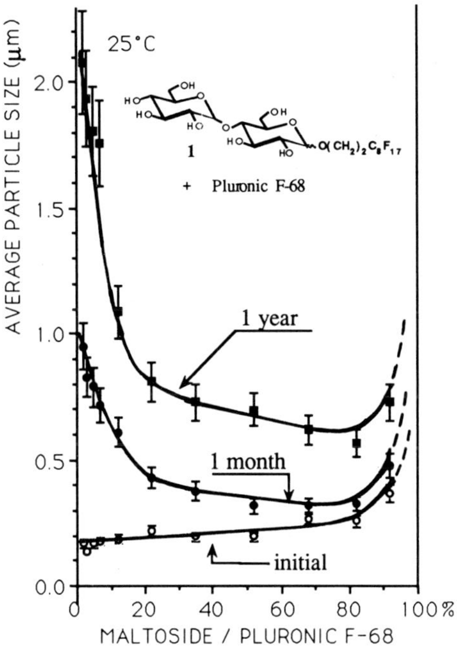

Fig. 3. Average particle size, initial (○), after 1 month (●), and 1 year (■) at 25 °C, in a 50% w/v perfluorodecalin emulsion as a function of the ratio of 2-(*F*-octyl)ethyl maltoside **1** to Pluronic F-68 in the surfactant mixture, the total amount of surfactant being held constant (5% w/v)

68. In this figure the average droplet sizes measured initially, then after 1 month and 1 year of storage at 25 °C are plotted as a function of the percentage of fluorinated compound in the surfactant mixture. The stability of the emulsion, which is rather poor when Pluronic is used alone, is significantly improved when 20% or more of the poloxamer is replaced by the perfluoroalkylated maltoside. Similar U-shaped stabilization effect curves were observed for the other members of this family and for other formulations [12]. When a series of *F*-hexyl derivatives, **2**, **3** or **4**, were investigated in a 20% w/v FDC, 2% w/v Pluronic F-68, and 1% w/v maltoside formulation, the average growth in particle size over 1 month at 25 °C is in the 1-(no change) to 1.6-fold range, far slower than for the reference emulsion prepared with Pluronic alone, for which a 6-fold increase was observed.

The important synergistic stabilization effect observed appears to support the hypothesis which had presided over the design of these surfactants. It should be noted that no significant stabilization effect was observed when these surfactants were used in conjunction with egg-yolk phospholipids.

Neutral perfluoroalkylated trehalose ester-derived surfactants

All the perfluoroalkylated trehalose monoesters tested were found to be efficient in stabilizing FDC emulsions when used as co-surfactants with Pluronic F-68. For example, 50% w/v FDC emulsions prepared by sonication with 5% w/v of a 1:1 mixture of the trehalose esters **7**, **8** or **10** and of Pluronic F-68 were monitored over a 60-day period at 50 °C (a rather high temperature chosen to accelerate the emulsion aging process). When the cube of the average diameter of the droplets was plotted against storage time, straight lines were found for all the preparations tested (Fig. 4; regression coefficient: 0.96–0.99), indicating that Ostwald ripening is likely to be the predominant aging process [13]. The particle size increase rate *S*, defined as the slope of these straight lines ($\times 10^5$, in μm^3/day) is, respectively, 530, 1950, and 6970 for the emulsions incorporating **7**, **8** or **10**, compared to 8300 for the reference emulsions prepared with Pluronic alone. This stabilization effect, when expressed as the ratio of the droplet volume increase rate in the test vs reference emulsions is 16, 4.3 and 1.2 for **7**, **8** and **10** respectively, indicating that the more fluorinated the tail, the more efficacious is the synergistic stabilization effect.

But this family differs markedly from the maltoside series in that highly stable emulsions could also be obtained with one of its members taken as the sole emulsifier. Thus, 6-*O*-[3′-(perfluorooctyl)propanoyl]-trehalose, **7**, allowed the preparation of stable 50% w/v FDC emulsions with 3% of **7**. Figure 5 shows the evolution of the particle sizes in such an emulsion as a function of time at 4°, 25°, and 50 °C, compared to the reference emulsion prepared with the same amount of Pluronic F-68; at 50 °C the average growth in 2 months is about 2-fold for **7**, compared to 5.5-fold for the reference emulsion. No synergistic effects with EYP were noted.

Anionic perfluoroalkylated phosphate esters of glucose as surfactant

Contrary to the neutral perfluoroalkylated sugar derivatives investigated above, all the phosphate

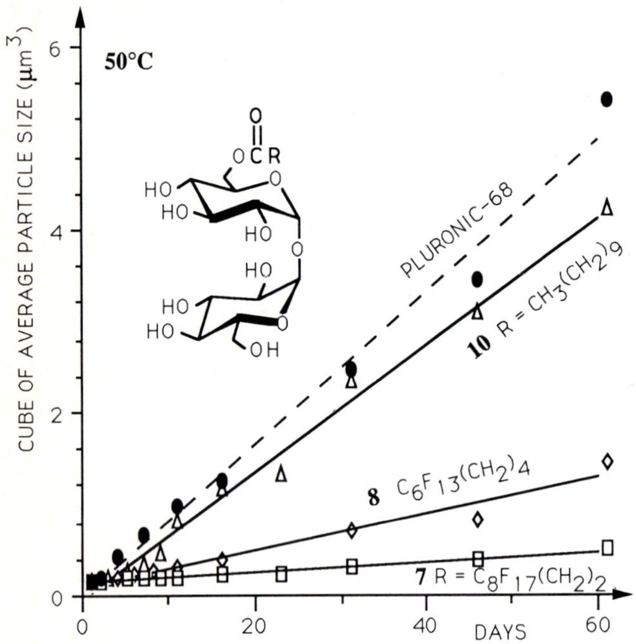

Fig. 4. Comparison of the aging curves, $d^3 = f(t)$, at 50°C for 50% w/v perfluorodecalin emulsions prepared with 5% w/v of a 1/1 mixture of Pluronic F-68 and of various trehalose fatty acid monoesters

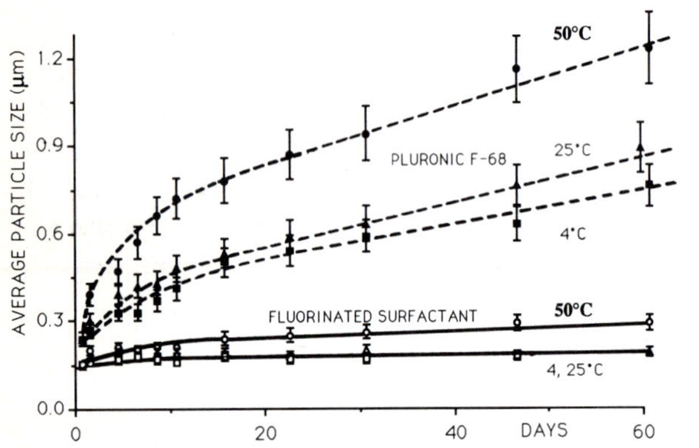

Fig. 5. Stability at 4°, 25°, and 50°C of 50% w/v perfluorodecalin emulsions prepared with 3% w/v of 6-O-[3′-(perfluorooctyl)propanoyl]-trehalose, **7**, (solid lines) compared to 3% w/v of Pluronic F-68 (dashed lines)

ester derivatives (**11–13**) we synthesized were found to allow the preparation of stable FDC or PFOB emulsions when used as the sole surfactant. They were also found to act synergistically with EYP, but not with Pluronic F-68.

Thus, for example, the sodium (6-D-glucosyl)[2-(perfluorooctyl)ethyl]phosphate **12**, a representative compound of this family on which more detailed studies were performed, allows the preparation of stable 90% w/v concentrated emulsions of PFOB with only 1% w/v of this surfactant in the formulation. It is noteworthy that the particle sizes of these emulsions remain as low as those of the emulsions prepared with 4% of EYP (Fig. 6).

A strong effect on the emulsion's characteristics was also observed when this surfactant was used in conjunction with EYP. Thus, 90% w/v PFOB emulsions in which the EYP: **12** ratio was decreased stepwise (the total amount of surfactant, 4% w/v, being held constant) were prepared by mechanical microfluidization (Fig. 7). It is noteworthy that the replacement of only one-eighth of the EYP by **12** is sufficient to obtain both a definite stabilizing effect and, interestingly, a decrease in particle sizes which is preserved, in comparison to the EYP-alone emulsions, during sterilization and aging at 40°C.

Figure 8 represents the slopes of the straight lines obtained by plotting the cube of the average diameter of the particles vs time (regression coefficient: 0.97–1.00) for emulsions prepared with EYP, or **12** alone, or with a 3.5% EYP:0.5% **12** mixture; the increase rates, S, were respectively 122, 18, and 34 ($\times 10^5$) μm^3/day.

Discussion

The drastic yet unpredictable differences in behavior observed for closely related amphiphiles are

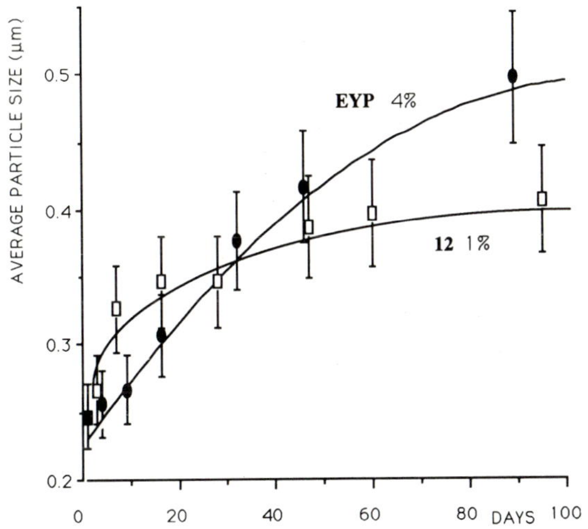

Fig. 6. Comparison of average particle size vs time for sterilized 90% w/v perfluorooctylbromide emulsions prepared with 1% w/v of the phosphate ester **12**, and with 4% w/v of EYP

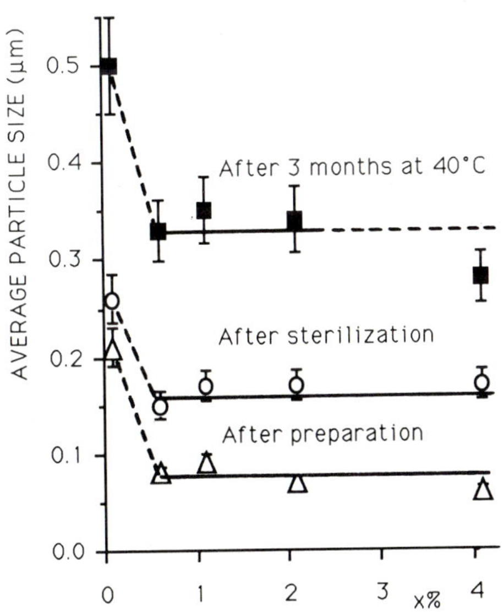

Fig. 7. Average particle sizes in 90% w/v perfluorooctylbromide emulsions as a function of the amount x% w/v of the perfluoroalkylated phosphate **12**, the remaining $(4-x)$% being EYP

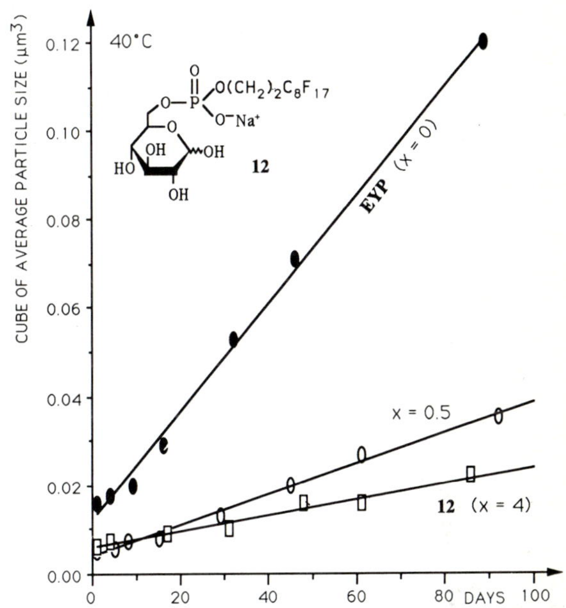

Fig. 8. Comparison of the aging curves $d^3 = f(t)$ at 40 °C for sterilized 90% w/v perfluorooctylbromide emulsions prepared with x% of **12** and $(4-x)$% of EYP

Table 4. Emulsifying and co-emulsifying capabilities of some typical perfluoroalkylated sugar-derived surfactants

Surfactant	As sole surfactant	As co-surfactant with Pluronic F-68	As co-surfactant with EYP
Maltoside **1**	no	yes	no
Trehalose ester **7**	yes	yes	no
Glucose phosphate ester **12**	yes	no	yes

probably the most remarkable result of this study; they are summarized in Table 4 for the most typical examples.

For the neutral perfluoroalkylated maltose and trehalose derivatives, the working hypothesis which directed this study appears to be borne out: a considerable synergistic stabilization effect has been observed for mixtures of these surfactants with Pluronic F-68. Failure to prepare emulsions with the maltoside-derived surfactants alone (while for a comparable hydrophobic chain a related trehalose derivative yields very stable emulsions) is striking. It can only be related to differences in conformation of the head due to the differences in the junction between the two cycles and in type and location of its junction with the tail, which in turn have an influence on the head surface. The fact that the phosphate glucose esters do not display the expected co-surfactant effect with Pluronic F-68 is noteworthy; it cannot be assigned only to its lower hydroxyl group content, as xylitol derivatives with similar hydrophobic chains have been shown to display very strong synergistic effects with Pluronic F-68 [12]. Instead, a co-surfactant effect with EYP is observed with this anionic amphiphile, which is not seen with the maltoside and trehalose derivatives; this effect results in both more stable and finer particle size emulsions.

These studies show that action on the characteristics of fluorocarbon emulsions (including action on their shelf stability and particle sizes) is possible, sometimes with only small amounts of perfluoroalkylated sugar-derived surfactants. However, the ability of a given molecular structure to produce a given effect is still unpredictable.

Acknowledgements

We thank the CENTRE NATIONAL DE LA RECHERCHE SCIENTIFIQUE and ATTA for financial support.

References

1. Riess JG (1991) Intl J Artif Org 14:255–258
2. Riess JG (1991) Vox Sanguinis 61:225–239
3. Riess JG (1988) Proc 2nd World Surfactant Congress (Paris, May 1988), ASPA, Paris 4:256–263
4. Greiner J, Manfredi A, Riess JG (1989) New J Chem 13:247–254
5. Milius A, Greiner J, Riess JG (1991) New J Chem 15:337–344
6. Abouhihale S, Greiner J, Riess JG (1991) Carbohydr Res 212:55–64
7. Milius A, Greiner J, Riess JG (1991) Carbohydr Res 213:xxx
8. Rosevear P, Van Aken T, Baxter J, Ferguson-Miller S (1980) Biochim 19:4108–4115. Koeltzow DE, Urfer AD (1984 J Am Oil Chem Soc 61:1651–1655. Jeffrey GA (1986) Acc Chem Res 19:168–173. Thiem J, Bocker Th, Hofmann P, Luders H, Schreier G (1988) Proc 2nd World Surfactant Congress (Paris, May 1988) ASPA, Paris 1:197–211
9. Rosen MJ (1985) CHEMTECH 15:292–298
10. Rosen MJ (1978) in "Surfactants and Interface Phenomena", J Wiley, New-York, Chichester, chap 2:26–82
11. Riess JG, Pace S, Zarif L (1991) Adv Mat 3:249–251
12. Zarif L, Manfredi A, Varescon C, Le Blanc M, Riess JG (1989) J Am Oil Chem Soc 66:1515–1523
13. Kabalnov AS, Pertzov AV, Shchukin ED (1987) J Colloid Interface Sci 118:590–597
14. Alpes H, Allmann K, Plattner H, Reichert J, Riek R, Schulz S (1986) Biochim Biophys Acta 862:294–302

Authors' address:

Professor Jean G. Riess
Laboratoire de Chimie Moléculaire
Associé au CNRS
Université de Nice-Sophia Antipolis
Parc Valrose
F-06108 Nice Cedex 2
France

Interactions between poly(styrene-allylalcohol) monolayers and surfactants. Correlations to water-in-crude oil emulsion stability

H. Ebeltoft, K. G. Nordli Børve, J. Sjöblom, and P. Stenius[1])

Department of Chemistry, University of Bergen, Norway
[1]) Institute for Surface Chemistry, Stockholm, Sweden

Abstract: The Π–A isotherms and relaxation behavior of monomolecular films at the air/aqueous solution interface of a copolymer of styrene and allylalcohol (PSAA, mean molecular weight 1500 g/mol) have been investigated as a function of the subphase surfactant concentration. The surfactants were sodium dodecylsulphate (SDS), cetyltrimethylammonium bromide (CTAB) or cetylpyridinium chloride (CPC). At low surfactant concentrations ($\lesssim$0.3 CMC), the PSAA film is penetrated by surfactant molecules and undergoes a relaxation consistent with conformation changes of the polymer without dissolution into the subphase. At higher concentrations ($\approx$CMC) the PSAA film is slowly dissolved into the subphase; Π–A isotherms show a plateau and the films relax by a gradually decreasing surface area. The dissolution of PSAA is probably brought about by adsorption of surfactant molecules onto the polymer chain, giving rise to a water-soluble polymer-surfactant "complex" of polyelectrolyte nature. At medium concentrations, hydrophobic interactions between the neutral polymer and the ionic surfactant increase the hydrophilicity of the film and, thereby, probably lead to the formation of loops of polymer segments into the subphase. All the surfactants seem to interact with the PSAA film in similar ways. PSAA monolayers are good model systems for emulsion-stabilizing monolayers in North Sea crude oils. The monolayers of both the PSAA and the interfacially active crude oil system respond similarly to the presence of ionic surfactants, indicating analogous dissolution mechanisms. Thus, specific surfactant-monolayer interactions of the type observed will be of importance for an understanding of water-in-crude oil emulsion stability.

Key words: Monolayers; air/water interface; polymer; surfactant; interactions; crude oil

Introduction

A prerequisite for predicting coalescence in emulsion systems is to understand the conditions at the water/oil interface. Depending on the type of emulsion, this interface may contain a variety of single constituents or mixtures of surfactants, cosurfactants, particles, and polymers. The water droplets in crude oil based emulsions are generally considered to be surrounded by an interfacial film of asphalthenes, porhyrines, and wax particles [1–5]. The mechanical properties of such films prevent the system from coalescence with a concomitant phase separation.

In a series of articles, studies on water-in-crude oil emulsions from the Norwegian Continental Shelf with regard to formation, characterization, and destabilization have been reported [5–11]. As a part of these studies, a method for separation of the interfacially active (i.e., emulsion stabilizing) fraction in the oils was developed [9]. In order to obtain a fundamental understanding of the film properties of this fraction, a Langmuir study of its monolayer properties on aqueous subphases was undertaken [10]. A correlation between the specific monolayer areas of the fractions and the stability of the corresponding crude oil-based emulsions was found. In order to make further progress with regard to a basic understanding of interactions in crude oil films, the monolayer behavior of model systems based on poly(styrene-allylalcohol) was investigated. The film properties of the indigenous crude oil fractions are, to a large extent, reproduced by this system [11].

It has been convincingly shown that surfactants destabilize water-in-crude oil emulsions [7, 12]. Thus, in order to understand the mechanisms behind destabilization, knowledge about the interplay between film constituents and single monomeric surfactant molecules seems crucial. In the present article, the influence of different ionic surfactants in the subphase on monolayers of the water-insoluble, neutral poly(styrene-allylalcohol) (PSAA) and on monolayers of the interfacially active fractions from a North Sea crude oil are investigated. From the comparison between the data on polymer and crude oil films, information about the mechanisms of emulsion destabilization by surfactants can be inferred.

Previous reports on interactions between polymers and surfactants mainly discuss water-soluble polymers (neutral or ionic) and ionic surfactants [13–16]. The interaction between surfactants and polymers is generally thought to involve the formation of micelle-like clusters adsorbed to the polymer chain. The formation of these clusters generally occurs at a rather well-defined surfactant concentration, denoted the critical aggregation concentration (CAC), which is frequently much lower than the critical micellization concentration (CMC) of the surfactant in bulk. The binding is usually cooperative and the same model has been extended to involve both neutral and charged water-soluble polymers. For polyelectrolytes and an oppositely charged surfactant, there is a large electrostatic contribution to the polymer-surfactant interaction, in addition to the hydrophobic interaction which is common to all polymers.

Only a few articles dealing with water-insoluble, neutral polymers and ionic surfactants have appeared. Isemura et al. have shown that uncharged, water-insoluble polymers such as polyvinylacetate, polyvinylformal, and polyvinylbutyral may be dissolved in highly concentrated solutions of sodium dodecylsulphate (SDS) [17, 18]. The driving force behind the solubilization is most likely nonspecific hydrophobic interactions. It is assumed that the surfactant molecules are adsorbed onto the neutral polymer with the non-polar part interacting with the polymer and with the ionic part directed to the aqueous phase. This structure, resulting in a negatively charged polymer-surfactant "complex" with polyelectrolyte properties has been verified experimentally [18].

Experimental

Materials

Poly(styrene-allylalcohol) (PSAA) was supplied by Polyscience Inc. The hydroxy content is 5.7% and the mean molecular weight is 1500 g/mol. Sodium dodecylsulfate (SDS) (> 99% purity) and NaCl (suprapure) were delivered by Merck, whereas cetyltrimethylammonium bromide (CTAB) (> 99% purity) and cetylpyridinium chloride (CPC) (99% purity) were from Janssen Chimica (Geel, Belgium). All these chemicals were used as supplied.

The surface-active fraction in a crude oil from the Norwegian Continental Shelf was separated and treated as described earlier [9]. The molecular weight average of the fraction is 1300 g/mol and the molecular range is 50–10 000 g/mol [10]. The elemental composition is 39.7 mol% carbon, 59.3 mol% hydrogen, 0.20 mol% nitrogen, 0.62 mol% oxygen, and 0.14 mol% sulfur [10]. The fraction was spread from a 1:1 (weight) mixture of redistilled toluene and hexane, both of p.a. quality from Merck. The spreading solvent in all the other experiments was a mixture of redistilled toluene and ethanol (4%).

The pH of the subphase solutions was adjusted by adding HCl (suprapure quality from Merck) to distilled water (pH = 5.6).

The water was first purified by decalcination, prefiltration, and reverse osmosis. Final purification was made by a modified MilliQ-unit which included two mixed bed ion exchangers, an activated charcoal cartridge, a 0.2-μm in-line filter, an Organex cartridge (all other cartridges were from Millipore), and a final 0.2-μm filter. All filters were Zetapore products from Cuno, France.

Surface balance experiments

The experiments were performed with a KSV 2200 Langmuir trough system (KSV Chemicals, Helsinki, Finland). The equipment includes automatic recording of Π–A isotherms at pre-determined compression rates, as well as recording of area losses at constant pressure (relaxation measurements). The temperature was kept at 294.5 ± 0.5 K.

A maximum change in surface pressure of 0.02 mN/m per minute was allowed for as the monolayer components were spread. The amount of material spread and the initial surface area were kept constant in all experiments for each subphase additive. The barrier speed was 10 mm/min and the compression started 30 min after spreading. In area relaxation experiments the films were initially compressed at 10 mm/min and then maintained at a predetermined surface pressure while the change in surface area was recorded as a function of time.

The fracture pressure Π_F is defined as the pressure where the film fractures or collapses at the given experimental conditions [19].

Results

Monolayers of PSAA

Π–A isotherms: The surfactants were added to the subphase prior to spreading the monolayers. The monolayer behavior of PSAA was studied as a function of the concentration of surfactant, as added to distilled water (pH = 5.6) in equilibrium with air, solutions of pH = 2.5 (corresponding to an ionic strength of 0.0032 M) or 0.0032 M NaCl solutions. This was done in order to separately observe the effect of electrostatic shielding of the ionic surfactant groups and of changes in the pH. The concentration of surfactant was in the range:

$$2.8 \times 10^{-5}\ \mathrm{M} - 1.4 \times 10^{-3}\ \mathrm{M} \quad \text{for CPC}\ .$$

$$5.6 \times 10^{-5}\ \mathrm{M} - 3.8 \times 10^{-3}\ \mathrm{M} \quad \text{for CTAB}\ .$$

$$1.0 \times 10^{-4}\ \mathrm{M} - 1.0 \times 10^{-2}\ \mathrm{M} \quad \text{for SDS}\ .$$

The CMC for CPC, CTAB, and SDS in distilled water at 298 K is 9.0×10^{-4} M, 9.2×10^{-4} M and 8.1×10^{-3} M, respectively [20]. The CMC decreases somewhat as the concentration of electrolyte increases [21].

The monolayer behavior of PSAA on distilled water and on subphases of different pH has been described [11]. PSAA is a completely water-insoluble polymer and forms a stable liquid expanded monolayer at specific surface areas below 0.94 m^2/mg on distilled water. At a surface area of about 0.45 m^2/mg, the isotherm starts to rise more steeply and the film fractures at a surface pressure of 33 mN/m and specific surface area $A = 0.30\ m^2/mg$.

Figure 1 shows the Π–A isotherms of PSAA on subphases of CPC in distilled water (pH = 5.6). The characteristic features of the isotherm change dramatically as the concentration of CPC (C_{CPC}) is varied. The surface pressures at $A = 1.4\ m^2/mg$ and 0.8 m^2/mg and the fracture pressure for the PSAA film are shown in Fig. 2 as a function of the

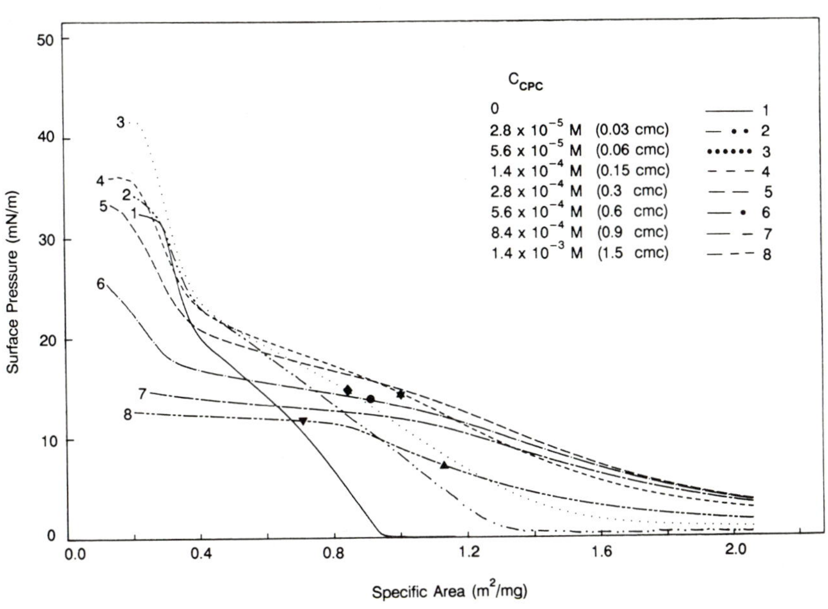

Fig. 1. Π–A isotherms of PSAA on subphases with different concentrations of CPC in distilled water (pH = 5.6). The symbols on the isotherms refer to constant pressure relaxation experiments shown in Fig. 5

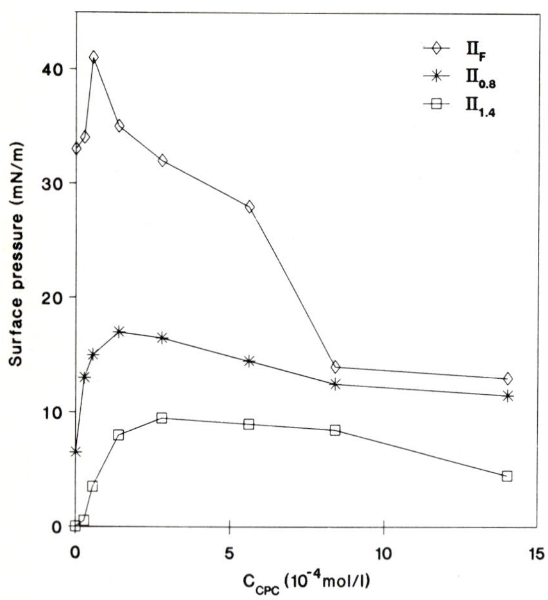

Fig. 2. The surface pressures at $A = 1.4\ m^2/mg$ ($\Pi_{1.4}$) and at $A = 0.8\ m^2/mg$ ($\Pi_{0.8}$) and the fracture pressure (Π_F) for the PSAA film as a function of the subphase concentration of CPC (C_{CPC}); pH = 5.6

Table 1. The surface pressures at $A = 1.4\ m^2/mg$ ($\Pi_{1.4}$) and at $A = 0.8\ m^2/mg$ ($\Pi_{0.8}$) and the fracture pressure (Π_F) for the PSAA film as a function of the subphase concentration of CPC (C_{CPC}); pH=5.6

C_{CPC} (10^{-4} mol/l)	$\Pi_{1.4}$ (mN/m)	$\Pi_{0.8}$ mN/m	Π_F (mN/m)
0	0	6.5	33
0.28	0.5	13.0	34
0.56	3.5	15.0	41
1.4	8.0	17.0	35
2.8	9.5	16.5	32
5.6	9.0	14.5	28
8.4	8.5	12.5	14
14.0	4.5	11.5	13

subphase concentration of CPC (the data are also given in Table 1). With increasing C_{CPC} four different concentration regimes may be identified:

1) For $C_{CPC} \leq 1.4 \times 10^{-4}$ M ($\approx$0.15 CMC), the surface pressure at large surface areas increases monotonously with increasing concentration. The fracture pressure is higher than on a pure water subphase, and attains a maximum of about 41 mN/m at $C_{CPC} = 5.6 \times 10^{-5}$ M ($\approx$0.06 CMC). The fracture pressure on distilled water is 33 mN/m.
2) For $1.4 \times 10^{-4}\ M < C_{CPC} \leq 2.8 \times 10^{-4}$ M ($\approx$0.15 CMC – $\approx$0.3 CMC), there is still an increasing expansion of the monolayer at large surface areas; the film area at which the film fractures becomes smaller.
3) For $2.8 \times 10^{-4}\ M < C_{CPC} < 8.4 \times 10^{-4}$ M ($\approx$0.3 CMC – $\approx$0.9 CMC) the surface pressure at large areas begins to decrease. The film becomes even more compressible and the fracture pressure decreases.
4) For $C_{CPC} \geq 8.4 \times 10^{-4}$ M (close to and above CMC in distilled water), the initial pressure decreases. A distinct plateau (almost constant pressure) is reached upon compression.

As the pH of the subphase is changed to 2.5, similar regimes for in the Π–A behavior of PSAA occur at lower concentrations of CPC (Fig. 3 and Table 2). The plateau of the isotherm occurs already at $C_{CPC} = 2.8 \times 10^{-4}$ M (in distilled water at $C_{CPC} \geq 8.4 \times 10^{-4}$ M). An almost identical effect on the isotherm is obtained by adding 0.032 M NaCl to the water (Fig. 3), resulting in an ionic strength corresponding to pH = 2.5.

The influence of CTAB on the monolayer behavior of PSAA is almost identical to that observed for CPC on all subphases investigated. Therefore, data for CTAB are not given here.

The addition of an anionic surfactant (SDS) to the subphase (Fig. 4 and Table 3) has, to a large extent, a similar effect on the PSAA monolayer as CPC and CTAB. However, higher concentrations of surfactant are required in the case of SDS to obtain the same effects. The effect of changing the pH and adding NaCl to the subphase in which SDS is present, follows the same trend as observed for CPC and CTAB.

Relaxation: Constant pressure relaxation of PSAA was performed on a subphase of CPC added to distilled water. The relaxation process was examined at four different concentrations of CPC; three below and one just above the CPC concentration which gives rise to a plateau on the Π–A curves (5.6×10^{-5} M, 1.4×10^{-4} M, 5.6×10^{-4} M and 1.4×10^{-3} M, respectively). At each of the lower concentrations only a single surface pressure was used. At the highest concentration the relaxation

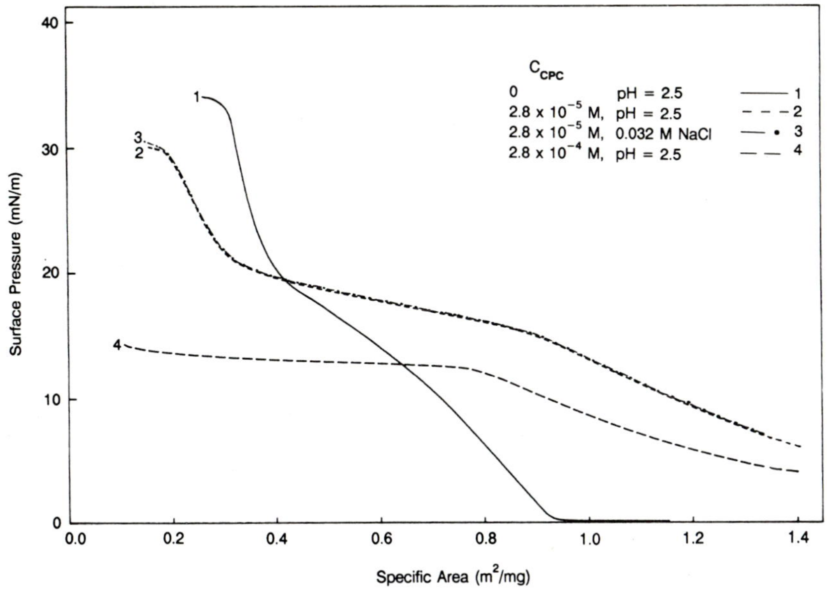

Fig. 3. Π–A isotherms of PSAA on subphases with different concentrations of CPC at pH = 2.5. An isotherm for the case of PSAA on a 0.0032 M NaCl subphase is also included

Table 2. The surface pressures at $A = 1.4\ m^2/mg$ ($\Pi_{1.4}$) and at $A = 0.8\ m^2/mg$ ($\Pi_{0.8}$) and the fracture pressure (Π_F) for the PSAA film as a function of the subphase concentration of CPC (C_{CPC}). pH = 2.5

C_{CPC} (10^{-4} mol/l)	$\Pi_{1.4}$ (mN/m)	$\Pi_{0.8}$ (mN/m)	Π_F (mN/m)
0	0	6.5	33
0.28	6.0	16.0	30
2.8	4.0	12.0	14

experiment was performed at two pressures, one below and one on the plateau.

The surface pressures for the relaxation experiments are marked on the Π–A curve in Fig. 1.

The relaxation process of PSAA is highly dependent on the concentration of CPC (Fig. 5). For $C_{CPC} = 5.6 \times 10^{-5}$ M and 1.4×10^{-4} M at $\Pi = 15$ mN/m and for $C_{CPC} = 5.6 \times 10^{-4}$ M at $\Pi = 14$ mN/m the relaxation is characterized by a rapid, initial area loss. The rate of relaxation then

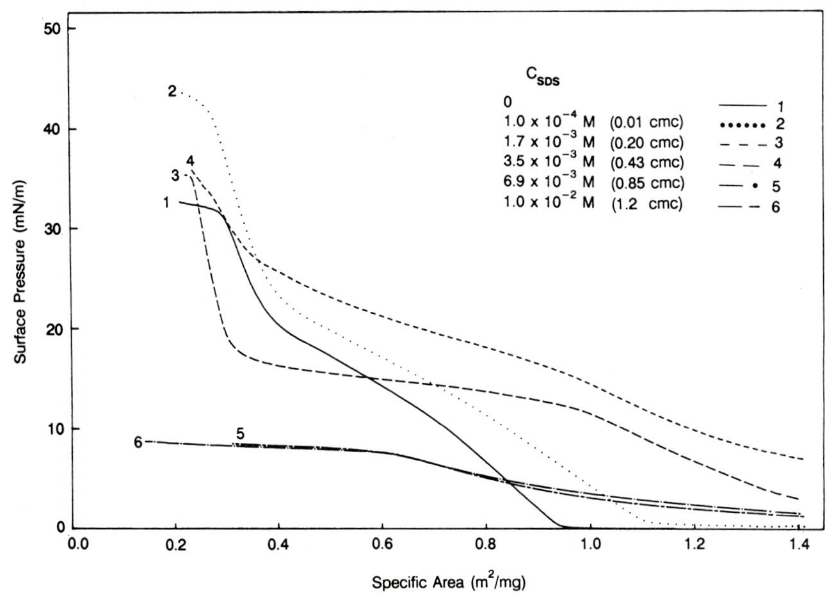

Fig. 4. Π–A isotherms of PSAA on subphases with different concentrations of SDS in distilled water (pH = 5.6)

Table 3. The surface pressures at $A = 1.4\ m^2/mg$ ($\Pi_{1.4}$) and at $A = 0.8\ m^2/mg$ ($\Pi_{0.8}$) and the fracture pressure (Π_F) for the PSAA film as a function of the subphase concentration of SDS (C_{SDS}). pH = 5.6

C_{SDS} (10^{-3} mol/l)	$\Pi_{1.4}$ (mN/m)	$\Pi_{0.8}$ (mN/m)	Π_F (mN/m)
0	0	6.5	33
0.1	0.5	11.0	42
1.7	7.0	18.0	36
3.5	3.0	13.5	35
6.9	1.6	5.5	8.5
10.4	1.4	5.0	8.5

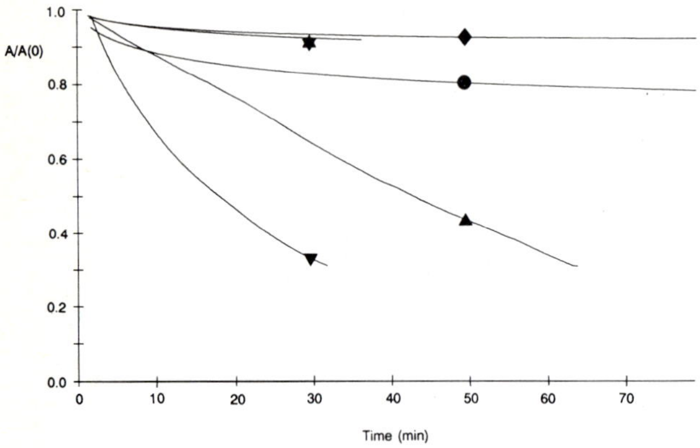

Fig. 5. Constant pressure relaxation of PSAA on subphases with different concentrations of CPC (C_{CPC}) in distilled water (pH = 5.6). ◆: $C_{CPC} = 5.6 \times 10^{-5}$ M (0.06 CMC), $\Pi = 15.0$ mN/m. ★: $C_{CPC} = 1.4 \times 10^{-4}$ M (0.15 CMC), $\Pi = 15.0$ mN/m. ●: $C_{CPC} = 5.6 \times 10^{-4}$ M (0.6 CMC), $\Pi = 14.0$ mN/m. ▲: $C_{CPC} = 1.4 \times 10^{-3}$ M (1.5 CMC), $\Pi = 7.0$ mN/m. ▼: $C_{CPC} = 1.4 \times 10^{-3}$ M (1.5 CMC), $\Pi = 11.6$ mN/m. The symbols also refer to the markings on the Π–A curves in Fig. 1

slows down and an almost constant surface area is obtained. The area loss is about 10% at $C_{CPC} = 5.6 \times 10^{-5}$ M and 1.4×10^{-4} M, and 20% at $C_{CPC} = 5.6 \times 10^{-4}$ M in a period of 60 min. The relaxation at $C_{CPC} = 1.4 \times 10^{-4}$ M is totally different. At both pressures employed a gradually decreasing surface area is observed with an area loss of about 35% at $\Pi = 7$ mN/m and about 70% at $\Pi = 11.7$ mN/m within 30 min. There is no indication of a flattening out of the relaxation curves.

Because of the very similar effects of CPC, CTAB, and SDS on the Π–A isotherms, relaxation was not investigated for the two latter surfactants.

Monolayers of the interfacially active crude oil fraction

Figure 6 shows the Π–A curves for the interfacially active crude oil fraction on subphases of CPC in distilled water (pH = 5.6). An expansion of the film at large surface areas followed by a gradually increasing compressibility with increasing subphase surfactant concentration is observed, as it is for PSAA.

The crude oil film contains small amounts of alkaline functional groups which become negatively charged at pH = 2.5 [10]. However, no specific effects are observed at pH = 2.5 with ionic surfactants present in the subphase (results not shown).

The effect on the crude oil film by replacing CPC with CTAB or SDS is similar to that observed for the PSAA monolayer (results not shown).

Discussion

Polymer-surfactant interactions

From the Π–A isotherms and the relaxation behavior of the PSAA film on surfactant solutions, it is clear that the concentration of surfactant is the essential parameter determining the properties of the PSAA film at the air/aqueous solution interface and the degree of polymer-surfactant interaction. The influence of surfactant concentration on the PSAA monolayer properties is very similar for CPC and CTAB. Both surfactants are cationic with saturated C_{16} hydrocarbon chains. The structural difference in the cationic part of the molecules does not seem to be of importance for their interactions with the PSAA monolayer. Higher concentrations are required in the case of SDS to obtain the same effects on the PSAA monolayer. This is probably due to the shorter hydrophobic part of SDS, which also results in an increase in the CMC of the surfactant [22]. In relation to the CMC of the surfactants, the concentration effects are almost equal for all surfactants investigated. Both the effect of adding electrolyte to the subphase for one specific surfactant and of increasing the hydrophobicity of the surfactant (the interactions are in both cases initiated at lower concentration of surfactant) are readily understood by considering corresponding effects on micellar systems [21, 22]. Also, in systems of *water-soluble* polymers and

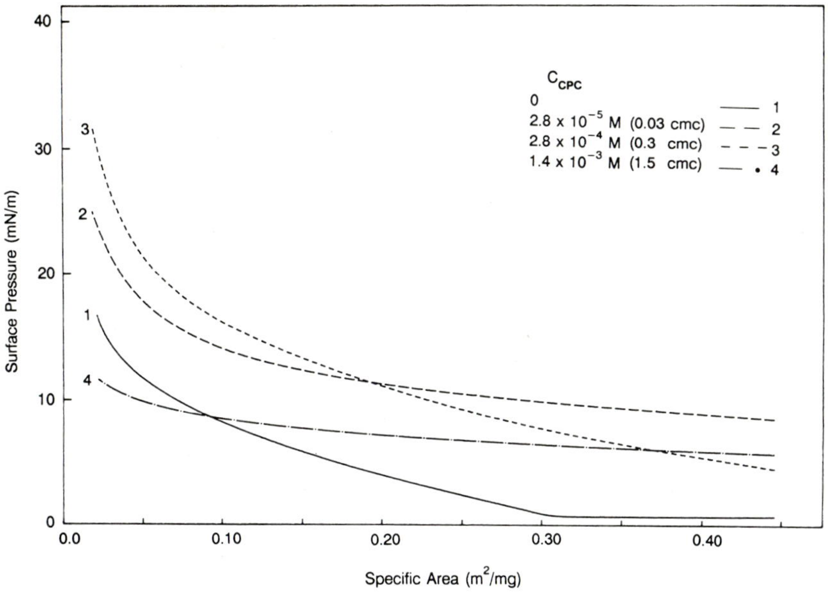

Fig. 6. *Π–A* isotherms of the interfacially active crude oil fraction on subphases with different concentrations of CPC in distilled water (pH = 5.6)

ionic surfactants similar trends are observed [23–25].

The following discussion concentrates on the monolayer behavior of PSAA on CPC subphase since this is the system we have examined most thoroughly. It is divided into two parts; one discussing CPC concentrations less than and one larger than $C_{CPC} = 2.8 \times 10^{-4}$ M ($\approx$0.3 CMC), since there seems to be different mechanisms of interaction between the polymer monolayer and the subphase surfactant molecules in these intervals.

For $C_{CPC} \leq 2.8 \times 10^{-4}$ M ($\approx$0.3 CMC) penetration of surfactant molecules into the PSAA film at large areas increases with increasing bulk concentration of surfactant (Fig. 1). This results in the formation of a mixed film of polymer and ionic surfactant. Such an expansion at large surface areas has been observed in many systems in which the subphase molecules interact weakly with the film molecules, which may be either polymers and proteins [26] or simpler molecules [27, 28]. The penetrating molecules are usually displaced from the mixed monolayer as the film is compressed and the surface pressure increases, which is similar to the observed behavior of the PSAA film on subphases with low surfactant concentration. The fracture pressure probably increases due to the force needed to completely displace the ionic surfactant from the PSAA monolayer. As the concentration of surfactant is further increased, the intermolecular forces between PSAA molecules in the monolayer will probably dominate, as seen in the subsequent decrease in fracture pressure and area [27, 28]. Intersections between the isotherms of a polymer on subphases with pure water and added surface-active material have also been observed in systems in which the subphase solute has caused the polymer to assume a more compact conformation [29].

The relaxation curves of the PSAA monolayers are identical at all concentrations of CPC less than 2.8×10^{-4} M($\approx$0.3 CMC) and are characterized by a small initial area loss and the rapid establishment of stable films (Fig. 5). The relaxation process is almost identical to that of PSAA on pure water [11] and is consistent with a rearrangement mechanism of the molecules in the film until a stable molecular conformation is reached [30, 31]. Although a stable polymer film is obtained also at 5.6×10^{-4} M CPC ($\approx$0.6 CMC), the initial relative area loss has increased to about 20% compared to 10% at the lower concentrations. A possible explanation for the larger area loss may be hydrophobic interactions between the neutral polymer and the ionic surfactant. This would lead to adsorption of surfactant molecules onto the neutral polymer in such a way that the non-polar part of the surfactant is directed towards the polymer and the ionic part is directed into the aqueous subphase, giving rise to a polymer with a slight polyelectrolyte character [17, 18]. This will increase the hydrophilicity of the film and, thereby, lead to a formation of loops of polymer segments penetrating into the subphase at high pressures. Partial solubilization of polyelectrolyte molecules into the subphase solution has earlier been observed for systems in which the solute is highly surface active and oppositely charged to the polymer [32].

The surface pressure of the PSAA monolayer at large surface areas increases with increasing surfactant concentration below $C_{CPC} = 2.8 \times 10^{-4}$ M ($\approx$0.3 CMC) and then starts to decrease. The onset of decreasing pressure with C_{CPC} coincides with the first signs of a plateau on the Π–A curve. It is not clear why the pressure should start to decrease at this concentration. As seen from Fig. 5, the monolayer is still completely insoluble. A further increase in surfactant concentration gives rise to a very distinct plateau on the isotherm. Similar isotherms are observed at all CPC concentrations close to and above CMC and may be explained by a gradual displacement of the PSAA film into the subphase as the available surface area decreases upon compression.

This conclusion is supported by relaxation at $C_{CPC} = 1.4 \times 10^{-3}$ M ($\approx$1.5 CMC) (Fig. 5) which gives additional information about the mechanism of interaction between the PSAA film and the surfactant at high subphase concentrations of the surfactant. The surface area decreases continuously at constant pressure at both pressures investigated. However, the rate of relative area loss is twice as large at the higher surface pressure (on the plateau of the isotherm) as that at the lower pressure (below the plateau). A relaxation behavior with a gradually decreasing surface area is usually observed for films which dissolve into the subphase solution [33]. At high concentrations of surfactant, the adsorption of surfactant onto the polymer probably gives rise to a polymer-surfactant "complex" of hydrophilic character which becomes dissolved in the surfactant bulk solution. This is, of course, entirely compatible with the formation of somewhat fewer hydrophilic and, hence, still insoluble complexes at very low surfactant concentrations.

The displacement of the polymer from the air/water interface first occurs at concentrations somewhat below CMC of the surfactant. Thus, it seems as if the polymer-surfactant interaction involves a gradual adsorption of individual surfactant molecules onto the polymer chain. The data do not allow us to conclude whether the adsorbed surfactant monomers act as sites for further aggregation [34–36], or if there is a continuous adsorption of monomeric surfactant molecules to the polymer chain [37].

Crude oil systems and emulsion destabilizing mechanisms

The Π–A isotherms of the interfacially active crude oil fraction are, in general, more featureless than those of the PSAA [11]. This is also the case when charged surfactants are introduced to the subphase. Nevertheless, the monolayers of both the crude oil fraction and the synthetic model polymer PSAA respond similarly to the presence of ionic surfactant. This indicates that the main mechanisms of interaction in the system of interfacially active crude oil fraction and ionic surfactant are analogous to those suggested for PSAA. The fact that the dissolution mechanism seems to be similar in the two systems, confirms the conclusion from our earlier paper [11] about PSAA being a good model substance for the crude oil fraction with respect to monolayer properties.

Earlier investigations on destabilizing water-in-crude oil emulsions by means of surfactants have clearly displayed the importance of the surfactant concentration [12, 38]. High concentrations (> CMC) do not give rise to an efficient release of water from the emulsion, while low concentrations do. Thus, concentrations exceeding the CMC in bulk are less efficient in destabilizing the emulsified systems. In the absence of specific interactions this is the way destabilization will be linked to the surfactant activity. The most plausible explanation of the efficiency at low surface concentrations is the lowering of the surface tension γ. Since the indigenous crude oil components are not especially efficient in lowering γ, the surfactant (or amphiphile) will replace these components and form a new interface which does not stabilize the water droplets against coalescence. Such a mechanism is observed, for example, for short-chain alcohols [7, 10, 11]. However, when specific surfactant-oil component interactions occur, the destabilizing mechanisms are different. The destabilizing effects should then not be looked for in the lowering of the surface tension, but in an enhanced surface charge density in the interfacial zone leading to instability [7, 8]. Other mechanisms behind instability may be changed interfacial rheological properties such as elasticity and viscosity [39]. In other cases the surfactants have been shown to change the wettability of stabilizing particles in the interfacial film [40].

On the basis of the present study on interactions between monolayers of polymer/crude oil fraction and ionic surfactants added to the subphase, we suggest a supplementary mechansim for destabilizing water-in-crude oil emulsions. If the water-soluble ionic surfactant adsorbs onto the emulsion-stabilizing polymeric crude oil components, the hydrophilicity of the indigenous polymer is gradually enhanced. This will eventually cause a destabilization of the emulsion as the hydrophilic polymer-surfactant "complex" is displaced from the water/oil interface by dissolution into the water phase. This mechanism is supported by the result from the PSAA model system as well as from the true crude oil system.

Acknowledgements

This work was sponsored by Statoil A/S, Norsk Hydro A/S and Saga Petroleum A/S. The Norwegian Research Council for Science and the Humanities is acknowledged for a research grant (KGNB). The Norwegian Chemical Society (Norsk Kjemisk Selskap) is also thanked for a grant (HE).

References

1. Eley DD, Hey MJ, Symonds JD (1988) Colloids Surfaces 32:103
2. Papirer E, Bourgeois C, Siffert B, Balard H (1982) Fuel 61:732
3. Dunning HN, Moore JW, Denekas MO (1953) Ind Eng Chem 45:1759
4. Layrisse I, Rivas H (1984) J Dispersion Sci and Techn 5(1):1
5. Mingyuan L, Christy AA, Sjöblom J (1991) In Sjöblom J (ed.): Emulsions–A Fundamental and Practical Approach NATO ASI Series, Kluwer p 157
6. Johansen EJ, Skjærvø IM, Lund T, Sjöblom J, Söderlund H, Boström G (1988) Colloids Surfaces 34:353
7. Sjöblom J, Söderlund H, Lindblad S, Johansen EJ, Skjærvø IM (1990) Colloid Polymer Sci 268:398
8. Sjöblom J, Mingyuan L, Høiland H, Johansen EJ (1990) Colloids Surfaces 46:127
9. Sjöblom J, Urdahl O, Høiland H, Christy AA, Johansen EJ (1990) Progr Colloid Polymer Sci 82:131
10. Nordli KG, Sjöblom J, Kizling J, Stenius P (1991) Colloids Surfaces 57:83
11. Nordli Børve KG, Sjöblom J, Stenius P, Colloids Surfaces (in press)
12. Aveyard R, Binks BP, Fletcher PDI, Lu J-R (1990) J Colloid Interface Sci 139:128
13. Robb ID (1981) In: Lucassen-Reynders E (ed.) Anionic Surfactants in Physical Chemistry of Surfactant Action. Marcel Decker, New York
14. Goddard ED (1986) Colloids Surfaces 19:255 and 301
15. Saito S (1987) In: Schick MJ (ed.) Nonionic Surfactant, Physical Chemistry. Surf Sci Ser vol 23. Marcel Decker, New York, ch 15
16. Thalberg K (1990) Doctoral Thesis, University of Lund, Sweden
17. Isemura T, Kimura Y (1955) J Polym Sci 16:90
18. Isemura T, Imanishi A (1958) J Polym Sci 33:337
19. Gaines Jr GL (1966) Insoluble Monolayers at Liquid-Gas Interfaces. Wiley (Interscience), New York
20. Mukherjee P, Mysels KJ (1970) Critical Micellar Concentrations of Aqueous Surfactant Solutions. National Bureau of Standards, Washington DC
21. Corrin ML, Harkins WD (1947) J Am Chem Soc 69:683
22. Tanford C (1980) The Hydrophobic effect: Formation of Micelles and Biological Membranes. John Wiley & Sons, New York
23. Arai H, Murata M, Shinoda K (1971) J Colloid Interface Sci 37:223
24. Shirahama K, Ide N (1976) J Colloid Interface Sci 54:450
25. Saito S (1969) J Polymer Sci 7:1789
26. Giles CH, McIver N (1975) J Colloid Interface Sci 53:155
27. Adam NK, Askew FA, Pankhurst KGA (1939) Proc Roy Soc A122:485
28. Pankhurst KGA (1942) Proc Roy Soc A179:392
29. Giles CH, McIver N (1974) J Textile Res 44:587
30. Smith RD, Berg JC (1980) J Colloid Interface Sci 74:273
31. Honig EP, Hengst JH, den Engelsen DJ (1973) J Colloid Interface Sci 45:92
32. Cameron A, Giles CH, MacEwan TH (1958) J Chem Soc (London) 1224
33. Ter-Minassian-Saraga L (1956) J Colloid Sci 11:398
34. Saito S, Taniguchi T, Kitemura K (1971) J Colloid Interface Sci 37:154
35. Saito S (1967) Kolloid Z 215:16
36. Fishman ML, Eirich FR (1971) J Phys Chem 75:3135
37. Gilanyi T, Wolfram E (1981) Colloids Surfaces 3:181
38. Urdahl O, Sjöblom J (to be published)
39. Kizling J, Nordli Børve KG, Sjöblom J, Stenius P (to be published)
40. Menon VB, Wasan DT (1986) Colloids Surfaces 19:107

Authors' address:

K. G. Nordli Børve
Dept. of Chemistry
University of Bergen
N-5007 Bergen
Norway

Progress in Colloid & Polymer Science Progr Colloid Polym Sci 88:140–145 (1992)

Preliminary structural studies of liposomes and vesicles in formamide

A. Meliani, E. Perez, I. Rico, A. Lattes, C. Petipas[1]), and X. Auvray[1])

Laboratoire des IMRCP, UA CNRS No. 470, Université Paul Sabatier, Toulouse, France
[1]) Faculté des Sciences et des Techniques, URA CNRS No. 808, Mont St Aignan, France

Abstract: We describe the first examples of liposomes and vesicles in formamide, a structured non-aqueous medium. Bilayer membranes were obtained after dispersion of lecithin (for liposomes) and dodecylalkyldimethylammonium bromide (for vesicles) in formamide. The structures of these systems are discussed and compared to those obtained in water.

Key words: Formamide; non-aqueous; vesicles; liposomes

Introduction

Understanding of self-assembly of molecules in solution is likely to be advanced by studies on the structure, stability, and dynamic behavior of vesicles. This would also aid development of applications for these systems.

The most widely studied vesicle systems are prepared in water from lecithins (liposomes) or double-chain surfactants (vesicles), but it has been reported that phospholipids form bilayers in both glycerol [1] and N-ethylammonium nitrate [2]. In previous publications, we have described the state of association of ionic and non-ionic surfactants in formamide, which has similar structural properties to those of water. Many surfactants were observed to form micelles, microemulsions, and lyotropic liquid crystals in formamide, like those formed in water [3].

We present here the first evidence that both lecithins and dodecylalkyldimethylammonium bromide form bilayer systems in formamide. For dodecylalkyldimethylammonium bromide, this is the first example, to our knowledge, of the formation of vesicles in a polar, structured medium.

Materials and methods

Preparation of liposomes in formamide

We used the method normally employed for preparation of liposomes (of the MLV type) in water [4].

One gram of lecithin (Lipoid 80 purchased from Seppic and freeze-dried before use), 0.15 g of cholesterol (Sigma, purum > 99%) are dissolved in a mixture of 15 ml of ethanol (Prolabo, purum > 99%) and 5 ml of formamide (Aldrich, purum > 99%, dried over Na_2SO_4). The solution is added dropwise to 50 ml of formamide under vigorous agitation at room temperature. The ethanol is then evaporated under vacuum. The resulting solution is slightly opaque with a water content of less than 0.1% (measured in a Karl Fisher apparatus).

Preparation of vesicles in formamide with dodecylalkyldimethylammonium bromides $C_{12}C_n$

Preparation of surfactants

The dodecylalkyldimethylammonium bromides $(C_{12}H_{25})(C_nH_{2n+1})N^+(CH_3)_2$, Br^- or $C_{12}C_n$ ($n = 4, 8, 9, 10$) were synthesized by quaternization of amine with the appropriate alkyl bromides using a conventional method [5, 6]. The surfactants were purified by recrystallization from a mixture of ethyl acetate and ether, and then freeze-dried. Surfactant purities (> 98%) were checked by ^1H and ^{13}C NMR, and elemental analysis.

Preparation of vesicles

Twenty ml of the suspension of $C_{12}C_n$ (0.2 M) in formamide is sonicated (Bransonic 60) for 30 min using the sonication procedure described for preparation of $C_{12}C_n$ vesicles in water [7]. The resulting solution is slightly opaque with a water content of less than 0.1%, as measured in a Karl Fischer apparatus.

Examination of liposomes and vesicles

Liposomes and vesicles were observed by electron microscopy according to the following typical procedure. A few drops of this solution (of liposomes or vesicles) in formamide were applied to a 150-mesh copper grid coated with a carbon film, which was then dried in a dessicator. A 2% solution of uranyl acetate in formamide was applied in a similar way. The samples were examined in a Philips EM 301 electron microscope. The sizes of the vesicles in both water and formamide were verified by light scattering (Coulter N4MD).

Rate of decarboxylation of 3-carboxy-6-nitro-benzisoxazole

Synthesis of the substrate

3-carboxy-6-nitro-benzisoxazole was prepared according to literature methods [8, 9], and recrystallized from methanol to give a final purity of over 98% (checked by ^1H, ^{13}C NMR and elemental analysis).

Decarboxylation reaction

The rate of decarboxylation of 3-carboxy-6-nitro-benzisoxazole (10^{-4} M) in solutions of C_nC_{12} in formamide was followed by UV spectroscopy in a diode array spectrophotometer (HP 8451A) at various temperatures ($\pm 0.1\,°C$). The concentration of C_nC_{12} was chosen to be above the CMC [5] (4.6×10^{-2} mol·l^{-1} for C_4C_{12}, 3.5×10^{-2} mol·l^{-1} for C_8C_{12}, C_9C_{12} and $C_{10}C_{12}$).

Results and discussion

Liposomes in formamide

Figure 1 illustrates the electron micrographs of solutions of lecithin in formamide as described in the experimental part. Multilayered liposomes can be clearly discerned. Similar bilayer membranes were observed under the same conditions in water. The liposomes were found to be stable for more than 2 weeks in both solvents. Stability was checked by measuring the sizes of vesicles in both water and formamide by light scattering (see experimental part). Liposomes were found to be larger in formamide than in water (500 nm versus 100 nm). Apart from this size difference, it should be noted that the layer thickness was 4 nm in formamide and between 7 nm in water (measured on electron micrographs).

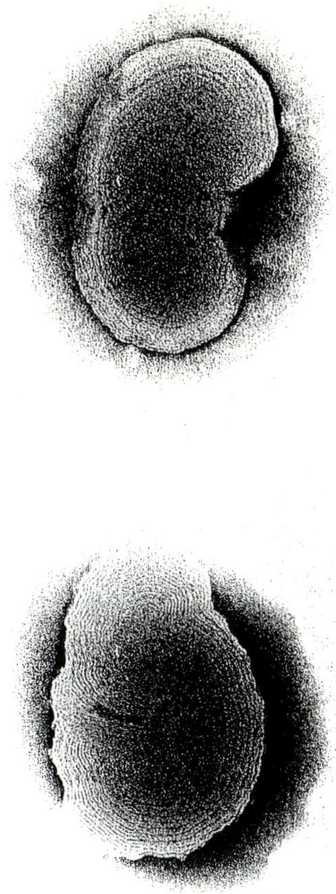

Fig. 1. Electron micrographs of multilamellar liposomes in formamide (X 196 000)

The fact that the liposomes were obtained under similar conditions in water and formamide is in line with the similarity of the binary phase diagrams of lipids in these two solvents described by Bergenstahl and Stenius [10]. These authors reported a similar layer thickness of the lamellar phase in the two solvents, with a tendency to higher thickness in formamide, which is in disagreement with our results. However, our findings are in line with those observed for conventional monocatenary surfactants (CTAB, SDS or alkylpyridinium halides) in formamide: In all cases, the structural parameters are smaller in formamide than in water [3]. Furthermore, the lamelles of lecithin bilayers in

N-ethylammonium nitrate are closer together in formamide than in water [2]. Interestingly, we observed an increase in layer thickness after addition of traces of water to the formamide-containing media. Above 1% water, the thickness of the lamelles approaches that observed in water. The results reported by Bergenstahl and Stenius could thus be due to traces of water in the media. A structural study of these systems by small-angle x-ray scattering is in progress.

Vesicles in formamide

Vesicles were observed in suspensions of $C_{12}C_n$ (0.2 M) in formamide for $n \geq 9$. Typical electron micrographs of vesicles of $C_{10}C_{12}$ in formamide are represented in Fig. 2. In contrast, vesicles were not observed for $n = 4, 8$. For $n = 9, 10$, results were similar to those obtained in water, but at a higher concentration (0.2 M vs. 0.1 M) and with longer sonication times than those required in water (30 min vs. 5 min). As for lecithin, multilamellar vesicles with a diameter of 1000–5000 Å and a layer thickness of 2 nm, rather smaller than those obtained in water (4 nm), were observed.

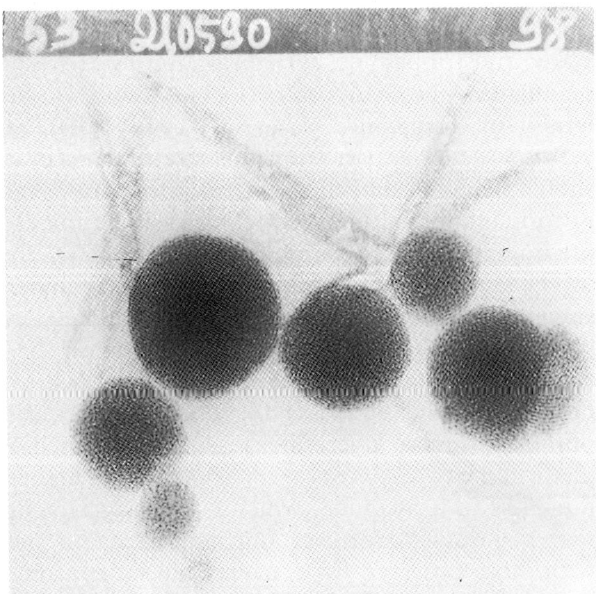

Fig. 2. Electron micrographs of multilamellar vesicles in the dimethyldecyldodecylammonium bromide (0.2 M) formamide system after sonication (X 196 000)

The vesicles grew rapidly, but were less stable than the liposomes. Polydispersion was observed in the light-scattering studies 24 h after preparation.

The binary systems, formamide/C_nC_{12}, have not yet been examined at various temperatures, but, in general, the behavior of systems with other cationic surfactants (CTAB, alkylpyridinium halides) is similar in water and formamide [3]. The only difference is the appearance of the lamellar phase at higher concentrations and temperatures in formamide. This could explain why higher concentrations were required to produce vesicles in formamide. Furthermore, the longer period of sonication enabled a higher temperature to be reached in the sample ($\approx 70\,°C$). Thus, the vesicles obtained with C_nC_{12} in formamide are likely to be in a metastable state at room temperature. A structural study of these vesicles and of the formamide/C_nC_{12} binary systems is in progress.

To account for the observation of vesicles only for $n \geq 9$, we employed a probe reaction in micellar solutions of dodecylalkyldimethylammonium bromide in formamide to determine the phase transition temperature Tc [11] and establish the minimum length of side chain required for formation of bilayers.

Decarboxylation of 6-nitrobenzisoxazole in micellar solutions of C_nC_{12} determination of phase transition temperatures Tc

The following reaction was chosen:

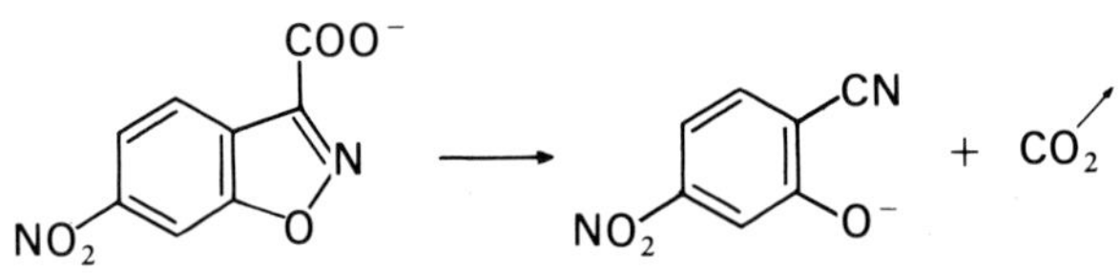

I II

It has been shown that the solvent plays an important role [12–14], depending on the effects of solvation of the initial state of delocalized charge or the less delocalized transition state. The rate constants of this reaction in various solvents are listed in Table 1 [12].

Minch and Bunton [15, 16] showed that the cationic micelles of cetyltrimethylammonium bromide (CTAB) lead to around a 100-fold increase in rate, whereas anionic micelles of sodium lauryl sulfate only have a weak kinetic effect. This difference was attributed to the fact that the cationic surfactants with a positively charged polar head facilitate penetration of anionic reagents into the micelles. On the other hand, the negatively charged anionic surfactants hinder approach of reagent by electrostatic repulsion.

Since this reaction has been employed to determine the transition phase temperature Tc [11] of dialkyldimethylammonium bromides in water, we applied it to the formamide system with a modification of the experimental conditions. In water it is generally carried out in the presence of paratoluene sulfonic acid and tetramethylguanidine to produce an acceptable duration of reaction. In formamide, we left out the acid and base, and just reacted the acid [1] in the presence of surfactants.

The reaction was followed at various temperatures by UV spectrophotometry (λ_{max} of final product = 410 nm) in pure formamide and in micellar solutions of C_nC_{12} in formamide. The results illustrated in Figs. 3 to 6 can be summarized as follows:

i) The rate is generally first order, as in the other solvents studied.
ii) The reaction is much faster in formamide (k_{obs} at 40 °C = $2.56 \cdot 10^{-3}\ s^{-1}$) than in water. This can be accounted for from differences in Kosower's parameter, expressing differences in polarity [17, 18]. Although formamide has both a high dielectric constant and dipole moment, the value of parameter E_T is lower than that of water (56.6 vs. 63.1 Kcal·mole^{-1}). Thus, for charge solvation, formamide can be considered

Table 1. Rate constants (k_{obs}) of the reaction in different solvents at 30 °C in the presence of paratoluene sulfonic acid and tetramethylguanidine

Solvent	Rate constant k_{obs} (s^{-1})
Water	7.4×10^{-6}
Ethanol	1.0×10^{-3}
Dioxan	4.0×10^{-2}
Nitromethane	5.8×10^{-1}
Acetone	2.4×10
Dimethylacetamide	1.6×10^{2}

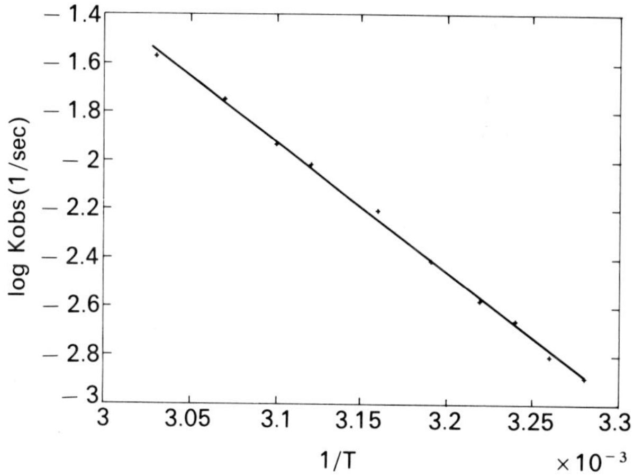

Fig. 4. Plot of log(k_{obs}) against $1/T$ for decarboxylation reaction in the dimethyloctyldodecylammonium bromide/formamide system

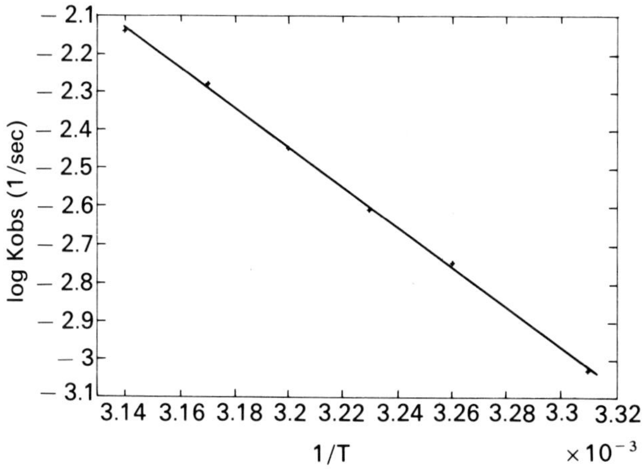

Fig. 3. Plot of log(k_{obs}) against $1/T$ for decarboxylation reaction in the dimethylbutyldodecylammonium bromide/formamide system

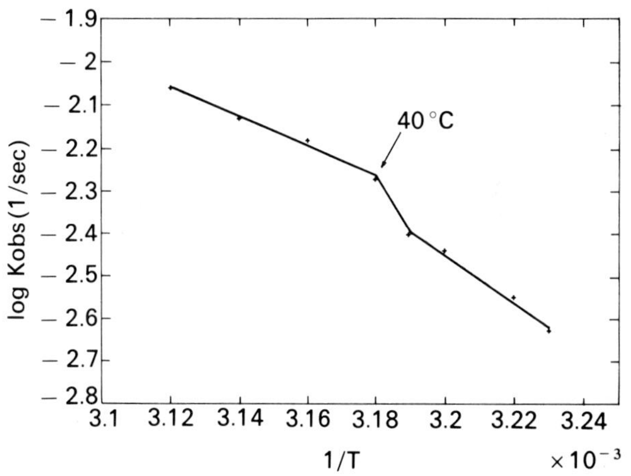

Fig. 5. Plot of log(k_{obs}) against $1/T$ for decarboxylation reaction in the dimethylnonyldodecylammonium bromide/formamide system

to be less polar than water, and so the decarboxylation reaction will be faster in formamide than in water.

iii) In formamide the surfactant concentration was set above the critical micelle concentration (CMC), and the reaction was carried out at temperatures above the Krafft point. The values of CMC and Krafft point determined in a previous study [5] are summarized in Table 2.

The micelle effect at 40 °C was modest, but significant for the various surfactants tested (the rate constant was increased by between 1.5 and 2-fold). This relatively small effect was probably due to the fact that the reaction is rapid in pure solvent, and thus only slightly susceptible to acceleration.

For the $C_{12}C_9$ and $C_{12}C_{10}$ surfactants, we observed a break in slope of the rate vs. temperature plots. This has been observed in water and is attributed to a phase transition. The *Tc* around 40 °C for C_9C_{12} and slightly over 41 °C for $C_{10}C_{12}$ are higher than those for similar surfactants in water. For longer chain surfactants, *Tc* was 45 °C for $C_{18}C_{18}$ and 28 °C for $C_{14}C_{14}$. An increase of 2 °C in *Tc* per carbon atom was observed in water, which is comparable to that obtained in formamide.

In contrast, we observed a linear change in rate constant with increase in temperature for $C_{12}C_4$ and $C_{12}C_8$. Considering the values of *Tc* for $C_{12}C_9$ and $C_{12}C_{10}$ and the Krafft temperature of 27 °C for $C_{12}C_8$, a phase transition temperature for $C_{12}C_8$ should have been observed over the temperature range (29–45 °C) tested. A value of *Tc* around 38 °C would be expected for a 2 °C fall in temperature per carbon atom like that observed in water. This indicated that a bilayer was not formed, as the second hydrophobic chain for both $C_{12}C_4$ and $C_{12}C_8$ was probably too short. In this case, only the micellar phase is formed. These results show that, in order to form a bilayer and hence vesicles, the side chain must be above a certain length; in our series, this was observed at C_9.

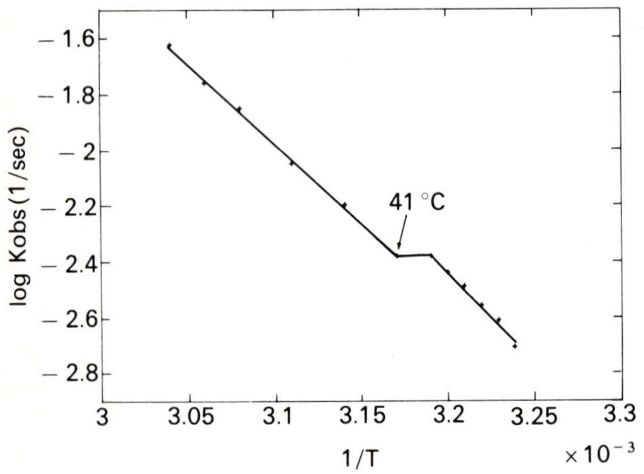

Fig. 6. Plot of $\log(k_{obs})$ against $1/T$ for decarboxylation reaction in the dimethyldecyldodecylammonium bromide/formamide system

Table 2. CMC (at 40 °C) and Krafft temperatures for $C_{12}C_n$ in formamide [5]

Compound	*T* Krafft (°C)	CMC ($mol \cdot l^{-1}$) (40 °C)
C_4C_{12}	27	4×10^{-2}
C_8C_{12}	30	3×10^{-2}
C_9C_{12}	34	2.5×10^{-2}
$C_{10}C_{12}$	36	2×10^{-2}

Conclusion

Taken together, these results indicate that double-chain surfactants are of particular interest in formamide since they can induce formation of large liposomes (for lecithin) or vesicles (for dialkyldimethylammonium bromides). The stable liposomes, which were larger than those obtained in water, could serve as microreaction environments in chemical synthesis. The structure of these vesicles and liposomes, only briefly described here, will be investigated in more detail by measuring the coefficient of elasticity of the membrane and by x-ray diffraction in the course of an analysis of solvent-polar head interactions of surfactants. The morphology of the structures obtained in different polar and structure solvents thus appears to be largely governed by such interactions.

References

1. McDaniel RV, McIntosh TJ, Simon SA (1983) Biochim Biophys Acta 731:97
2. Evans DF, Kaler EW, Benton WJ (1983) J Phys Chem 87:533
3. a) Auvray X, Petipas C, Perche T, Anthore R,

Marti MJ, Rico I, Lattes A (1990) J Phys Chem 94:8604
b) Auvray X, Perche T, Anthore R, Petipas C, Rico I, Lattes A (1991) Langmuir 7:2385
c) Lattes A, Rico I (1989) Colloids and Surfaces 35:221
d) Auvray X, Petipas C, Rico I, Lattes A (1989) J Phys Chem 93:7588
e) Rico I, Lattes A, Das KP, Lindman B (1989) J Am Chem Soc 111:7266
f) Gautier M, Rico I, Lattes A (1990) J Org Chem 55:1500 (and references therein)
4. We thank Société P Fabre (Castres. France) for supplying us with details for the preparation of liposomes in water
5. Fernandez de Sousa M, Perez E, Rico I, Lattes A (1991) Jorn Com Esp Deterg 22:397
6. Jada A, Lang J, Zana R (1990) J Phys Chem 94:381 (and references therein)
7. Kunitake T, Okahata Y (1977) J Am Chem Soc 99:3800
8. Borsche W (1909) Chem Bers 42:1310
9. Lindemann H, Cissee H (1929) Justus Liebi Ann Chem 464:44
10. Bergenstahl BA, Stenius P (1987) J Phys Chem 91:3944
11. Kunitake T, Okahata Y, Ando R, Shinkai S, Hirakawa S (1980) J Am Chem Soc 102:7877
12. Kemp GA, Paul KG (1975) J Am Chem Soc 97:7305
13. Hall GA, Verhoek FH (1947) J Am Chem Soc 69:613
14. Masamitsu S, Smid JJ (1970) J Am Chem Soc 92:2553
15. Bunton CA, Minch MJ (1970) Tetrahedron Lett 44:3881
16. Bunton CA, Kamego AA, Minch MJ, Wright JL (1975) J Org Chem 40:9, 1321
17. Kosower EM (1958) J Am Chem Soc 80:3253
18. Reichardt C (1988) Solvents and Solvent Effects in Organic Chemistry Ed by VCH Verlag gesellschaft Germany

Authors' address:

I. Rico
Laboratoire des IMRCP
UA CNRS No. 470
Université Paul Sabatier
118 route de Narbonne
F-31062 Toulouse Cédex
France

 Progr Colloid Polym Sci 88:146–153 (1992)

ESCA studies of solution/vacuum interfaces

H. O. G. Siegbahn

Department of Physics, University of Uppsala, Sweden

Abstract: The application of electron spectroscopy for chemical analysis (ESCA) to liquid phase samples is described. The surface sensitivity of this technique allows structures within the first few molecular layers to be investigated. In particular, the study of electron spectra in angular resolved mode provides a means to perform depth profiling of the surface region. Applications are presented for solutions of fatty acid salts and phase transfer catalysts. In these solutions details of the electric double layer at the surface have been investigated.

Key words: ESCA; solution surfaces; electric double layer

Introduction and experimental background

ESCA (or XPS) is now a commonly available technique in surface physics and chemistry research. Its versatility is, to a large extent, based on the comparatively straightforward interpretation of the measurement results in terms of elemental abundance and chemical state. The typical surface sensitivity of ESCA is around 25 Å, due to the inelastic mean free path of the excited photoelectrons. Although mainly applied in the investigation of solid surfaces, the sample studied may be also in the gaseous and liquid state. We have during recent years developed the techniques to perform measurements on the latter state [1–6]. The present article is intended to describe the experimental means to perform such measurements and to present some recently obtained results.

A schematic picture of the experimental set-up used by us in performing ESCA measurements on liquid-phase samples is shown in Fig. 1. The sample is irradiated with monochromatic photons and the expelled electrons are then analyzed by means of an electron spectrometer. The number of electrons is thus recorded as a function of position in the spectrometer focal plane (i.e., kinetic energy), in our case by means of a multichannel technique. As discussed below, such spectral recordings are performed for a number of different take-off angles of the electrons with respect to the surface plane of the sample. The excitation source used in the spectra presented here is soft x-rays, viz. AlKα-radiation at 1487 eV, which allows us to study deep core levels.

The use of electrons as information agents requires that the residual gas pressure be reduced to at least high-vacuum conditions before the entrance to the electron energy analyzer region in order to avoid excessive intensity losses. For liquid samples whose vapor pressures are of the order of 0.1 mbar or more, this implies the use of at least one stage of differential pumping between sample and analyzer. This demands, in turn, a continuous renewal of the liquid sample during the recording of the ESCA spectrum. Figure 2 shows schematically the sample arrangement developed by us for these purposes. The basic feature is the use of a moving metal surface as a backing for the liquid. The metal backing is a stainless steel disc immersed halfway in a liquid sample container and rotatable around two different axes. Rotation around the first axis, which is coincident with the centrum normal of the disc, implies that the disc is continuously covered with fresh sample from the container. The other rotation is performed around the cross-section line between the focus of the AlKα x-ray monochromator on the liquid sample surface and the acceptance to the electron spectrometer. This latter rotation allows the take-off angle of the electrons with respect to the sample surface plane

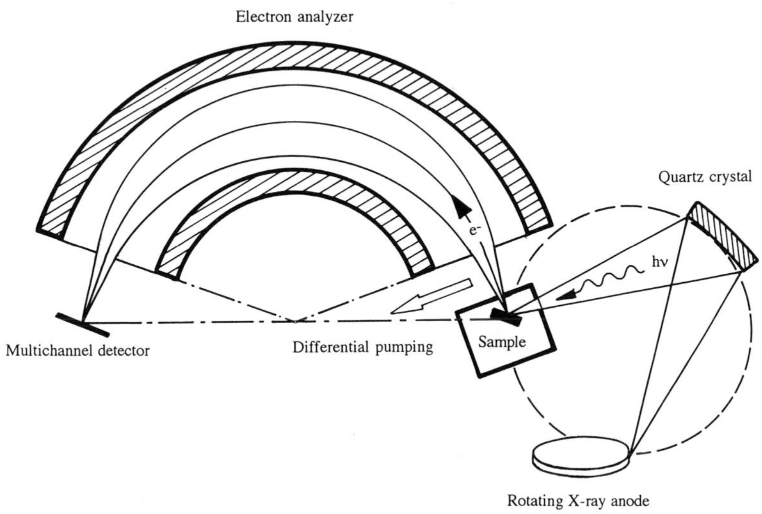

Fig. 1. Set-up of electron spectrometer for studies of liquid phase samples ($h\nu$ = 1487 eV)

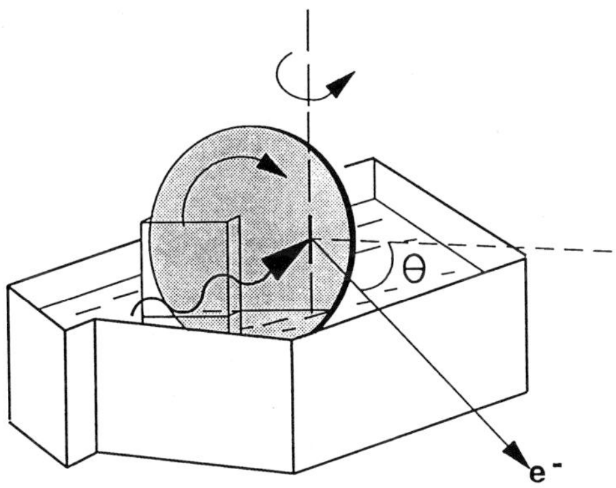

Fig. 2. Schematic picture of the arrangement used for recording of angular resolved electron spectra from solutions

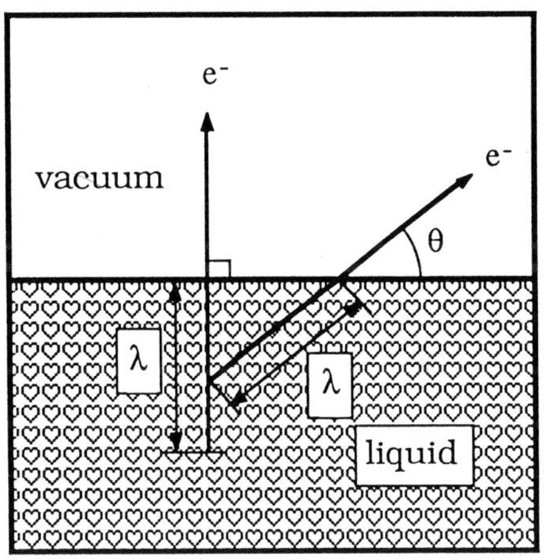

Fig. 3. Illustration of the variation in information depth obtained by means of angular resolved electron spectroscopy (λ = inelastic mean free path of the photoelectrons ≈ 25 Å at electron kinetic energies = 1000 eV)

to be varied continuously between 0° and 50°. This feature of the experimental arrangement has important advantages for the investigation of surface structures to be discussed below. As illustrated in Fig. 3, it implies that the surface sensitivity can be conveniently varied between a bulk-dominated mode at high take-off angles to a surface-dominated one at grazing angles.

The results presented below pertain to electric double layer structures at solution surfaces. Although the electric double layer phenomenon has been thoroughly studied experimentally in terms of macroscopic manifestations (surface pressure, surface potential, electrochemical studies, etc.) its microscopic structure (i.e., charge distribution as function of distance through the layer) has been subject mainly to theoretical modeling based on statistical mechanical methods as well as large scale simulations (e.g. [7, 8]). X-ray [9] and neutron [10] scattering techniques have recently been developed to obtain microstructural data on double layer systems. As exemplified by the present paper, we believe that electron spectroscopy provides another fruitful approach to the investigation of such systems.

The structure of the double layer is a function of several parameters which may be broadly divided into nonspecific electrostatic interactions (i.e., those that do not depend on the internal structure of the ions and may be treated within a classical Poisson–Boltzmann type of double layer description or development thereof) and those interactions that depend specifically on the chemical character of the ions involved. The latter give rise to short-range ion-pairing effects in the topmost surface layers of the solution. The results that we have obtained using angular resolved electron spectroscopy on two classes of double-layer systems (fatty acid salts and phase transfer catalysts) show distinctly different characteristics in terms of such short-range interactions. As a polar substrate medium formamide has been used; it has several properties akin to water, e.g., high dielectric constant as well as high surface tension.

Results and discussion

Fatty acid salts

Figure 4 shows the C1s binding energy region of a 0.5 m solution of potassium octanoate in formamide. The spectra, recorded at different take-off angles, show the well-resolved lines from the three relevant species in the solution; cation (K2p, left lines), solvent (C1s in $HCONH_2$, middle line), and anion (C1s in hydrocarbon moiety of octanoate, right line; the carboxyl carbon signal is hidden under the solvent line). Comparison between bulk concentration and relative line intensity shows that the potassium octanoate is strongly surface segregated within the effective sampling depth of the photoelectrons (< 25 Å). At the lower take-off angles the intensities of both the cation and the solvent lines are substantially reduced with respect to the anion line compared to the relation between the lines at 40°. This relative intensity reduction is also observed to be larger for the solvent line than for the cation line. This unambiguously shows the relative ion distribution to be that characteristic of an electric double layer with the hydrophobic anions in the topmost layer of the solution and the charge-compensating counterions distributed between this layer and the bulk of the solution. The anion coverage is estimated from the spectra to be of the order of half a monolayer at this concentration [5].

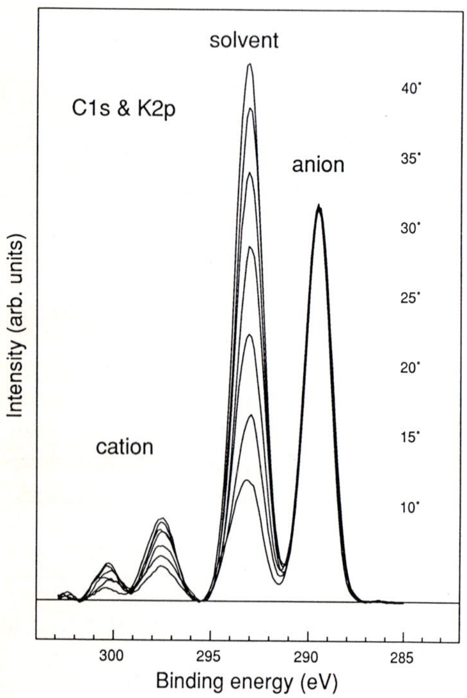

Fig. 4. C1s and K2p spectra obtained from a 0.5 m solution of potassium octanoate in formamide at different photoelectron take-off angles. The spectra are normalized to the solute anion peak

It may be concluded from the observed angular variations of Fig. 4 that the counterion distribution has a pronounced diffuse character within the photoelectron escape depth. It is possible to invert angular distribution data (as in this figure) to obtain experimental depth concentration profiles of the different species at the solution surface. Alternatively, the angular variation of the photoelectron intensities may be calculated on the basis of theoretical models of the electric double layer. When an analysis of the latter type is performed [5], based on a simple Poisson–Boltzmann description of the double layer, cation intensities are found to be underestimated. The concentration of counterions thus seems to be larger near the surface than predicted by such a model, which is also indicative of the existence of Stern-layer-type states.

An alternative confirmation of the formation of double layers in these systems may also be obtained by comparison with spectra recorded for the solid phase compound. Figure 5 shows the Cs4d and I4d binding energy region of the salt cesium 8-iodooctanoate. Spectra are presented both for the solid compound and a solution in formamide at a concentration of 0.5 m and take-off angles of 40° and 20°. Normalization of the spectra is performed with respect to the I4d peak regarding both energy

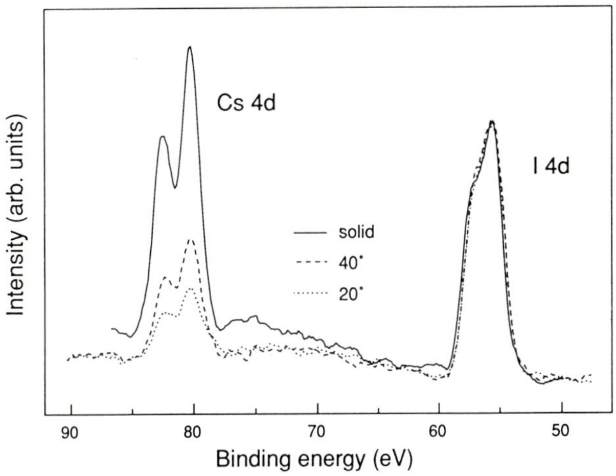

Fig. 5. The Cs4d and I4d spectral region for solid cesium 8-iodooctanoate and a 0.5 m solution in formamide (recorded at 20 ° and 40 ° take-off angles of the photoelectrons). The spectra are normalized to the I4d peak

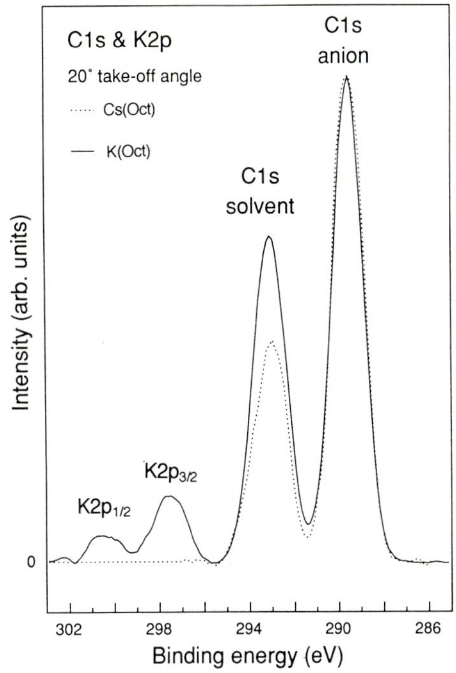

Fig. 6. Comparison of C1s spectra obtained from 0.5 m solutions of potassium octanoate (K(Oct)) and cesium octanoate (Cs(Oct)), respectively, in formamide at a take-off angle of 20 °. Normalization to the solute anion peak

and intensity. As seen in this figure the intensity ratio between the anion and cation peaks is substantially larger in the solution than found for the solid. This shows that the cations and anions are unevenly distributed at the solution surface with the latter ions preferentially residing in the topmost molecular layers of the solution. The relation between the cation and anion distributions is again manifested in the angular variation of the spectrum. The cation peak is thus further reduced in relative intensity at lower take-off angles, which shows that the distribution of these ions has a spread deeper into the solution than that of the anions in consistency with the potassium octanoate case of Fig. 4.

With reference to Figs. 4 and 5, it is of interest to compare the double layer structures for a given anion, but with different cations. Since the fatty acid salt systems are generally assumed to be dominated by long-range electrostatic forces in the formation of the double layer, such a comparison might be expected to yield only a small dependence on cation. The relative surface adsorption for potassium and cesium as counterions can be estimated from the C1s spectra of Fig. 6. Here, spectra are shown for 0.5 m solutions of potassium octanoate and cesium octanoate (normalized to the anion peak). As observed, there is indeed a substantial difference in anion/solvent intensity ratio between the cesium and potassium cases. Thus, with cesium as counterion the anion relative intensity increases, indicating a higher coverage than for the potassium case. There seems, consequently, to be specific interactions involved in the formation of the double layer in these systems. We interpret this as a result of the larger polarizability of the cesium ion compared to that of the potassium ion [11].

Discussing next the variation of the photoelectron relative intensities with respect to bulk concentration of the fatty acid salt, Fig. 7 shows a plot of the normalized intensity ratios for the anions vs. concentration (i.e., measured anion/solvent intensity ratios divided by the number of carbon atoms in the anion hydrocarbon chain [8]). Data are obtained for potassium and cesium octanoates, and values are, in each case, shown for two take-off angles of the electrons and, in addition, the variation expected on the basis of bulk concentration (dashed line) is displayed. Some interesting observations can be made from this plot. First, we note that all the curves have slopes that deviate strongly from the bulk variation, in particular, at low concentrations. This is the trend expected with the present surface-sensitive mode of investigation in

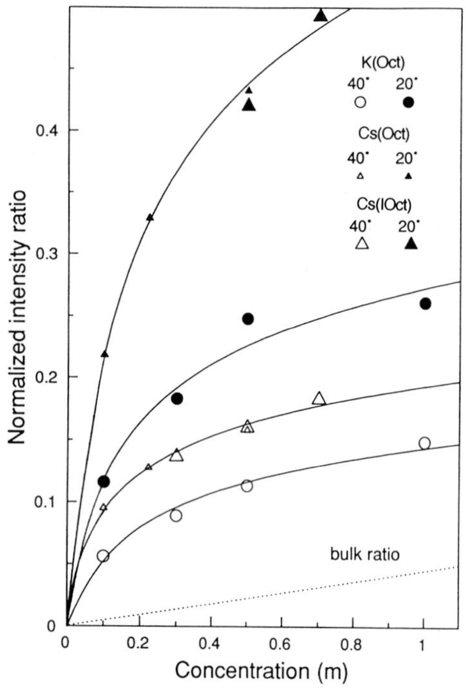

Fig. 7. Anion: formamide normalized intensity ratios obtained for the octanoate of cesium (Cs(Oct)) and potassium (K(Oct)) and 8-iodooctanoate of cesium (Cs(IOct)) vs. concentration at take-off angles of 20° and 40°. Curves are visual fits to the data

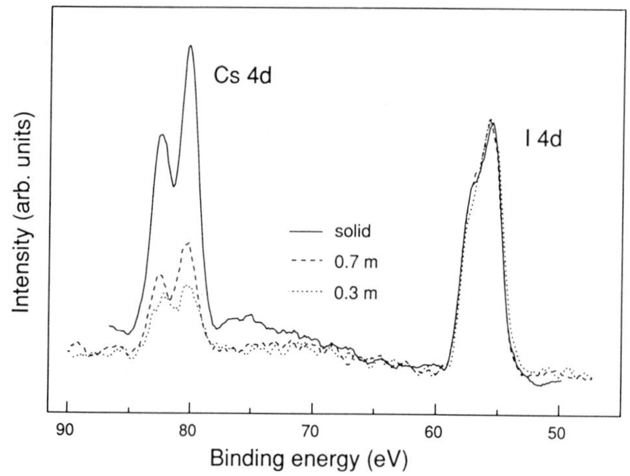

Fig. 8. The Cs4d and I4d spectral region of solid cesium 8-iodooctanoate and solutions of 0.7 m and 0.3 m concentration obtained at a take-off angle of 30°. Normalization to the I4d peak

which the sampling depth is comparable to the size of the anion overlayer. Thus, the data reflect, to a large extent, the overlayer growth which will be fastest at low bulk concentrations and then gradually level off as the overlayer approaches a saturated monolayer. Asymptotically, at high bulk concentrations, one expects a slope of the curves akin to the bulk variation which is also clearly observed for the higher take-off angles where the information is more bulk-dependent. In moving to lower take-off angles the slopes of the curves are larger for a given bulk concentration and show that the anion overlayers for the cases shown are still growing up to at least 0.5 m bulk concentration. Indeed, measurements of the surface tension of potassium octanoate formamide solutions that we have performed do not give evidence of a CMC point within the concentration range investigated here.

As an example of corresponding cation data, Fig. 8 shows the same spectral region as Fig. 5 of cesium 8-iodooctanoate, but with a comparison of the solid phase spectrum with solution spectra at two different bulk concentrations (0.7 m and 0.3 m, take-off angle 30°). These spectra show that the cation-anion ratio increases with bulk concentration, as also observed [5] for potassium octanoate. We believe that there are mainly two reasons for this, viz. the decrease in effective thickness of the cation diffuse layer with bulk concentration (according to Poisson-Boltzmann theory, inversely proportional to $(\text{concentration})^{1/2}$) and the formation of contact pairs at the higher concentrations.

Phase transfer catalysts

Phase transfer catalysis using tetraalkylammonium or phosphonium salts is an extremely powerful method in organic synthesis and has been employed in a multitude of applications [12–14]. Key features are: 1) the surface activity of these compounds in aqueous media, and 2) contact ion pairing at the interface to an organic phase, properties which together lead to efficient transport of reactants into the organic phase for subsequent reaction. The tetrabutylammonium ion is one of the most efficient catalytic agents in this group. Figure 9 shows spectra recorded from a solution of tetrabutylammonium perchlorate mixed with potassium chloride in formamide. The concentration of these two solutes was the same (0.5 m). The high binding energy Cl2p peak (due to the perchlorate ion) is, nevertheless, seen to be of substantially higher intensity (by about a factor of four at 40°)

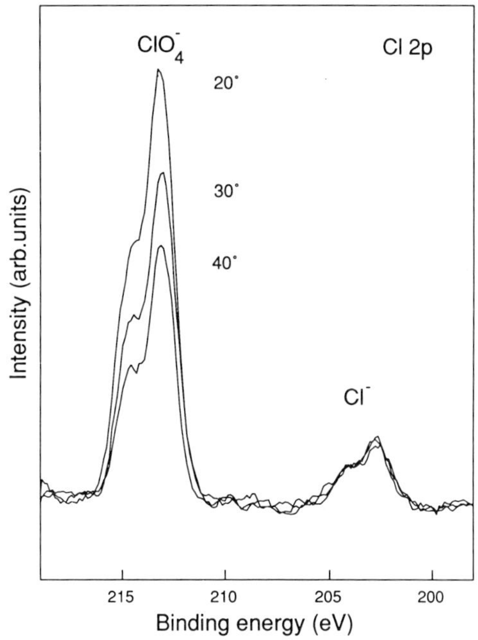

Fig. 9. Cl2p spectra recorded from a solution of tetrabutylammonium perchlorate (0.5 m) and potassium chloride (0.5 m) in formamide at three different take-off angles of the photoelectrons. Normalization to the chloride peak

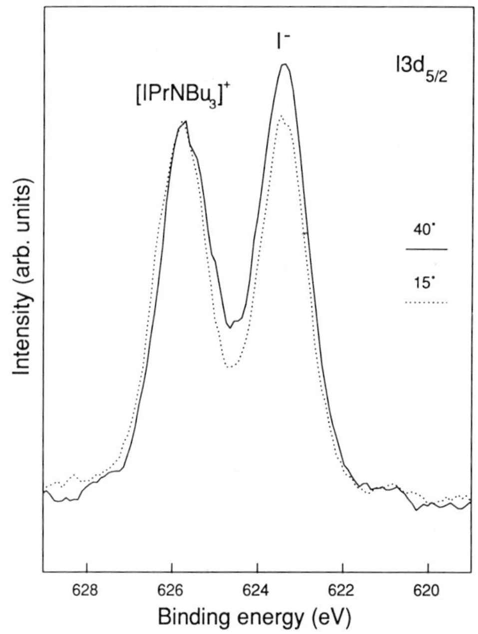

Fig. 10. $I3d_{5/2}$ spectra obtained from a 0.5 m solution of $\{(IPr)NBu_3\}^+I^-$ in formamide at photoelectron take-off angles of 15° and 40°. Normalization with respect to the positive ion peak. This peak is somewhat broader than the iodide peak which leads to the observed peak height ratio

than that of lower binding energy (due to the chloride). This clearly indicates formation of a surface active complex involving the tetrabutylammonium and the perchlorate ions. This complex is formed while the chloride ion remains essentially in the bulk of the solution. The nature of the surface-active perchlorate complex may be inferred from the fact that the two negative ions do behave differently, hence, specific chemical interactions have to be involved in the interaction with the cation. Moreover, we find that, in comparison with corresponding spectra obtained from the fatty acid salts, the enhancement observed for the perchlorate signal is high. Both of these factors indicate that the perchlorate ions form a layer with direct contact to the cation overlayer without solvent separation.

In order to study in more detail the relation between cations and anions in these systems, we have also synthesized a compound, $(IPrNBu_3)I$, which contains iodine as common element in both ions. Iodine is advantageous also from the point of view of containing photoelectron lines of substantial intensity (I3d). Figure 10 shows the $I3d_{5/2}$ spectrum from a 0.5 m solution of this compound at two different take-off angles. As can be seen, there is a substantial binding energy shift between the two chemical states of iodine. Also, the intensity (peak area) ratio is very close to 1, which strongly indicates the presence in this case of contact ion pairing at the surface which is similar to the perchlorate case above. The small, but clearly measurable, angular variation in the ratio can be interpreted in terms of a structure where the anions lie underneath the cations at the surface.

When used in phase transfer systems, the tetraalkylammonoim iodide and perchlorate salts show by far the highest transfer numbers between aqueous and organic phase [15]. This is in line with the formation of contact ion pairs also at the solution/vacuum interface in these cases, as observed in Figs. 9 and 10. It is thus of interest to study the effect of competition between the iodide and perchlorate ions on the surface activity in these systems. Figure 11 shows the $I3d_{5/2}$ spectrum obtained from a solution of $(IPrNBu_3)I$ (0.5 m) and $KClO_4$ (0.5 m) in formamide at 40° take-off angle. In comparison with the corresponding spectrum in Fig. 10 (shown as a solid line in Fig. 11), there is now a pronounced

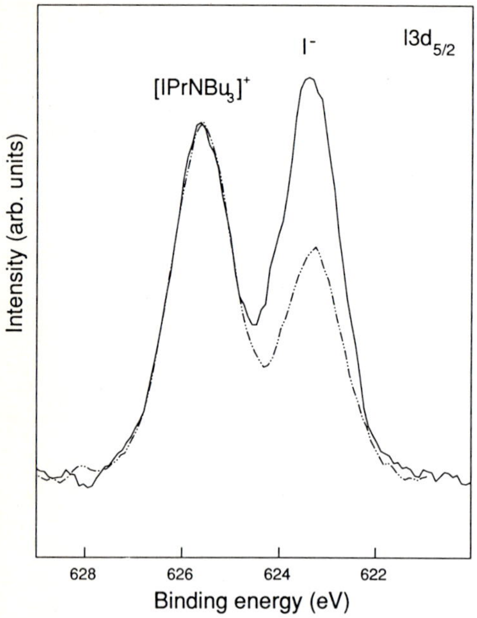

Fig. 11. Dashed-dot spectrum: $I3d_{5/2}$ spectrum obtained from a solution of $\{(IPr)NBu_3\}^+I^-$ (0.5 m) $KClO_4$ (0.5 m) in formamide at a take-off angle of 40°. For comparison, we show (solid line) the $I3d_{5/2}$ spectrum obtained from a formamide solution of only $\{(IPr)NBu_3\}^+I^-$ (0.5 m) at 40° take-off angle. Normalization is with respect to the positive ion peak

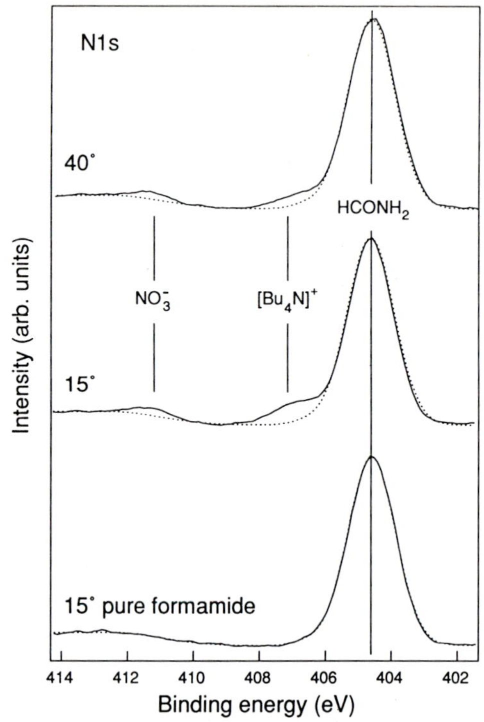

Fig. 12. N1s spectra of tetrabutylammonium nitrate (0.5 m) in formamide at take-off angles of 40° (top) and 15° (middle). The background shown in the spectra was obtained by fitting to the corresponding spectra of pure formamide (identical for 40° and 15°; bottom)

increase in the relative intensity of the positive ion peak with respect to the iodide peak. A corresponding relative intensity increase is found also for the positive ion C1s peak relative to that of the solvent. Our interpretation of this result is that of a salting-out effect, where the perchlorate ions participate at the surface in forming additional contact ion pairs with the tetraalkylammonium ions. Corroborative evidence for this interpretation is provided by a significant enhancement of the Cl2p relative intensity with respect to that expected from the bulk concentration of the salt. A similar effect has also been observed by us for corresponding bromine-substituted analogues [16].

In contrast to the iodide and perchlorate tetra-alkylammonium salts, the corresponding nitrate ion salts, in general, show comparatively low transfer numbers [15]. Figure 12 shows the N1s spectra obtained from a 0.5 m solution of $(Bu_4N)NO_3$ in formamide at two different take-off angles, along with the spectrum of the same region obtained from pure formamide. The solute peaks at 40° take-off angle are enhanced by ≈ 1.8 (NO_3^--peak at 411.5 eV) and ≈ 2.5 (Bu_4N^+-peak at 407 eV) with respect to expected bulk intensities. Changing the take-off angle to 15° clearly shows that the cation peak increases relative to the anion peak. Subtractions of the pure N1s spectrum from the 40° and 15° spectra of the solution are displayed in Fig. 13 to demonstrate the effect in more detail. This observed angular variation is significantly larger than that observed in the iodine case of Fig. 10. We thus conclude that the nitrate compound does represent a different kind of surface activity, whereby the anions do not form close ion pairs with the cations at the surface, but reside deeper in the bulk. The structure of the surface double layer is thus more diffuse in character, and akin to the fatty acid salt cases. This is indicative of the relatively larger importance of bulk solvation forces in this case.

Conclusion

In this paper, we have presented the technique of angular resolved core electron spectroscopy for

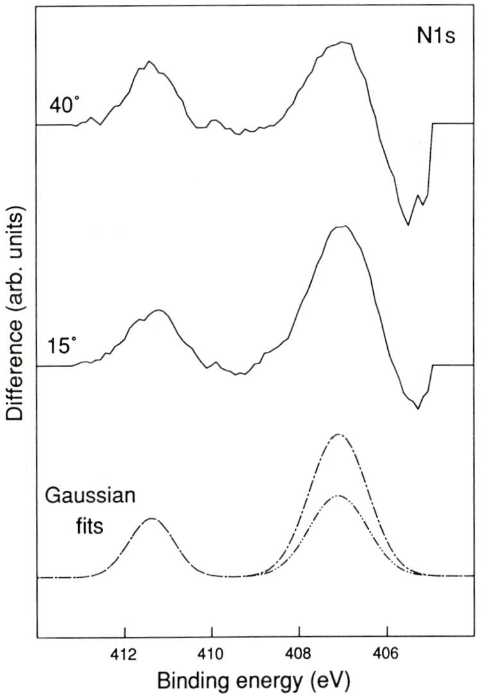

Fig. 13. N1s spectra obtained by subtraction of the pure formamide spectrum from the 40°- and 15°-spectra in Fig. 12. Bottom shows Gaussian fits to the respective spectrum

liquid phase samples. With this technique, we have investigated a number of electric double-layer systems at solution surfaces. For fatty acid salt systems, diffuse counterion distributions generally prevail, which is in line with the predominance of long-range electrostatic forces in deciding the double layer structure. Significant differences have, however, been observed between potassium and cesium as counterions, which shows that specific forces may also be of importance in these systems.

For phase transfer catalyst compounds the structure of the surface double layer has been found to be strongly dependent on the anion. The occurrence of closely bound anion layers in the cases of iodide and perchlorate is in agreement with high interphase transfer numbers for these compounds. Conversely, a low transfer number for nitrate is found to correlate with the absence of contact pairs at the surface and, hence, a more diffuse character of the anion layer.

We conclude that electron spectroscopy provides an interesting future possibility for the investigation of structures at the solution/vacuum interface. Further studies involving other ions, concentration dependencies, as well as different types of solvents are in progress.

Acknowledgements

This work was supported by the Swedish Natural Science Research Council (NFR) and the Bank of Sweden Tercentenary Foundation.

References

1. Siegbahn HOG (1985) J Phys Chem 89:897
2. Holmberg S, Moberg R, Yuan ZC, Siegbahn H (1986) J Electron Spectrosc 41:337
3. Holmberg S, Yuan ZC, Moberg R, Siegbahn H (1988) ibid 47:27
4. Moberg R, Bökman F, Bohman O, Siegbahn HOG (1991) J Am Chem Soc 113:3663
5. Moberg R, Bökman F, Bohman O, Siegbahn HOG (1991) J Chem Phys 94:5226
6. Moberg R, Bökman F, Hagfeldt A, Lindquist S-E, Bohmån O, Siegbahn HOG (1990) UUIP (Uppsala University Institute of Physics Report) 1218
7. Croxton T, McQuarrie DA, Patey GN, Torrie GM, Valleau JP (1981) Can J Chem 59:1998
8. Patey GN, Torrie GM (1989) Chem Scr 29A:39
9. a. Braslau A, Persham PS, Swislow G, Ocko BM, Als-Nielsen J (1988) Phys Rev A38:2457
   b. Yun WB, Bloch JM, Rev Sci Instr 60:214 (1989)
10. Penfold J, Thomas RK (1989) RAL-89-057 Rutherford Appleton Lab Rep)
11. Bökman F, Bohman O, Siegbahn HOG Surf Sci (in press)
12. a. Makosza M, Serafinowa B (1965) Rocsz Chem 39:1223
    b. Makosza M, Wawrzyniewicz W (1969) Tetrahedron Lett 4659
13. Starks CM (1971) J Am Chem Soc 93:195
14. Brändström A (1974) Preparative Ion Pair Extraction, an Introduction to Theory and Practice. Apotekarsocieteten, Hässle Läkemedel
15. a. Gustavii K (1967) Acta Pharm Suec 4:233
    b. Gustavii K, Schill G (1966) ibid 3:259
16. Bökman F, Bohman O, Siegbahn HOG (1992) J Phys Chem 96:2278

Author's address:

H. O. G. Siegbahn
Dept. of Physics
University of Uppsala
Box 530
S-751 21 Uppsala
Sweden

Scanning force microscopy – examples of applications to surface chemistry

R. Erlandsson, H. Elwing, M. Eriksson, L. Olsson, P. Tengvall, R. Wigren, S. Welin-Klintström, and I. Lundström

Laboratory of Applied Physics, University of Linköping, Sweden

Abstract: Some recent results from the scanning force microscopy activity at our laboratory are presented. A brief description of attractive mode force microscopy is followed by a discussion of the following examples: O_2/H_2-induced morphology changes in thin palladium films, structure of spin cast polysulfone films, fibrinogen adsorption on hydrophobic SiO_2, and force measurements on hydrophobic/hydrophilic substrates.

Key words: Scanning force microscopy; atomic force microscopy; palladium; polysulfone; fibrinogen

Introduction

Scanning force microscopy (SFM, also known as AFM, atomic force microscopy) is one of the new scanning probe techniques which holds promise for becoming an important tool for the study of surface chemical phenomena. Since the original demonstration of the SFM principle by Binnig et al. [1], several interesting applications of the technique related to organic systems and surface chemistry have occurred in the literature [2–11]. In this contribution, we describe the development of SFM in our laboratory with a special emphasis on examples of applications which demonstrate the usefulness and versatility of the technique for the investigation of surface chemical phenomena. After a brief description of the instrumentation and the different modes of operation, we discuss experimental observations related to i) structural changes in thin catalytic metal films, ii) the crystallinity of thin polymer films (on silicon), and iii) the structure and behavior of fibrinogen molecules adsorbed on a hydrophobic silicon (dioxide) surface. Furthermore, we discuss how the details of the tip-surface interaction can lead to ultra-high resolution of the border between a hydrophobic silicon surface and an adsorbed fibrinogen molecule. The sharpness of this structure is much higher than would be expected from the size of the probe (tip) used in our SFM.

Instrumentation

The principle of a scanning force microscope is shown in Fig. 1. The sample is mounted on a scanner consisting of a piezoelectric ceramic tube as described by Binnig and Smith [12]. With such a tube it is possible to move the sample in all three dimensions, x, y, and z, with a resolution better than 1 Å and with maximum scan lengths on the order of a few μm. The movements are controlled by voltages applied over the different segments of the piezoelectric tube. A small tip, mounted at the end of a lever (a spring element), is placed close enough to the sample surface to feel the interaction forces with the surface. The motion of the lever is measured with a deflection sensor, which in our instrument consists of a laser interferometer [13]. The proximity to the surface can be measured either by the instantaneous deflection of the lever, which gives the force when multiplied with the spring constant, or by a resonance method that uses the modulation of the lever resonance frequency that takes place when a force is acting on the tip (see below). In either case, a topographic

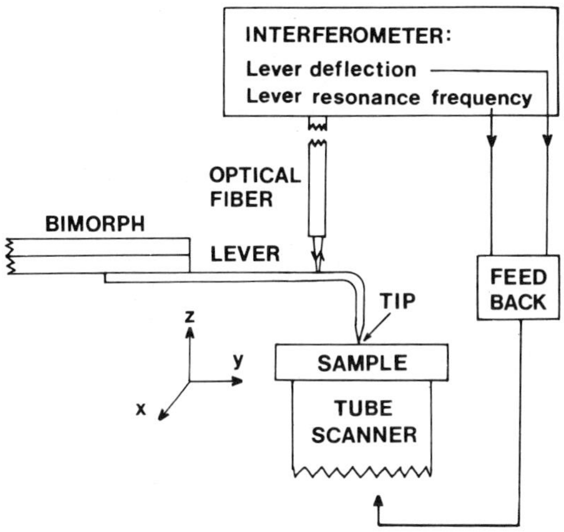

Fig. 1. Schematic illustration of an SFM. The laser interferometer is used to measure either lever deflection or the shift in lever resonance frequency. See text for further details.

image is obtained as the sample is raster scanned laterally under the tip while the tip-sample separation is kept constant by feedback using one of the methods above. By plotting the z-position of the sample (derived from the voltage on the piezo) as a function of the (x, y) coordinates, a representation of the sample topography is obtained.

As the interaction forces between the tip and the surface are either attractive (van der Waals, electrostatic, capillary, magnetic) or repulsive (contact forces due to wave function overlap), the force microscope can be operated in two distinctly different modes with different characteristics. To use the instrument in the attractive mode is an advantage when studying fragile structures (for example biological molecules), as the disturbance of the sample is smaller. The instruments used in this mode are also better suited to measure the interaction forces far from the surface. Repulsive force imaging, on the other hand, is capable of obtaining higher resolution, in many cases atomic, and can be used in liquids. We have specialized in using attractive mode imaging, which is advantageous for many of the problems we are addressing.

As the overall force on the tip is normally very small when working in the attractive mode, it is advantageous to use the resonance method mentioned above. In this case, the lever is oscillated close to its resonance frequency by the use of a bimorph which is a piezoelectric device capable of exciting oscillations of the lever. The resonance frequency is then modified by the derivative of the force between the surface and the tip, F', according to:

$$\frac{\delta\omega}{\omega} = F'/2k \,, \tag{1}$$

where ω is the resonance frequency of the lever without force interaction, k is the spring constant of the lever, and $\delta\omega$ is the induced frequency shift [14]. In this case, the z-coordinate of the sample is controlled in order to keep the frequency shift constant. We thus get traces of a constant force derivative during lateral scanning, which also follows the topography of the sample surface if the interaction properties do not vary. The resonance frequency shift is measured by detecting how the oscillation amplitude varies as the resonance curve shifts while the oscillation frequency is kept constant. If the shape of the resonance curve is known and no damping occurs, the change of amplitude will give the frequency shift. When working in ambient atmosphere, however, the gap between tip and surface is bridged by a liquid phase for small separations, which introduces damping in the system. A stable distance regulation can often still be obtained, as the damping varies continuously with separation, but the interpretation in terms of force derivative is no longer possible. The characteristics of the damping is, however, critically depending on the surface state and can be used to investigate lateral variations of the tip-surface forces as will be described below. All measurements presented here were made in air which means that liquid bridge formation normally takes place at some separation distances.

The tip/lever unit can be micromachined in silicon [15] or be made out of a metal wire. In our instrument, they were made out of approximately 1-mm-long and 20-μm-thick tungsten wires which were bent 90° close to one end to form the tip. The final tip radius obtained through electrochemical etching was about 20 nm, at best, as revealed by scanning electron microscopy (SEM). These levers had resonance frequencies of about 20 kHz.

Force microscopy data (as well as data from other scanning probe techniques) are normally presented in one of three different ways: In line scan images each line represents a profile of the surface along the x direction, and the subsequent lines are displaced as the y coordinated is incremented stepwise during the measurement (Fig. 7a is an example of this). In top-view images, the intensity (or color) represents the height of the surface at each point (Figs. 3a and 5a). In three-dimensional (3D) rendered images, the topographic data is used to construct a perspective view of the surface where the intensity at each point corresponds to the "reflection" from a hypothetical light source, giving a realistic looking representation of the surface topography (Figs. 2, 3b, 4, 5b, and 7b).

A detailed description of our instrument for measurements in air has been published elsewhere [16]; a review of the applications of SFM can be found in [17].

Structural changes in thin catalytic metal films

Heterogeneous catalysis and chemical sensors are among the areas where we have applied SFM. The studies concern the properties of thin films of metals like Pd and Pt. Figure 2 shows SFM images of how structural changes occur in a 10-nm-thick Pd-film supported on SiO_2 substrates due to catalytic reactions on the metal surface. Figure 2a is a 3D representation of a 10-nm-thick Pd-film after thermal evaporation at room temperature at a pressure of 10^{-7} Torr, showing a roughness of less than 1 nm. The sample was subsequently transferred to a UHV chamber where cyclic oxygen/hydrogen exposures, consisting of O_2 pulses in the 10^{-8} Torr region and H_2 pulses at half the pressure were made at 473 K. Subjecting the Pd film to this treatment for a couple of hours leads to a reconstruction of the film, giving the topography shown in Fig. 2b. The Pd film is now completely changed and consists of individual, drop-shaped islands of 20–30-nm height with a bare SiO_2 surface in between. A more thorough description of this work can be found elsewhere [18]. We believe that the reconstruction taking place at the temperature used here is not caused by hydrogen interacting with the Pd film, or that it is initiated by the exothermicity of the water forming reaction. Instead, we propose the following explanation. The as-prepared Pd film, with a roughness of less than 1 nm, contains a number of grain boundaries in which impurities like carbon and oxygen are enriched. The film is in a metastable state stabilized by these impurities. Upon exposure to cyclic H_2/O_2-pulses in a UHV chamber these impurities are catalytically combusted. When they disappear it is energetically more favorable for the metal to form small droplets. The possibility to form 3D images of these surfaces is very advantageous for the studies of this phenomenon, as the previously available TEM data only give the lateral extension of the metal islands [18].

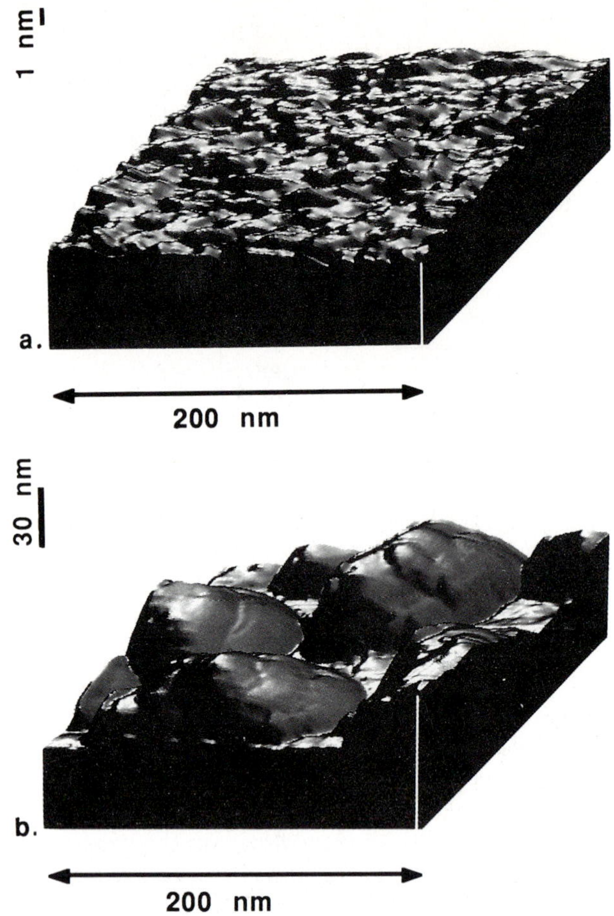

Fig. 2. Three-dimensional representations of SFM-results on a 10-nm-thick Pd-film evaporated on SiO_2 (on Si); a) before and b) after cyclic exposures to O_2 and H_2 at 200 °C in a UHV-chamber at a base pressure of 10^{-10} Torr. The oxygen pulses were about 10^{-8} Torr and the hydrogen pulses about 0.5×10^{-8} Torr. (From [18])

Crystalline regions in thin polymer films

One of the interesting properties of SFM is that it can be applied to any type of material without any special sample preparation prior to the microscopy. SFM therefore seems well suited for the study of polymers and polymer films. Dietz et al.

[9] have recently published some SFM-images of commercial ultrafiltration membranes made of amorphous polysulfone which show that the polymer surface is amorphous with no crystalline structure. It is therefore of interest that in a preliminary study we found crystalline structures in polysulfone films prepared by dripping a solution of the polymer onto a hydrophilic silicon substrate spinning at 2000 rpm. The SFM-pictures in Fig. 3 were obtained in air after the evaporation of the solvent. The top view (Fig. 3a) indicates that the polymer has formed a hexagonal lattice with a repeat distance of about 50 nm. A 3D representation of the results shows the crystalline regions as cone-shaped structures with a height of about 40 nm (Fig. 3b). However, the true shape of the individual units can very well be spherical as tip-imaging effects distort the image. It should be noted, however, that we have not yet more closely examined the nature of these ordered structures. A scanning force microscope operating in the attractive mode appears to be well suited for the analysis of structures having rather large dimensions perpendicular to the surface, like in this and the previous example. It is, however, important to realize that the geometry of the tip will always be convoluted with the sample topography. The arrangement, the average separation, and the height of the structures should, however, be rather insensitive to the tip shape for structures of these dimensions.

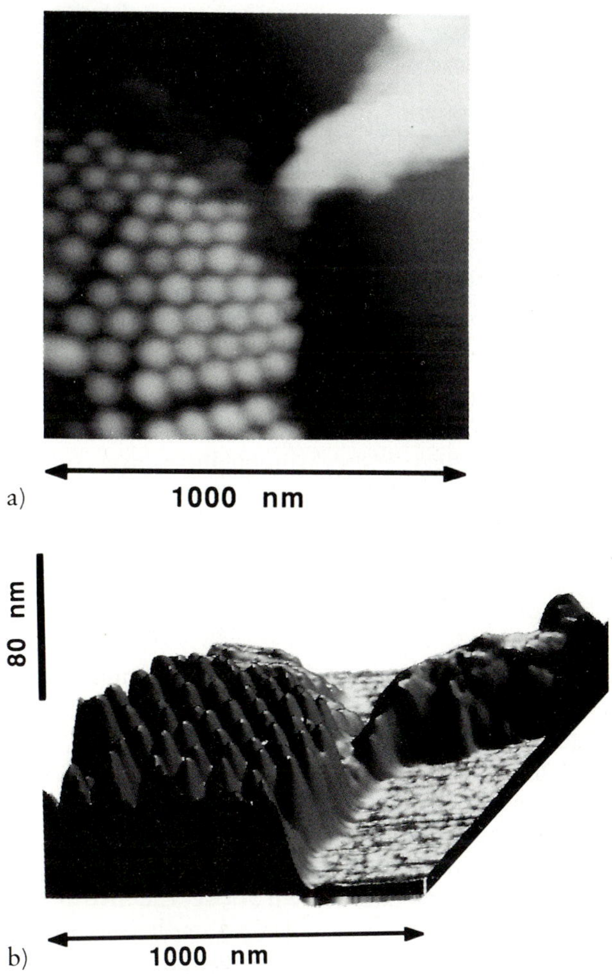

Fig. 3. SFM image of a part of a polysulfone film cast on a silicon surface with a spinning technique: a) top view, b) three-dimensional representation.

Structures of adsorbed fibrinogen molecules

SFM has been used to image adsorbed proteins in buffer and to observe the formation of fibrin networks on surfaces [3, 5, 6]. The details of the internal structure of individual molecules were, however, not revealed in these experiments. Recently, however, molecular resolution images of Langmuir–Blodgett films and DNA in aqueous buffer have been presented for an SFM using repulsive contact forces [8]. In another work, it was argued that, for biomolecules adsorbed on surfaces, the repulsive mode imaging may lead to interaction forces strong enough to give rise to a distortion or destruction of the adsorbed molecules [5]. It is therefore of interest that it is possible to obtain high-resolution images of adsorbed protein mo' :-cules in the attractive mode in air, where the tip moves further away from the surface and there are no repulsive contact forces involved [19, 20]. Figure 4 shows a 3D-rendered image of a fibrinogen molecule adsorbed on a hydrophobic SiO_2 substrate. The SFM revealed some extra features in addition to the generally accepted tri-nodular elongated form of (adsorbed) fibrinogen molecules. We also observed that the molecules could be adsorbed in two different ways, either as the shown tri-nodular form, but also as more globular structures, approximately 40 nm in diameter. These two forms have also been observed in recent TEM studies by Nygren and Stenberg [21]. The possibility to obtain three dimensional pictures of adsorbed molecules, as shown in Fig. 4, is perhaps the most unique feature of scanning force microscopy. During the studies of adsorbed fibrinogen molecules, we also observed a slow polymerization of adsorbed fibrin-

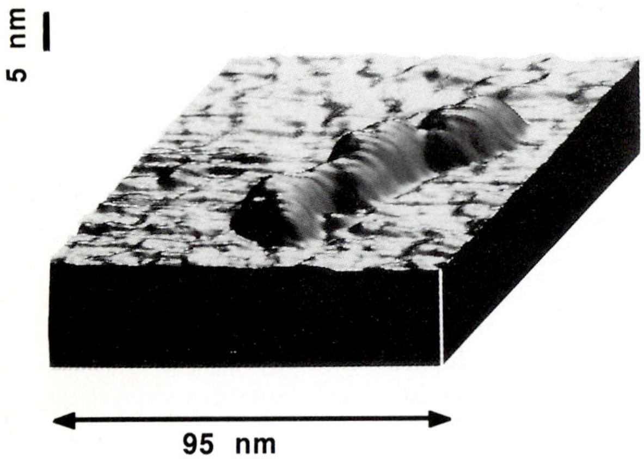

Fig. 4. Three-dimensional representation of an adsorbed fibrinogen molecule on a hydrophobic silicon dioxide surface. The protein was adsorbed during 4 h from a solution (PBS-buffer) containing 0.2 μg/ml fibrinogen. The surface was then rinsed in distilled water and dried in flowing nitrogen. (From [19])

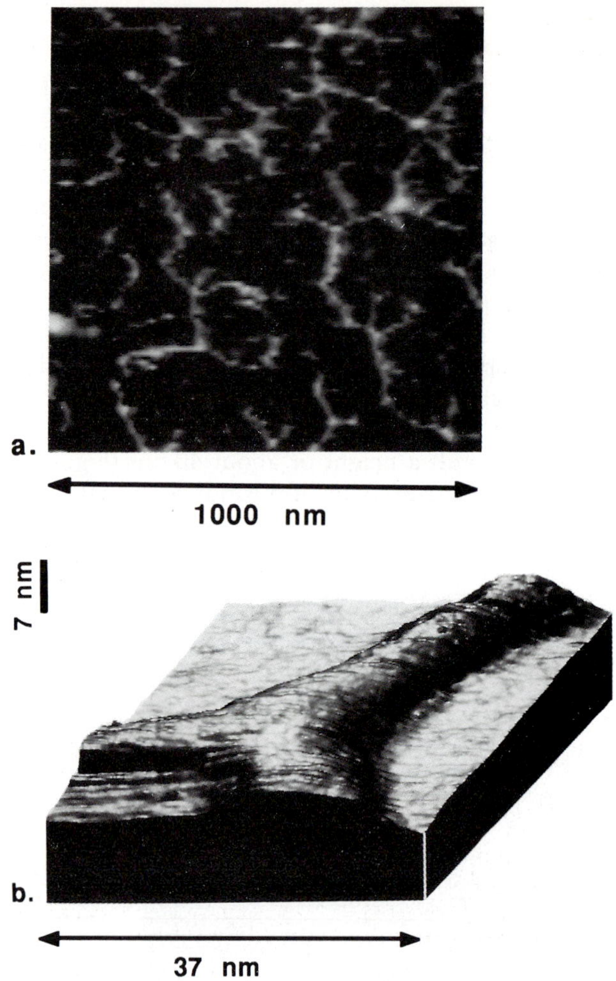

Fig. 5. a) Top view of a polymeric network of fibrinogen formed on a hydrophobic silicon surface stored in clean PBS buffer for 11 days after adsorption of fibrinogen for 5 min from 1.0 μg/ml protein (From [19]). b) High magnification 3D rendered image of an intersection of the network.

ogen molecules during storage of the sample in clean buffer after the adsorption step. Figure 5a shows a top view of a polymeric network formed after 11 days incubation in clean buffer. Detailed studies of the polymeric network indicated that the polymeric strands are about 12-nm-wide and 4-nm-thick (see Fig. 5b). Furthermore, the branching points almost always showed take-off angles close to 120°, suggesting a symmetry in the polymer formation [19].

Observation of surface properties related to hydrophobic interactions

Our experience from the investigations of biomolecules adsorbed on SiO_2 substrates tells us that the details of the tip-sample interaction forces are very important for the interpretation of the images. For large tip-sample separations (> 5 nm), the interaction is normally due to a van der Waals force, while capillary forces due to a liquid bridge between the tip and the sample dominate for smaller distances [10, 11, 22]. The force microscope used for attractive mode imaging is well suited to study these phenomena, as the resonance method described above is extremely sensitive and can measure the interaction force at large separations. At separations where no liquid bridge formation has occurred, we observe a force derivative increase when decreasing the tip-surface separation that is consistent with a van der Waals interaction. As the liquid bridge formation takes place, there is a strong signature in the resonance amplitude due to damping which can be used to investigate the properties of this liquid bridge.

For high-resolution imaging of biomolecules, as in the fibrinogen data shown above, the instrument is operated with a small tip-sample separation where these capillary forces are dominating. It is clear that these kind of interactions are strongly dependent on the state of the surface and, in particular, on its hydrophobicity. In order to study these phenomena, we made an investigation of forces versus tip-sample separation on SiO_2 samples treated to give different surface free energies [23]. The details of the force interactions are given by the "approach curves", as shown in Fig. 6a. The upper curve gives the static force (derived from the deflection of the lever), while the lower curve gives the resonance amplitude which gives the force derivative (before liquid bridging occurs at point A) and damping properties (after liquid bridging occurs). On a hydrophilic SiO_2 sample, these curves are identical over the surface, while a methylated sample (hydrophobic) gives two distinctly different behaviors. As is shown in Fig. 6b, it is possible to operate the instrument in a mode that gives images in which the contrast is related to interaction properties. The hydrophobic surface has islands of approximately 100-nm-diameter where the interaction properties are different from the surrounding area. The interpretation of these images is complex and is the subject of our present work. The observation could indicate that the methylated silicon surfaces is non-uniform, with islands of CH_3 groups. It can, however, not be ruled out that the effect is due to micro-droplet formation due to condensation from the atmosphere. In a study of fundamental wetting properties in UHV by Dubois et al. [24], it is suggested that water molecules will cluster on a hydrophobic surface, in analogy with the non-wetting behavior observed in contact-angle studies.

Figure 7b shows an expanded view of the border between an adsorbed fibrinogen molecule of the globular form (Fig. 7a) and the hydrophobic silicon dioxide surface. A phenomenon taking place over only a few Å is thus easily visible, although we are operating in the attractive mode with a tip radius which is about 20 nm at best. The significant feature in Fig. 7 looks like a very sharp downwards step from the hydrophobic surface to the protein molecule. This observation indicates that the image formation (in the attractive mode) is strongly dependent on the force-inducing properties of the surface under study. In Fig. 7, we make the assumption that the protein molecule presents a more hydrophilic surface towards the tip than the hydrophobic silicon surface. A tentative explanation of the phenomenon in Fig. 7 could then be a difference in microcapillary effects between the tip and the surface in the two regions. On the hydrophobic surface there is a water bridge between the tip and the surface. On the hydrophilic surface the proper-

a)

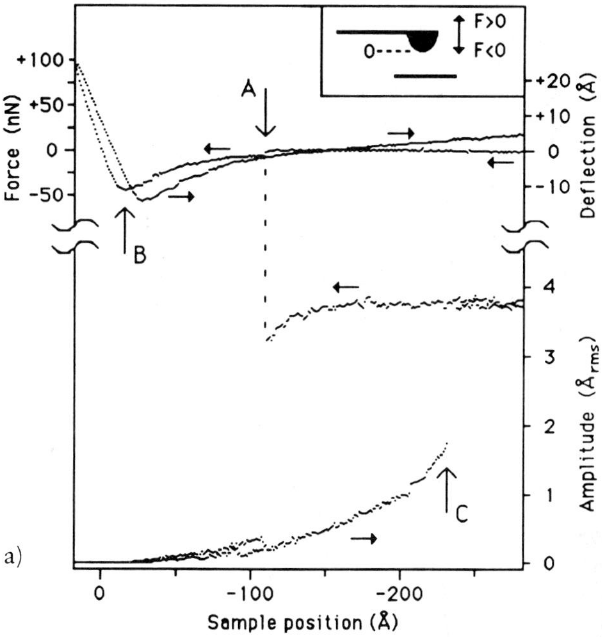

b)

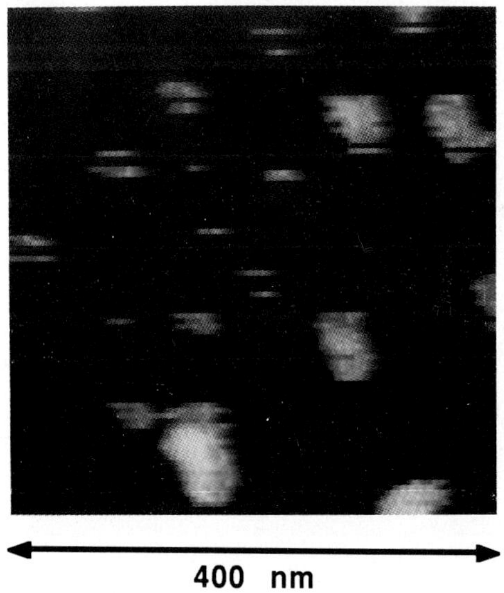

Fig. 6. a) A typical "approach curve" showing the force (upper curve) and oscillation amplitude (lower curve) as the sample is moved in and out of the interaction region with the feedback turned off. At point A a liquid bridge is formed and the resonance is damped, which also gives a small response in the force curve. Point B indicates when the sample comes in repulsive contact with the tip. At point C the liquid bridge disappears when retracting the sample. (In this particular case, a bias voltage of 1.5 V was applied to the tip, giving an electrostatic long-range attraction.) b) Image showing the lateral extension of areas with modified interaction properties on a methyl-silanized SiO_2 surface. See text for further details. (From [23])

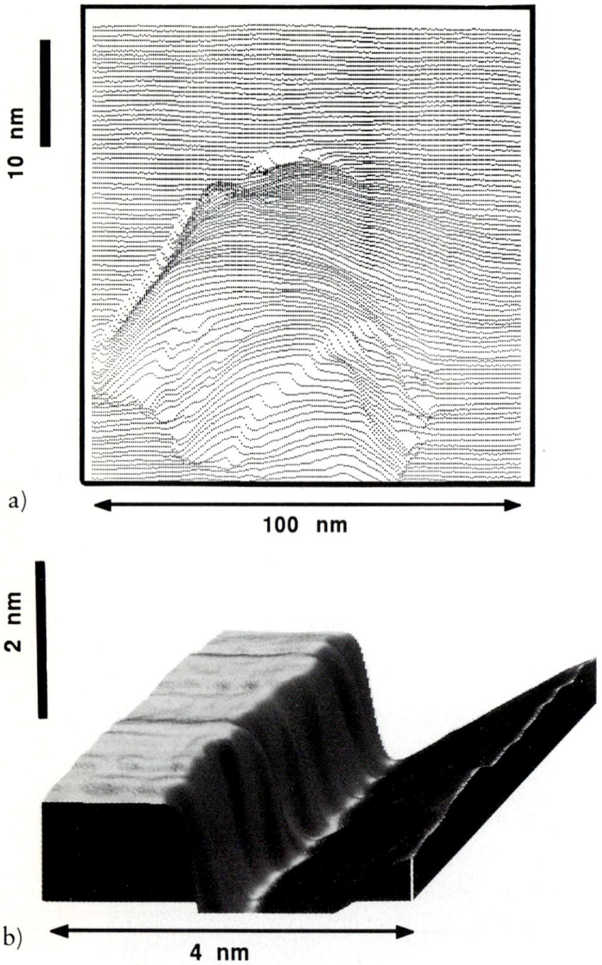

Fig. 7. a) A line scan image of a globular fibrinogen molecule adsorbed on a hydrophobic SiO_2 surface. Observe the depression at the edge of the molecule. b) A high-resolution 3D-rendered image of the edge between the substrate and the molecule.

ties of this bridge change, leading to a weaker interaction. The surface will therefore approach the tip, which will appear as a step downward, although the molecule sticks out from the surface. It has been suggested by Wiggins [25] that the internal structure of water is influenced by the hydrophobicity of an adjacent surface, and that an ice-like, low-density phase will occur close to a hydrophobic surface. An effect like this could account for the observed behavior, but much more work is needed to clarify the situation. One interesting aspect is the sharpness of the step in Fig. 7, being only a few percent of the tip radius. Even though the detailed mechanisms behind this effect

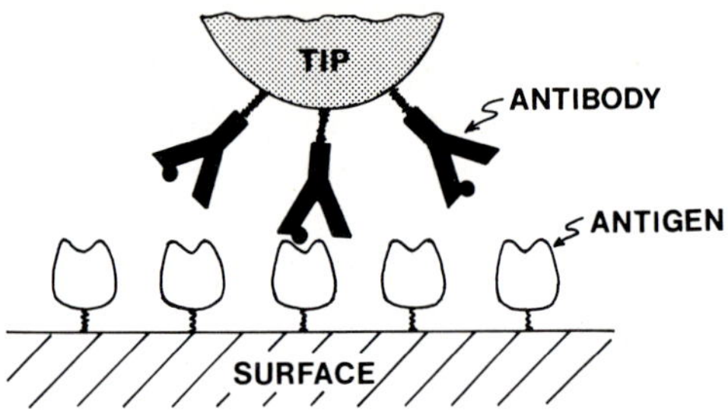

Fig. 8. The figure shows a possible future use of SFM to study molecular interactions, like that between antigen and antibodies.

are not understood at present, it clearly indicates that variations in surface properties linked to hydrophobicity can be imaged with very high resolution.

Discussion

We have demonstrated the use of SFM for surface chemical problems by a few examples. In the examples chosen we have observed structural changes of thin catalytic metal films due to chemical reactions on the film, crystalline regions in thin layers of polymers, high-resolution images of adsorbed protein molecules and polymer formation from adsorbed molecules. Furthermore, it was pointed out that (in the attractive mode) the details of the interaction between the surface and the SFM-tip may give rise to interesting phenomena, such as a resolution of features close to the atomic scale, although the tip diameter is several tens of nanometers. This effect which, at present, is under study suggests that the SFM is of interest to surface chemistry, not only as an imaging device, but also for detailed studies of surface physical phenomena such as the properties of (micro) hydrophilic regions on a hydrophobic surface and vice versa. The study of interaction forces can be further extended by modifications also of the surface of the tip. One may speculate about the possibility to bind one of the molecules in a molecular interaction pair on the surface and the second on the tip to facilitate the direct measurement of binding forces like those between antigens and antibodies, as suggested in Fig. 8.

Acknowledgments

The work presented in this paper was supported by several grants from the Swedish National Board for

Technical Development (STU) and the Swedish Research Council for Engineering Sciences (TFR).

References

1. Binnig G, Quate CF, Gerber Ch (1986) Phys Rev Lett 56:930–933
2. Albrecht TR, Dovek MM, Lang CA, Grütter P, Quate CF, Kuan SWJ, Frank CW, Pease RFW (1988) J Appl Phys 64:1178
3. Drake B, Prater CB, Weisenhorn AC, Goulder SAC, Albrecht TR, Quate CF, Cannell DS, Hansma HG, Hansma PK (1989) Science 243:1586–1589
4. Gould SAC, Drake B, Prater CB, Weisenhorn AC, Manne S, Hansma HG, Hansma PK, Massie J, Longmire M, Elings V, Dixon Northern B, Mukargee B, Patersson CM, Stoeckenius W, Albrecht TR, Quate CF (1990) J Vac Sci Technol A8:369–373
5. Lin JN, Drake B, Lea AS, Hansma PK, Andrade JD (1990) Langmuir 6:509–511
6. Weisenhorn AL, Drake B, Prater CB, Gould SAC, Hansma PK, Ohnesorge P, Egger M, Heyn SP, Gaub HE (1990) Biophys J 58:1251–1258
7. Weisenhorn AL, MacDougall JE, Gould SAC, Cox CD, Wise WS, Massie J, Maivald P, Elings VB, Stucky GD, Hansma PK (1990) Science 247:1330
8. Weisenhorn AL, Egger M, Ohnesorg G, Gould SAC, Heyn S-P, Hansma HG, Sinsheimer RL, Gaub HE, Hansma PK (1991) Langmuir 7:8–12
9. Dietz P, Hansma PK, Herrman K-H, Inacker O, Lehmann H-D (1991) Ultramicroscopy 35:155–159
10. Blackman GS, Mate CM, Philpott MR (1990) Phys Rev Lett 65:2270
11. Mate CM, Novotny VJ (1991) J Chem Phys 94(12):8420
12. Binnig G, Smith DPE (1986) Rev Sci Instr 57:1688–1689
13. Rugar D, Mamin HJ, Erlandsson R, Stern JE, Terris BD (1988) Rev Sci Instr 59:2337
14. Martin Y, Williams CC, Wickramasinghe HK (1987) J Appl Phys 61:4723
15. Albrecht TR, Akamine S, Carver TE, Quate CF (1990) J Vac Sci Technol A8(4):3386
16. Erlandsson R, Wigren R, Olsson L (1991) Microscopy Microstructures Microanalysis 1:471–480
17. Rugar D, Hansma P October 1990, Physics Today, 23–30
18. Erlandsson R, Eriksson M, Olsson L, Helmersson U, Lundström I, Petersson L-G (1991) J Vac Sci Technol 89:825–828
19. Wigren R, Elwing H, Erlandsson R, Welin S, Lundström I (1991) FEBS Lett 280:225–228
20. Wiegräbe W, Nonnenmacher M, Guckenberger R, Wolter O, J of Microscopy (in press)
21. Nygren H, Stenberg M (1988) J Biomed Mater Res 22:1–11
22. Erlandsson R, McClelland GM, Mate CM, Chiang S (1988) J Vac Sci Technol A6:266–270
23. Olsson L, Tengvall P, Wigren R, Erlandsson R (1992) Ultramicroscopy
24. Dubois LH, Zegarski BR, Nuzzo RG (1990) J Am Chem Soc 112(2):571
25. Wiggins PM Dec (1990) Microbiological Reviews 432–449

Authors' address:

R. Erlandsson
Laboratory of Applied Physics
University of Linköping
S-581 83 Linköping
Sweden

Progress in Colloid & Polymer Science Progr Colloid Polym Sci 88:162–168 (1992)

Force feedback measurements of surface forces

J. L. Parker and A. M. Stewart

Department of Applied Mathematics, Research School of Physical Sciences and Engineering, The Australian National University, Canberra, Australia

Abstract: The surface force apparatus developed by Israelachvili, and variants thereof, rely on the measurement of the deflection of a weak spring in order to detect the force between two surfaces. Mechanical instabilities occur when the gradient of the surface force overcomes the spring constant and regions of the force curve become inaccessible. We report here a technique which partly overcomes thisrestriction. It is based on a combination of the bimorph force sensor (Parker, to be published) and a magnetic force transducer (Stewart and Christenson, 1990) which allows the deflection of the sensor to be both controlled and measured with a resolution of 0.05 nm and a force sensitivity of 10^{-8} N. A force is applied to the spring with the magnetic transducer in order to compensate the surface force. This has been achieved with a simple feedback system which allows the spring constant to be increased and strong attractive and repulsive forces to be measured easily.

Key words: Surface force; force feedback; elasticity; deformation

Introduction

The last 10 years has seen a large number of important results emerge from the direct measurement of surface forces. DLVO [1], structural [2] and hydration [3] forces have been measured and the anomalous long-range nature of the interaction between neutral hydrophobic surfaces has been discovered [4]. One tool used for surface force measurement is the surfaces force apparatus, undoubtedly the most successful of which is that of Israelachvili [1] and later variants [5]. These devices use an optical interferometry, fringes of equal chromatic order (FECO) technique to determine the separation between the surfaces [6]. The refractive index of the medium between the surfaces can also be measured and the instrument is capable of studying a number of properties of thin films. Despite the success of the technique, there are a number of important restrictions. The number of surfaces available for study is severely restricted because of the requirements imposed by FECO [7], equilibrium and dynamic surface forces are easily confused and strongly attractive surface forces are impossible to measure. These problems all arise because, with the simple spring-based apparatus there is no method of controlling and measuring the spring deflection independently of the surface separation.

Some of these restrictions have been removed by replacing the force measuring spring with a force sensor [8]. The force sensor, based on a piezoelectric bimorph, produces an electric charge proportional to the deflection applied. The device is capable of remarkable sensitivity, it can detect displacements as small as 0.05 nm and forces less than 10^{-8} N. Unfortunately, the device differs from a spring only in that it produces a charge, and so it is subject to the same mechanical instabilities as a normal spring. Further problems complicate the use of the device. The bimorph is used as a single cantilever spring, and so at high forces or with large deflections the surfaces may both roll and shear across each other. It has been shown that this type of unwanted movement between the surfaces can dramatically affect the values obtained for adhesion between those surfaces [9].

Mechanical instabilities can be understood with reference to Fig. 1, where one possible surface force profile is shown. At infinite separation there is no

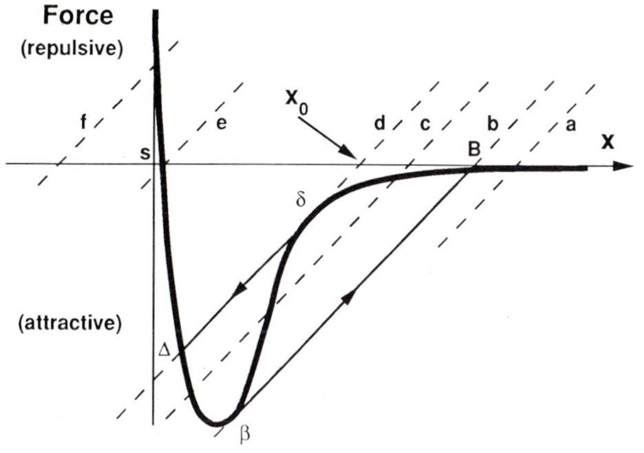

Fig. 1. A typical surface force law is shown schematically. The force law of the force measuring spring is shown as the straight lines. The surface separation is determined by point of intersection of the force laws of the measuring spring and the surface force. If the distance between the surfaces is reduced from infinity, then at surface separation δ a mechanical instability occurs and the surfaces jump to the position Δ. Similarly, on separating the surfaces a mechanical instability occurs at position β and the surfaces jump to position B. As a result of these instabilities it is impossible to measure forces between points δ and β

force present between the surfaces, on decreasing the separation the force becomes attractive and this is followed by a short-range repulsion. The force law of the spring is shown by the straight lines. The surfaces are stable at the positions where the spring force law intersects the force law being measured. At some point, however, the slope of the force law exceeds the spring constant and the surfaces jump from the stable position to another (δ to Δ in Fig. 1); the same thing occurs on separating the surfaces (b to B). The region between δ and β is therefore inaccessible to measurement.

The total force balance for the surfaces is given by:

$$F_{tot} = F_{surf} - k(x - x_0) + F_{ext} \; ; \qquad (1)$$

F_{tot} equals zero when the system is in mechanical equilibrium, k is the spring constant and $x - x_0$ the spring deflection. With a suitable technique we can choose to add an external force F_{ext} to the force balance. If we adjust this so that $F_{ext} = -Gk(x - x_0)$, where G is a positive number, we find that the surface force is given by $F_{surf} = k(x - x_0)(1 + G)$, demonstrating that the effective spring stiffness has been increased by a factor of $(1 + G)$.

The bimorph force [8] sensor provides a convenient method of measuring $(x - x_0)$; the external force can also be generated very conveniently with the magnetic force transducer developed by Stewart and Christenson [10]. A small magnet is attached to the end of the cantilever with its magnetic moment m parallel to the force between the surfaces. A magnetic field B is applied in the same direction, so the force experienced by the magnet is $F = m\, dB/dx$. Two coils mounted outside the chamber of the apparatus are used to generate the magnetic field, one mounted on either side of the magnet with their windings in series opposition. In this way the applied magnetic field at the position of the magnet (and the surfaces) is small but its gradient is large. The effect of magnetic field on a paramagnetic electrolyte solution has been investigated and no differences were detected in the force curves [10].

With the bimorph and the magnetic coil system we have a suitable method for both controlling and measuring the spring deflection. To implement a constant deflection scheme we need only amplify the signal from the bimorph and apply this to the coils in a feed-back loop. A negative feed-back loop has been employed, similar to that in the instruments developed by Derjaguin et al. [11]. Figure 2 shows a schematic arrangement for the device. The feedback amplifier consists of several stages, including an integrator for reducing the steady-state error. The design and dynamic behaviour of this device is beyond the scope of this article and will be dealt with in another publication. The feedback amplifier compares the voltage generated by the bimorph with a set point (usually zero) and generates a voltage that is applied to the coils to minimize this difference. In this paper we report examples of repulsive and attractive forces measured with the system. Perhaps the ultimate test that the system could be put to is to measure the oscillatory force profile present between mica surfaces in a non-polar liquid [2]. The force oscillates with surface separation with a periodicity of the size of the molecule. To measure such an interaction a distance resolution of better than 0.2 nm is required. With the constant deflection system we find that we can measure the interaction with a separation resolution of better than 0.05 nm. The optical interferometry technique is used only for calibration.

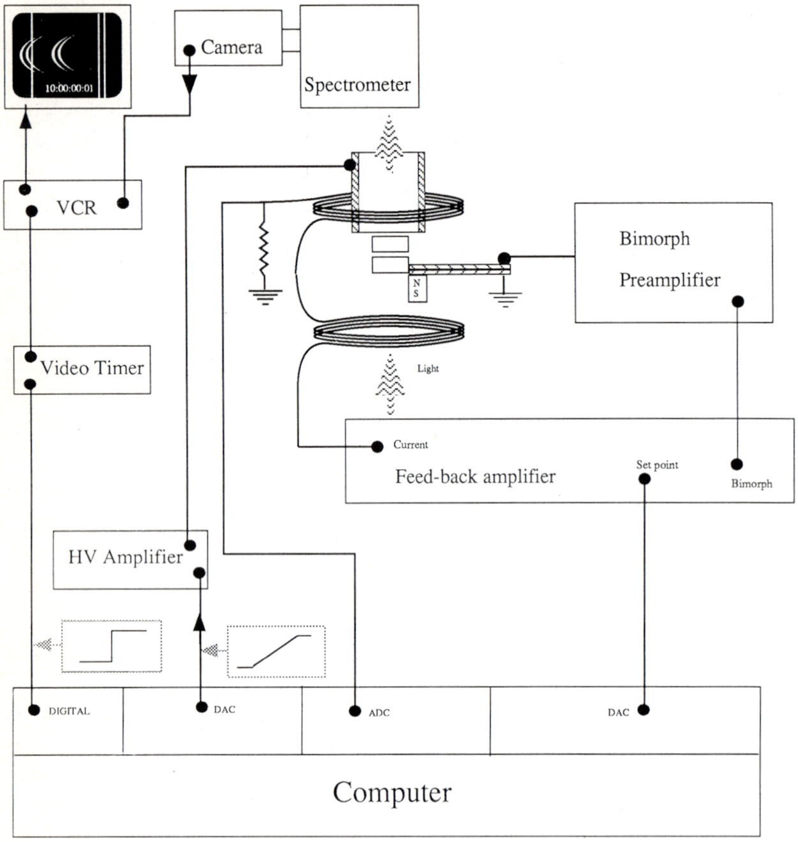

Fig. 2. Schematic diagram of the force feedback system. The system employs the Mk IV surface force apparatus with the bimorph force sensor and magnetic force transducer. The force between the surface is detected with the bimorph force sensor. The signal from this device is amplified and applied to a set of coils mounted outside the apparatus. A small magnet is mounted at the end of the force-measuring spring. The force on this magnet is determined by the applied magnetic field gradient which is, in turn, dependent on the current which flows through the coils. During operation the force experienced by the magnet almost exactly cancels the force between the surfaces, resulting in a nominally null deflection. The position of the bimorph can be controlled by adjusting the set point of the feedback loop

Materials and methods

The surface force apparatus used in these experiments has been described elsewhere [6]. Freshly cleaved mica surfaces were silvered on one side and glued (silver side down) to silica discs. One disc is mounted at the end of a piezoelectric actuator and the other at the end of the bimorph force sensor. All solutions were freshly distilled prior to use. Water was purified by reverse osmosis, distillation and final purification with a Milli-pore Milli-Q system.

A computer controls all aspects of the force-measuring system. The current through the coils is encoded by an analogue to digital convertor and the piezo displacement is controlled with a HV amplifier driven by a digital to analogue convertor. The images of the fringes (used for calibration purposes) are recorded on video tape along with a time signal which is initiated at the start of each measurement by the computer. With this, it is possible to synchronize the position of the surfaces with the other parameters from the apparatus.

Calibration

The deflection of the spring is directly proportional to the current I in the coil, and this is, in turn, related by Hooke's law to the force:

$$x - x_0 = \alpha I/k \ . \tag{2}$$

The proportionality constant α depends on the strength of the magnet, the position of the coils and the number of windings. In order to convert the measured coil current into the spring deflection we need to determine α. The most convenient way to obtain this quantity is to use FECO to calibrate the displacement of the bimorph. This is done by altering the set-point of the feedback loop and measuring the resulting change in distance between the surfaces at separations well beyond the range of any surface force present in the system. Results of this calibration are shown in Fig. 3. A linear voltage ramp, generated by the computer, is used to control the set point, and the coil voltage and FECO separation are determined as a function of

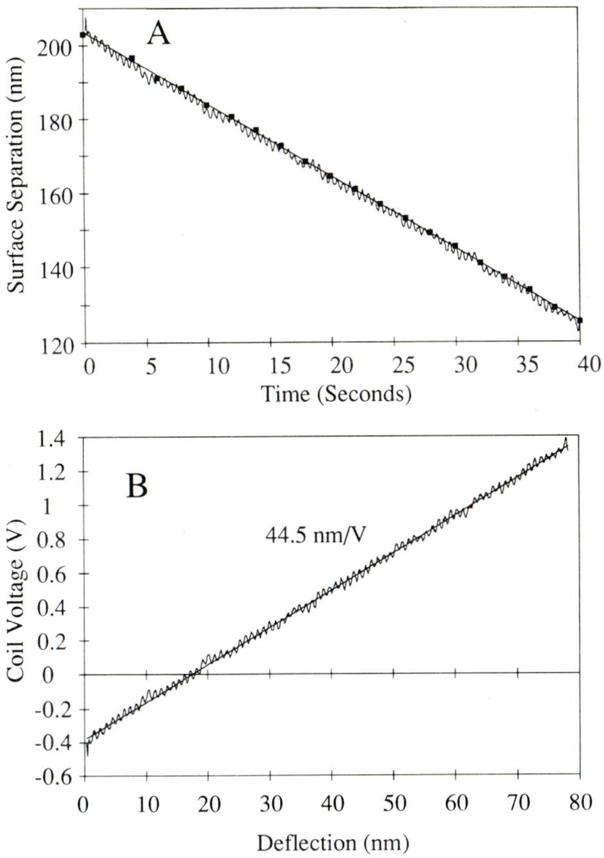

Fig. 3. The force sensitivity of the system is determined by applying a linear voltage ramp (generated by the computer) to the set point of the feedback loop; the linear change of distance with time is then recorded with FECO. a) Resulting separation of the surfaces as a function of time. b) The voltage across a 55 Ω resistor in series with the coils as a function of bimorph displacement. The slope of this line gives the deflection sensitivity of the system. During force measurement the deflection of the bimorph is held constant with the feedback system and the coil current varies according to the force between the surfaces

time. The spring constant of the bimorph is obtained by placing weights on the end of the device and measuring the deflection with the feedback turned off. The calibration of the sensor and coils remains constant to a few percent in repeat experiments. Having obtained the calibrations, it is possible to perform subsequent experiments without FECO interferometry so long as the bimorph mounting position is unchanged.

With the feedback system operating the spring remains nominally undeflected regardless of the force between the surfaces. As a result, the motions of the two surfaces are uncoupled from each other and the separation between the surfaces depends only on the motion of the upper surface (this is true only to a point; see later on). The motion of the upper surface is controlled by a voltage applied to a piezoelectric actuator; the displacement of this actuator is highly linear with applied voltage. FECO was used to calibrate the motion of the device. This can be done either by recording the fringe pattern during the surface force measurement or by measuring the displacement of the actuator as a function of the applied voltage prior to surface force measurement.

Force measurements and discussion

Attractive and repulsive interactions have been measured with the system. Results for the measurement of an attractive force profile are shown in Fig. 4. For this experiment the mica surfaces have been hydrophobed by adsorption of N-(α-(trimethylammonio)acetyl)-O,O-bis-(1H,1H,2H,2H-perfluorodecyl)-*L*-glutamate chloride (fluorocarbon surfactant) from a 2×10^{-5} M solution in ethylene glycol. The ethylene glycol was then diluted to a 50% water and ethylene glycol mixture. Figure 4(a) shows the data recorded as the surfaces are driven together at a constant rate. The distance between the surfaces is plotted as a function of time and the points indicate the separation measured by FECO interferometry. The surface separation changes linearly with time down to a separation of about 20 nm, where the surfaces spontaneously jump together. The lower line in the figure is an estimate of how the distance varies with time with the feedback system turned off. The fact that the distance response is linear with time and the bimorph force signal remains at zero up to the jump point indicates that the feedback system is maintaining the spring deflection at a constant value. Moreover, the bimorph force signal remains at zero through the measurement, even during the jump, indicating that the jump is not due to instability in the feedback system.

The surface separation is measured between the backs of two mica plates by FECO. The reflections which give rise to a fringe pattern arise at the silver-mica interface. The mica sheets are glued with a relatively thick layer of glue to two silica discs and it is the separation and force between the silica discs which is measured with this system. The glue

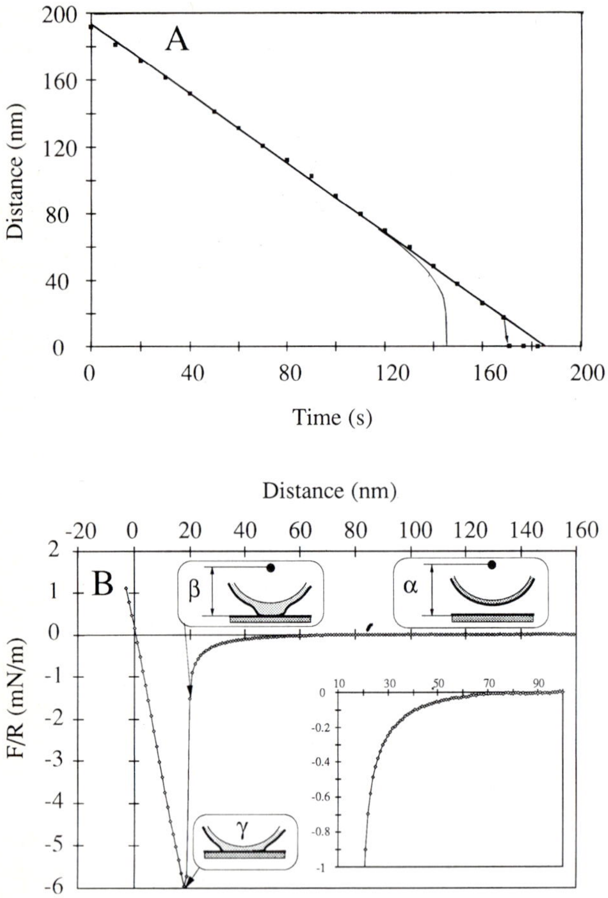

Fig. 4. An example of an attractive force curve measured with the force feedback system. The surfaces were hydrophobed by adsorption of a fluorocarbon surfactant and the measurements were performed in a 50% mixture of water and ethylene glycol. The raw data collected by the instrument is shown in A). The separation obtained from FECO (squares) and the calibrated driving function (straight line) are shown as a function of time. The lower solid line is an estimate of the distance between the surfaces as a function of time with the feedback loop open. The lower figure B) shows the force distance curve computed from the data in the upper figure. The surfaces show a mechanical instability at strongly attractive forces due to the elasticity of the glue. The surface deformations responsible for this instability are illustrated schematically (see text)

used has a relatively low Young's modulus ($Y = 10^9$ to 10^{10} Nm^{-2}) and the flattening of the fringe pattern which occurs when the surfaces are in strong adhesive contact is due to the elastic deformation of the glue. With the constant deflection arrangement we are able to control the total force on the surfaces; however for an attractive interaction there must come a point where the gradient of the interaction force between the surfaces exceeds the effective spring constant for the glue. At this point the surfaces jump together as observed by FECO but the spring deflection still remains constant. This process is illustrated schematically in Fig. 4b. At a separation of 20 nm the glue deforms and the surfaces bulge out to come into contact (β). As the separation is decreased still further this bulge is progressively flattened out until the equilibrium flattened state is reached (γ). After this point the glue comes into compression, and pushing the surfaces together results in an increasing contact area and an increasing force, but during all this the surface separation, as measured between the mica sheets, remains constant.

The effect of the glue elasticity is observed for the interaction of two mica surfaces in a non-polar liquid. The force between two surfaces in octamethylcycloterasiloxane (OMCTS) oscillates with surface separation with a periodicity close to the size of the molecule. The force curve reflects the free energy of packing of molecules into the gap between the surfaces and has been the subject of extensive study. Figure 5A shows the raw data for a single inward run in such an experiment. The voltage applied to the piezo actuator, which is linearly related to the surface separation, is plotted against the current in the coils (which is proportional to the force). The oscillatory nature of the interaction is clearly evident in the response; the force law oscillates from attractive to repulsive, but the measured coil voltage only drops below zero for the first few oscillations. The feedback set point remains constant during the measurement and the surfaces jump together from one maxima to a position part way down the next oscillation due to the elasticity of the glue. During compression of the surfaces along a repulsive branch of the oscillatory force curve FECO shows an immeasurable change in the surface separation. The slope of the voltage applied to the piezo against the coil current curve is due to the deformation of the glue as it is compressed. The compressibility of the glue adds an elastic term to the force law. The surface separation and force can be calculated from the raw data by assuming that this elastic deformation adds a linear spring constant to the system. At first sight, this appears to be a questionable approximation. In the simplest case the load applied to an elastic sphere (where interactions between the spheres is ignored) is given by the well known Hertz formulae [12] which, for two interacting elastic

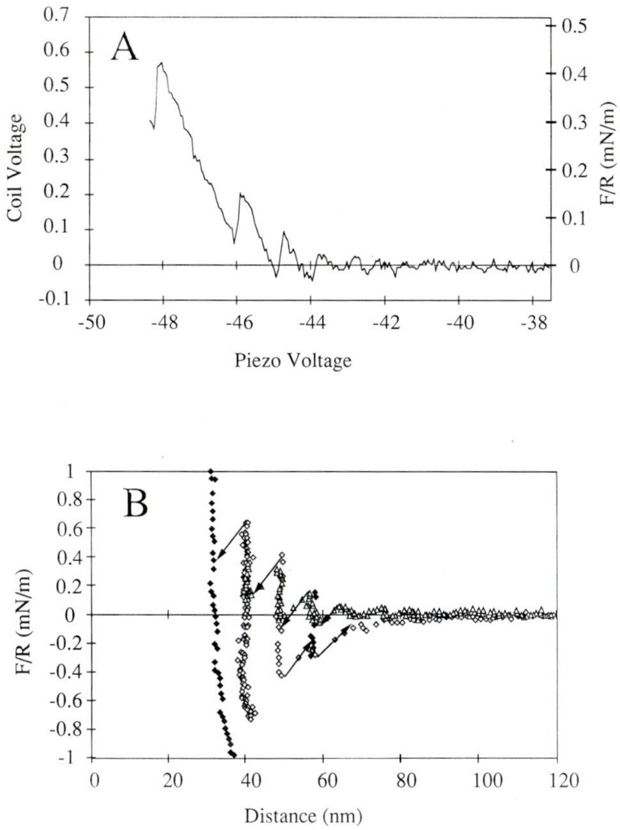

Fig. 5. An oscillatory force curve measured with the force feedback system between mica surfaces in OMCTS. The upper plot a) shows the raw coil voltage as a function of the potential applied to the piezo actuator. The lower figure shows the computed force curves for a number of both inward and outward force runs

spheres, is:

$$\frac{\varepsilon}{R^{1/3}} = \left(\frac{WD}{R}\right)^{2/3}, \tag{3}$$

where ε is the change in separation between the centre of the two spheres for an applied load of W, $1/R$ is given by $1/R_1 + 1/R_2$, the respective curvatures, and

$$D = \frac{3}{2}\left(\frac{1-\sigma^2}{E}\right), \tag{4}$$

where E is the Young's modulus for the sphere and σ is Poisson's ratio.

Clearly, from the above equation we do not expect the force, measured as a function of the compression of the two spheres, to be linear with the distance of compression over a large range. However, for very small forces and small compression distances, this is certainly true, as evidenced by the extraordinary linearity in the compression shown in Fig. 5B. The fact that the slopes of the repulsive branches of the interaction profiles measured in OMCTS all have the same values also indicates this. The total compression range is only of the order of at most a few tens of nanometers, and this is very small by comparison to the overall thickness of the glue layer which is several microns. The forces are small enough so that there is no measurable change in the contact area on compression, so we are safe in assuming that the glue adds only a simple linear elastic term to the system.

We can convert the measured profile shown in Fig. 5A into a force-distance curve by simply correcting the slope, i.e. subtracting a linear elastic term ($F = kx$) from the measured curve. The result of this is shown in Fig. 5B along with several repeat force runs with both increasing and decreasing surface separation. The resolution and repeatability of the measurements are quite remarkable. FECO interferometry was used only to calibrate the motion control and to determine the separation for one oscillation in each run relative to the mica-mica contact in air. There is insufficient range of feedback force available in the present system to measure the innermost three oscillations and so the surface separation cannot be decreased below a value of 3.4 nm. The arrows in Fig. 5B indicate the positions of the instabilities due to elastic deformation of the glue. Note that on separating the surfaces from the third oscillation the surfaces actually jump to the position of the next local minima and continue jumping from one minima to another as the surfaces are separated. On separation from the next most inward oscillation the surfaces jump to a position of zero force because this oscillation is sufficiently deep so that the jump skips over the outer oscillations.

Summary and outlook

A combination of a bimorph force sensor and magnetic force transducer has been used to measure and control the deflection of the spring in a surface force apparatus. With this setup forces can be measured with unparalleled separation and force resolution over a wide range of attractive forces. The rolling induced between the surfaces on separation from a deep attractive well in the interaction force precludes the accurate measurement of the well depth when the bimorph is used alone (see introduction). However, if the magnetic transducer has a sufficiently large enough range such rolling cannot occur because the spring remains nominally undeflected and accurate measurements of the depth of attractive minima are obtained for all but the most attractive force profiles.

The possibility for future work with this system is enormous. FECO interferometry is required for calibration only, so other surfaces become accessible for study. Surface contamination is instantly recognizable from the FECO fringes and by eliminating the use of FECO it would appear that valuable insight is lost to the experimentalist. However, with the force feedback system the data is analysed while the experiment is running and, in fact, the results appears (unscaled) on an X-Y chart recorder and so contamination is also instantly detected.

With the addition of extra hardware a number of new types of experiment can be performed. For instance, one can envisage the addition of another feedback loop to control the voltage applied to the piezoelectric actuator and hence the separation between the surfaces. In this scheme the feedback signal from the constant deflection loop is applied as the input to the second feedback loop which controls the piezo actuator. With both feedback loops closed the condition of constant force between the surfaces can be maintained. A constant force device has obvious applications in the measurement of frictional [13] and non-equilibrium forces.

Perhaps one of the most exciting possibilities that the feedback scheme offers is in the control of the force between a tip and surface in an atomic force microscope. These devices suffer from the same spring instabilities present in the surface force apparatus, so imaging can be done only at certain tip-surface distances. Feedback schemes utilizing the application of electrostatic forces have been proposed to overcome this difficulty [14, 15]. Implementation of the scheme described here in an atomic force microscope should allow imaging of surfaces at any point in the force-distance profile.

References

1. Israelachvili JN, Adams GE (1978) J Chem Soc Faraday Trans 1 74:975
2. Christenson HK (1988) J Dispersion Sci Technol 9:171
3. Pashley RM (1981) J Colloid Interface Sci 80:528
4. Pashley RM, McGuiggan PM, Ninham BW, Evans DF (1985) Science 229:1088; Christenson HK, Claesson PM (1988) Science 240:189
5. Parker JL, Christenson HK, Ninham BW (1989) Rev Sci Instrum 60:3135
6. Israelachvili JN (1973) J Colloid Interface Sci 44:259
7. Clarkson MT (1989) J Phys D 22:917
8. Parker JL Langmuir (in press)
9. Christenson HK (1988) J Colloid Interface Sci 121:170
10. Stewart AM, Christenson HK (1990) Meas Sci Technol 1:1301
11. Derjaguin BV Rabinovich YI, Churaev NV (1978) Nature 272:313
12. Fogden A, White LR (1990) J Colloid and Interface Sci 138:414
13. Israelachvili JN, McGuiggan PM, Homola A (1988) Science 240:189
14. Miller GL, Griffith JE, Wagner ER, Grigg DA (1991) Rev Sci Instrum 62:705
15. Joyce SA, Houston JE (1991) Rev Sci Instrum 62:710

Authors' address:

J. L. Parker
Dept. of Applied Mathematics
Australian National University
G.P.O. Box 4
Canberra, A.C.T. 2601
Australia

Author Index

Subject Index